Advances in Sustainability Science and Technology

The book series aims at bringing together valuable and novel scientific contributions that address the critical issues of renewable energy, sustainable building, sustainable manufacturing, and other sustainability science and technology topics that have an impact in this diverse and fast-changing research community in academia and industry.

The areas to be covered are

- Climate change and mitigation, atmospheric carbon reduction, global warming
- Sustainability science, sustainability technologies
- Sustainable building technologies
- Intelligent buildings
- Sustainable energy generation
- Combined heat and power and district heating systems
- Control and optimization of renewable energy systems
- Smart grids and micro grids, local energy markets
- Smart cities, smart buildings, smart districts, smart countryside
- Energy and environmental assessment in buildings and cities
- Sustainable design, innovation and services
- Sustainable manufacturing processes and technology
- Sustainable manufacturing systems and enterprises
- Decision support for sustainability
- Micro/nanomachining, microelectromechanical machines (MEMS)
- Sustainable transport, smart vehicles and smart roads
- Information technology and artificial intelligence applied to sustainability
- Big data and data analytics applied to sustainability
- Sustainable food production, sustainable horticulture and agriculture
- Sustainability of air, water and other natural resources
- Sustainability policy, shaping the future, the triple bottom line, the circular economy

High quality content is an essential feature for all book proposals accepted for the series. It is expected that editors of all accepted volumes will ensure that contributions are subjected to an appropriate level of reviewing process and adhere to KES quality principles.

The series will include monographs, edited volumes, and selected proceedings.

Mohammad Aslam · Sanjeev Mishra ·
Jorge Arturo Aburto Anell
Editors

Production and Biorefining of Biocrude Oil: Current Status and Future Developments

Springer

Editors
Mohammad Aslam
National Institute of Technology Srinagar
Srinagar, Jammu and Kashmir, India

Sanjeev Mishra
Sardar Swaran Singh National Institute of Bioenergy
Kapurthala, Punjab, India

Jorge Arturo Aburto Anell
Biomass Conversion Division
Mexican Institute of Petroleum
Mexico City, Distrito Federal, Mexico

ISSN 2662-6829 ISSN 2662-6837 (electronic)
Advances in Sustainability Science and Technology
ISBN 978-981-96-5197-9 ISBN 978-981-96-5198-6 (eBook)
https://doi.org/10.1007/978-981-96-5198-6

This Springer imprint is published by the registered company Springer Nature Singapore Pte Ltd.
The registered company address is: 152 Beach Road, #21-01/04 Gateway East, Singapore 189721, Singapore

Preface

Most of energy requirements of world are currently fulfilled by the petroleum-based products. World's petroleum reserves are dwindling day by day, and products derived from them are one of the major causes of GHG emissions that adversely affect the environment. As we all are aware about the consequences of petroleum-based fuels and their related world's challenging issues like climate change and global warming, biocrude oil derived from the renewable resources is only the viable option for reducing GHG emissions and to enhance the nature's green cover. The global energy scenario of twenty-first century clearly unveils the fact of transition of fuel market and economy from fossil fuels to carbon-neutral fuels or to a decarbonized fuel system. Current focus of the world's scientific community is to develop second/third-generation biorefineries to produce more efficient sustainable fuels and value-added products. A number of biorefineries based on the second-generation feedstocks like lignocellulosic biomass, agro-wastes, non-food crops, lake's biomass, algal biomass, etc., are currently in operation in the Europe, USA, and Asia. Global market for biorefinery estimated at US\$553.7 Billion in the year 2020 is projected to reach a revised size of US\$979.5 Billion by 2026.

Motivated by the literature, there is not a single specific book available on the proposed subject in international markets. Keeping in view of significant contribution in the field of green energy/biorefinery, the editors are highly enthusiastic to fill the existing market gaps and step forward for an initiative to edit a specific book on the subject. Biocrude oil is converted to green fuels and value-added products in a biorefinery mode. Unlike petroleum crude oil, biocrude oil does not consist of undesirable sulphur and nitrogen derivatives and is very less harmful as compared to that of petroleum crude oil. In addition, fuels derived from biocrude oils are carbon negative in nature and significantly contributed to save the humankinds and environment. Now, the modern society is moving from fossil-based fuels to more sustainable greener fuels.

This book deliberates a comprehensive discussion on the production and biorefining of biocrude oil derived from second- and third-generation biomass feedstocks. Chapters 1 and 2 discuss about the concept of biorefinery, biorefining raw materials/feedstocks, and its associated challenges. Chapter 3 emphasised on the third-generation biorefining feedstock (Algae) as an emerging feedstock for the biorefineries. In addition, co-feedstocks employed in biorefinery is also deliberated in Chap. 4. Chapters 5 and 6 are based on the various thermochemical conversion approaches, such as hydrothermal liquefaction, pyrolysis, and hydroprocessing, employed for biocrude oil production from a variety of biomass feedstocks including solid residues, organic and inorganic wastes, lignocellulosic biomass, and algal biomass. Chapter 7 is about the biocrude production and subsequent upgradation via hydrothermal liquefaction, distillation, and hydrotreatment of agro-forestry biomass in a Canadian context. Chapter 8 deals with the sustainable production of biocrude oil from algal biomass and its upgradation for enhanced fuel properties. Chapter 9 deliberates on current scenario and future developments in the area of biorefining catalysts. Chapter 10 is based on the comparative assessment of properties and downstream upgradation of biocrude oil through hydrothermal and pyrolysis processes. In addition, Chap. 11 deals with biocrude oil by Fischer-Tropsch (FT) process. Chapter 12 is focussed on the life cycle approach for the sustainability assessment of biocrude oil from pyrolysis process, while Chap. 13 is about the current status and future developments in biorefining of biocrude oil. Finally, sustainable production of biofuels and value-added products from biorefinery and biocrude upgrading technology and challenges are deliberated in Chaps. 14 and 15. This book will be of great interest to professionals, post-graduate students, researchers, and policymakers involved in biorefineries for sustainable development. This book makes a forceful foundation for the establishment of biocrude oil-based biorefineries and hence contributes towards a sustainable, cleaner, and greener blue planet for next generations. Biocrude oil will likely to be a "Green Gold" for the future's biorefinery, and this book will be marked as the milestones for biorefineries.

Srinagar (J&K), India — Mohammad Aslam
Kapurthala, India — Sanjeev Mishra
Mexico City, Mexico — Jorge Arturo Aburto Anell

Contents

About the Editors

Dr. Mohammad Aslam is Assistant Professor at the Department of Chemistry, National Institute of Technology, Srinagar (India). Dr. Aslam holds master's degree in "Industrial Chemistry" from Aligarh Muslim University, Aligarh (India). Dr. Aslam also completed M.Tech. in "Fuels and Combustion" from Birla Institute of Technology, Mesra, Ranchi (India). Later, Dr. Aslam has awarded Ph.D. in Chemistry from National Institute of Technology, Jalandhar (India). The thesis entitled "Hydroprocessing of Seed Oils into Transportation Fuels". Dr. Aslam has worked as Post-Doctoral Fellow/Research Associate/Senior Research Fellow in Chemical Conversion Division, Sardar Swaran Singh National Institute of Bio-Energy (an autonomous institute of Ministry of New and Renewable Energy, Government of India), Punjab (India). Dr. Aslam holds more than 17 years of academic and research experience. His research is focussed on the sustainable fuels production from 2-G and 3-G biomass feedstocks under biorefinery concept.

Dr. Sanjeev Mishra is working as Scientist-D at Sardar Swaran Singh National Institute of Bio-Energy (an autonomous institute of Ministry of New and Renewable Energy, Government of India), Punjab, India. Prior to this, Dr. Sanjeev worked as Research Fellow (Postdoc) at the Asian School of the Environment, Nanyang Technological University, Singapore. His research focuses on "Production of 3rd generation biofuels and high-value compounds from municipal, agricultural, and industrial waste via algal biorefinery approach". Dr. Sanjeev has done bachelor's and master of technology in industrial biotechnology and has over two years of pre-Ph.D. research experience working at national labs. During this, his research focussed on lab-scale to large-scale microalgal cultivation and bioenergy production. Dr. Sanjeev did his Ph.D. in energy science and engineering from the Indian Institute of Technology Guwahati (India). The thesis entitled "Microalgal Wastewater Treatment, Enhanced Biomass Productivity, and Biofuel Conversion Under a Biorefinery Approach".

Dr. Jorge Arturo Aburto Anell holds a B.Sc. in "Food Chemistry" from the National Autonomous University of Mexico (UNAM), the diplôme d'études approfondies (DEA), and Ph.D. from National Polytechnic Institute of Toulouse (INPT), France, and was Visiting Researcher in the Agrobiotechnology Institute (IFA-Tulln) from the

Agricultural University of Vienna (Boku), Austria. Dr. Jorge Aburto is Researcher at the Mexican Petroleum Institute (IMP) since 2000, and he is Project Leader and Head of biomass conversion division and In Charge of the division of energy efficiency and sustainability. Dr. Aburto's main research areas are bio-based chemicals and materials for applications in the petroleum industry, biomass deconstruction and conversion, production of lignocellulosic ethanol, road and aviation fuels, enzyme kinetics and applications to the oil industry, petroleum biorefining, precipitation inhibitors of asphaltenes, and petroleum pipelining through o/w emulsions. He is Author of several international articles, book chapters, and patent applications and patents.

Chapter 1
An Introduction to the Biorefinery

Antônio Luthierre Gama Cavalcante, Rafael Leandro Fernandes Melo, and José C. S. dos Santos

Abstract This chapter examines the production and biorefining of biocrude oil, highlighting the differences and potentials between traditional petroleum refineries and biomass biorefineries. Biorefineries utilize renewable sources, such as biomass, to produce fuels and chemicals, unlike petroleum refineries that rely on non-renewable fossil sources. This transition to biorefineries faces significant challenges, including high operational costs, technological needs for efficient biomass processing, and integration of new processes. However, biorefineries offer promising possibilities to reduce dependence on fossil fuels, lower carbon emissions, and generate sustainable products, marking a crucial evolution in the search for greener energy alternatives. Therefore, the biorefinery industry represents a vibrant technological development and innovation field, requiring supportive policies and continuous advancements to achieve its full potential.

Keywords Biocrude · Biorefineries · Biomass · Petroleum refineries · Renewable sources

A. L. G. Cavalcante
Departamento de Química Analítica e Físico-Química, Universidade Federal do Ceará, Fortaleza, Brazil

R. L. F. Melo
Instituto Federal de Educação, Ciência e Tecnologia, Campus Tabuleiro do Norte, Tabuleiro do Norte, Brazil

J. C. S. dos Santos (✉)
Instituto de Engenharias e Desenvolvimento Sustentável, Universidade da Integração Internacional da Lusofonia Afro-Brasileira, Redenção, Brazil
e-mail: jcs@unilab.edu.br

M. Aslam et al. (eds.), *Production and Biorefining of Biocrude Oil: Current Status and Future Developments*, Advances in Sustainability Science and Technology,
https://doi.org/10.1007/978-981-96-5198-6_1

1 An Introduction to Biorefinery

In recent decades, oil refineries and derivatives have served as the leading suppliers of fuels, chemicals, and various compounds necessary to meet the energy demand, which today is classified as a fundamental good for life [1–3]. Notably, fossil fuels supplant more than 80% of global energy demand. Furthermore, this energy source is scarce and non-renewable, causing some long-term concerns on the part of the energy industry [1, 2, 4]. In this sense, many countries have been searching for new, less polluting, renewable energy sources as sustainable alternatives that balance energy demand and optimized use of sources [1, 3, 4].

In recent years, using renewable resources such as solar, hydroelectric, wind, geothermal, and tidal energy is a promising alternative to replace non-renewable sources [5–9]. In this trend, natural in recent years, biorefineries have emerged as a potential alternative to pollution caused by refineries. Biorefineries are classified as structures in which any type of waste/biomass is used ideally to produce multiple products. They seek to be self-sustainable and not harmful to the environment [2, 6, 8, 10, 11]. However, the need for chemicals and materials derived from the petrochemical industry will continue until biorefineries can compete with petrol refineries [1, 2, 10–12]. Therefore, many research laboratories have invested efforts in developing biorefineries, which they point to as the main alternative to meeting market needs in the long term, offering products with more excellent added value due to the intrinsic renewable capacity in the process [5, 6, 10].

The possibilities in this field are gigantic. However, no diverse political drivers invest continuously in developing new processes and optimizing existing ones [4–6, 12, 13]. This is one of the main limiting factors for the exponential growth of this field of energy and chemical production [6, 7, 9]. On the other hand, the accelerated growth in using lignocellulosic raw materials will require the market to make structural changes to the supply chain. It may also require more state investment [7, 8, 10].

From this perspective, the refinement of lignocellulosic biomass, as by-products of agriculture and industry such as straw, wood, textile industrial waste, and others, can generate a variety of high-value renewable fuels, chemicals, biomaterials, and others. Notably, hemicellulose, lignin, and cellulose are the main compounds found in the composition of lignocellulosic biomass. Table 1 presents the chemical composition of various lignocellulosic biomasses of different origins [14–18]. These biorefinement industries are also known as bio-based refineries due to their biological versatility and sustainability [17–21]. Therefore, many researchers point out that replacing chemical products based on fossil fuels with those derived from renewable biomass is essential to sustain the development of chemical industries and the expansion of the global economy [5, 6, 19, 20, 22].

On the other hand, it is necessary to state several considerable challenges in bioenergetic and biochemical production from a large-scale perspective from biomass regarding product yield and energy input into the process [2, 5, 7–10]. The complex structure of lignocellulosic biomass makes it difficult for the various enzymes to hydrolyze cellulose into single-chain sugars, thus reducing product yield [2, 12],

Table 1 Chemical composition of different lignocellulosic biomasses from different origins

Lignocellulosic biomass	Percentage composition (%)				References
	Hemicellulose	Lignin	Cellulose	Others	
Agricultural residues/residues					
Corn husk	39.78	11.07	29.34	19.81	[23–25]
Oat straw	33.11	25.41	17.28	24.2	[26, 27]
Rice straw	35.45	14.58	37.52	12.45	[28, 29]
Sugarcane bagasse	27.85	23.25	46.32	22122.58	[30, 31]
Bamboo	26.53	27.15	40.74	5.58	[32, 33]
Wheat straw	43.22	22.14	31.56	3.08	[34, 35]
Industrial/municipal waste					
Recycled paper	11.98	1.12	46.11	40.79	[36–38]
Brewer's spent grains	16.8	29.36	27.58	26.26	[39–41]
Newspaper	25.12	18.91	55.52	0.45	[36–38]
Solid cattle manure	2.3	1.3	2.7	93.7	[42, 43]
Forest residues					
Aspen	20.3	25.5	44.5	– 9.7	[37, 44]
Eucalyptus	25.3	18.2	50.34	– 6.16	[42, 45]
Japanese beech	28.4	24.25	43.46	– 3.89	[46, 47]
Pinho	14.3	27.29	53.71	– 4.7	[42, 45]

Table 1. Biomass pre-treatment is a possibility that has been analysed. However, it is known that this treatment makes the process more costly and may result in more economic disadvantages than commercial viability [1, 3, 4].

However, the advent of industrial biotechnology approaches has played a catalytic role in the flourishing of lignocellulosic biorefineries worldwide [2, 4, 48]. Some limitations still exist, but enzymatic and materials engineering fields sought to develop efficient biocatalysts and catalysts to accelerate the process [2, 3, 9, 10, 12, 49].

In summary, this chapter's fundamental objective is to present the basic concepts of biorefinery and the main possibilities for overcoming the challenges encountered in the energy balance, which until now was insufficient for large-scale work [5, 7, 21]. Furthermore, we will seek to expose the current situation of biorefineries in the global context, explicitly citing some success cases [17, 19, 20].

2 Comparison Between Petroleum Refineries and Biomass Biorefineries

The continuous depletion of fossil resources, the energy crisis faced in several countries, causing a collapse in some sectors, and accelerated environmental pollution in emerging countries have become the main factors that force some countries to transition from the traditional oil refinery to a new alternative that appeared recently and is classified as one of the most promising alternatives, the sustainable biorefinery [11, 50–52]. In this sense, these world powers have increasingly adopted biofuels as a sustainable alternative to the indiscriminate use of fossil fuels to resolve this environmental issue [14, 15, 53, 54]. The biorefinery, which has sustainable biomass processing with a range of marketable products and energy, is compared to petrochemical refineries with a balance of respect for environmental issues [15–19]. Figure 1 presents a brief comparison between oil refineries and biomass biorefineries.

The lower generation of polluting gases, the reuse of agricultural and industrial waste, and the possibility of scaling encourage large countries to invest in this new technology [19–21, 55]. Regarding waste, biomass is an essential renewable source for biorefineries. Environmental policies and strategic plans use this factor to determine the possibility of large-scale production of biofuels, biomaterials, and important chemical derivatives [5–8, 22]. It is worth mentioning that biomass is an important energy source representing between 10 and 14% of the world's energy supply [5–8,

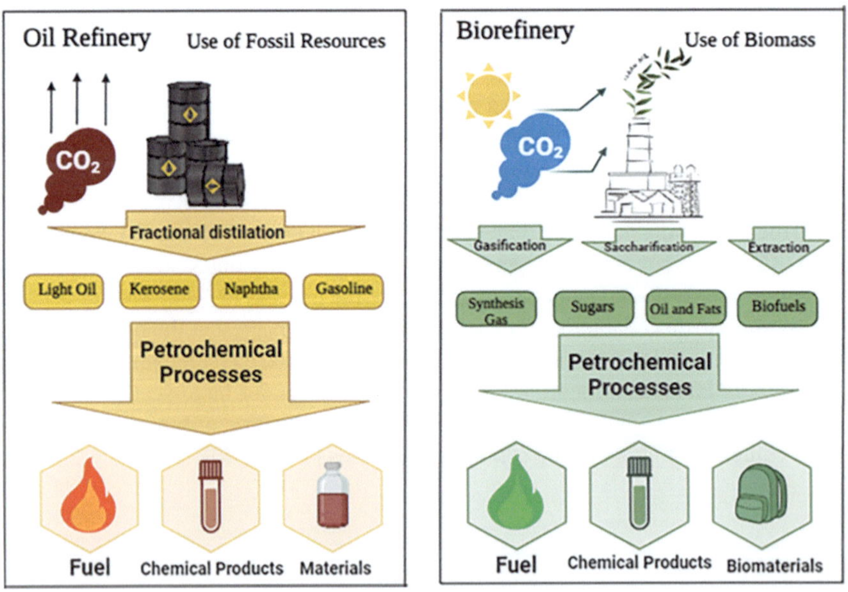

Fig. 1 Comparison between oil refineries and biomass biorefineries

22]. As previously mentioned, biomass is mainly formed by cellulose, hemicellulose, lignin, and extractives in small quantities [10, 49].

Lignocellulosic biomass is a promising source for generating value-added chemicals, such as biofuels, platform chemicals, and biomaterials production [11, 49–52]. Notably, cellulose and hemicellulose are the two main carbohydrates in lignocellulosic biomass and are strongly linked to lignin, forming a robust, complex, and unmanageable structure [51–53]. This factor dramatically favours the production and development of new products due to the operational stability of this type of biomass. However, pre-treatments are required to break recalcitrance and use the sugars for later stages [11, 52, 53].

To this end, despite all the advantages of biorefineries, some challenges are associated with their implementation, including the availability of raw materials, technological advances, and market competitiveness [14, 52–54]. Therefore, political support from the world's great potential is necessary to promote the development of the biorefinery sector and facilitate a sustainable transition to a sustainable environmentally based economy. In summary, biorefineries have the potential to act in a similar way to oil refineries and as an essential component of the renewable energy panorama [14, 15, 52–54].

3 Challenges and Possibilities of the Biorefining Industry

Subsequently, several countries from all socioeconomic contexts designed national policies, plans, and actions to mitigate their dependence on conventional energy sources (generally non-renewable and polluting) and design technologies for the production of clean and renewable energy [53, 56–58]. Lignocellulosic biorefineries entered with full force this process of structuring new sustainable alternatives [59–63]. As a form of standardization and planning, some global potentials have stipulated steps to be followed to successfully develop efficient biorefineries [11, 50, 51, 63, 64]. One of the critical points in this process is objective planning, which enables the production of biofuels and chemical derivatives [51–53]. This is followed by partial or total exemption from sales taxes for the biofuels market, which is still in the structuring phase and requires more care (benefits). Furthermore, the release of capital for promising projects and investment in customs benefit strategies will be fundamental in consolidating the biorefining market [14–16, 19, 54].

In this vein, the European Union initiated an action plan in 2012 to establish a low-cost carbon economy that minimizes gas emissions that intensify the greenhouse effect and cause alarming climate changes [14, 16, 18, 19, 53]. The "European Union Renewable Energy Directives" were revised, and some essential modifications were proposed to be carried out in the period 2020–2030 [11, 50–52]. The main one is a 40% reduction in emissions of gases harmful to the environment, focusing mainly on the transport sector, which is one of the largest generators. From this perspective, producing biofuels from biorefineries is one of the most promising alternatives perceived by the market [59–62].

The United States of America and China have also committed to reducing their emissions and have invested part of their income in developing new biofuel production technologies [53, 56–58]. It is worth remembering that the production of electric cars is on the rise and that many investors are moving into the market to improve this sector. However, many researchers warn of the possibility of collapse in hydroelectric, thermal, and other plants due to the exacerbated demand in the transport market. Automobiles [53, 57–62]. In this sense, China has realized that it is necessary to direct its attention to implementing cellulosic ethanol in the transport sectors. Therefore, investment in biorefineries is fundamental to the feasibility of this bold project [11, 50, 63, 64].

In this sense, as a promising alternative for several countries, biotechnological platforms that are characterized by the enzymatic conversion of biomass into products with high-added value stand out [60–62]. The process can occur under mild temperature and pressure conditions, enabling lower energy consumption for biomass refining [60–62]. It is worth remembering that these platforms direct their efforts towards preparing high-value chemical products of biofuels [56, 58, 61, 62].

Therefore, the pharmaceutical, food, and biofuel industries will benefit most from this technological advance that generates new sources of products necessary to maintain these industries [53, 57, 58]. Notably, microorganisms and enzymes are applied to accelerate and prepare traditional products and good-grade bio preservative nutraceutical substances [56–58]. In this sense, Fig. 2 presents biorefineries' applicability and sustainability, driven by a circular economy that connects interactively.

In contrast to the evolutionary bias of biorefineries, some challenges are encountered while building this new technology. One of the main challenges biorefineries face is the depolymerization of lignin [58, 59, 61, 62]. Many challenges remain after countless efforts, where research laboratories have dedicated themselves to optimizing efficient and expanded lignin depolymerization processes for years [11, 50–52, 63, 64]. It is noteworthy that lignin is a randomly cross-linked achiral polymer, mainly composed of three units with a phenolic nucleus, that is, p-coumaric alcohol, coniferyl alcohol, and sinapyl alcohol, also known as monolignols [2, 6, 10, 18, 53–55]. Furthermore, there is a need to valorize lignin, which requires recondensation, which occurs during depolymerization.

In this sense, the expansion of second-generation cellulosic bioethanol production remains limited due to high costs and uncertainty in the European Union's policy-making process, which is one of the challenges for second-generation biorefineries [2, 11, 53, 56, 60, 65]. Among several future challenges, the main one is the need to improve biomass transformation processes to make them economically viable in continuous, large-scale operations [59–61].

Finally, authorization for producing aromatic molecules on an expanded lignin platform is probably one of the most complex challenges for the coming years due to the risks encountered in production [22, 53, 55–60, 64]. However, to achieve this objective, biorefineries must be updated, using only renewable energy (solar, wind, and other sources) to reduce costs and make the process even more sustainable and cyclical.

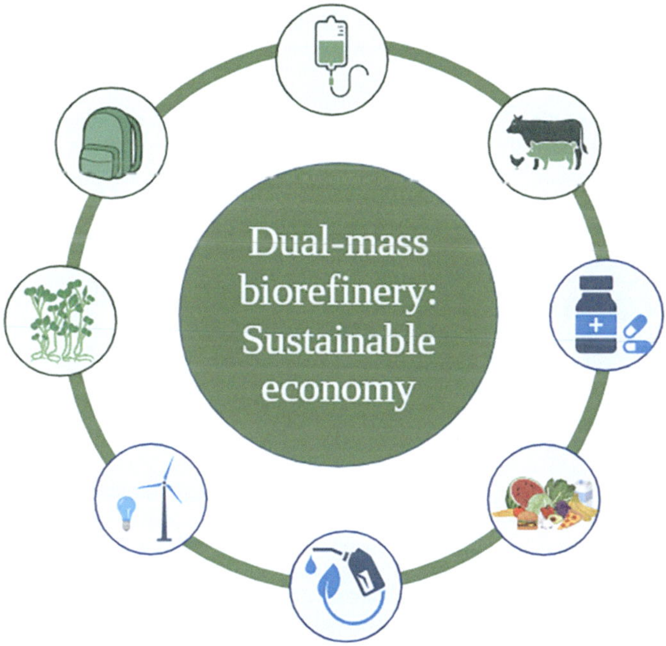

Fig. 2 Cyclization of products obtained by the biomass biorefinery

4 Conclusion

In conclusion, the production and biorefining of crude oil are crucial areas for sustainable development and energy transition. Biorefineries, significantly different from traditional petroleum refineries, use biomass as raw material, reducing the environmental impact of producing fuels and chemicals. Despite offering greener solutions, biorefineries face considerable challenges, including the need for technological advancements to improve the efficiency and scalability of processes, as well as high initial costs and complexity in raw material management. However, the prospects for the biorefinery industry are promising, driven by continuous innovation and the integration of new technologies that promise to optimize production and align with global sustainability goals. Ongoing development in this sector is vital to overcome current obstacles and maximize the potential of biorefineries, playing an essential role in reducing global greenhouse gas emissions and promoting a cleaner, more sustainable energy future.

Acknowledgements The authors would like to express their gratitude for the support that was provided through the Instituto de Engenharias e Desenvolvimento Sustentável (IEDS) at the Universidade da Integração Internacional da Lusofonia Afro-Brasileira (UNILAB). We gratefully acknowledge the following Brazilian Agencies for Scientific and Technological Development: Fundação Cearense de Apoio ao Desenvolvimento Científico e Tecnológico (FUNCAP) (PS1-0186-00216.01.00/21; 31052.000417/2023; UNI-0210-00537.01.00/23), Conselho Nacional de Desenvolvimento Científico e Tecnológico (CNPq) (311062/2019-9;440891/2020-5; 307454/2022-3), and Coordenação de Aperfeiçoamento de Ensino Superior (CAPES) (finance code 001).

Declaration of Competing Interest The authors declare that they have no known competing financial interests or personal relationships that could have appeared to influence the work reported in this paper.

Data Availability Data will be made available on request.

References

1. Narayanan M (2024) Biorefinery products from algal biomass by advanced biotechnological and hydrothermal liquefaction approaches. Discov Appl Sci 6:146. https://doi.org/10.1007/s42452-024-05777-6
2. Chaudhary G, Chaudhary N, Saini S et al (2024) Assessment of pretreatment strategies for valorization of lignocellulosic biomass: path forwarding towards lignocellulosic biorefinery. Waste Biomass Valor 15:1–36. https://doi.org/10.1007/s12649-023-02219-z
3. Zhang W, Luo Y, Yuan X (2024) Optimization of biofuel supply chain integrated with petroleum refineries under carbon trade policy. Front Chem Sci Eng 18:34. https://doi.org/10.1007/s11705-024-2397-1
4. Taghizadeh-Alisaraei A, Tatari A, Khanali M, Keshavarzi M (2023) Potential of biofuels production from wheat straw biomass, current achievements and perspectives: a review. Biofuels 14:79–92. https://doi.org/10.1080/17597269.2022.2118779
5. KN Y, TM MU, SK et al (2023) Lignocellulosic biorefinery technologies: a perception into recent advances in biomass fractionation, biorefineries, economic hurdles and market outlook. Fermentation 9:238. https://doi.org/10.3390/fermentation9030238
6. Bader AN, Sanchez Rizza L, Consolo VF, Curatti L (2023) Bioprospecting for fungal enzymes for applications in microalgal biomass biorefineries. Appl Microbiol Biotechnol 107:591–607. https://doi.org/10.1007/s00253-022-12328-9
7. Gong C, Meng X, Thygesen LG et al (2023) The significance of biomass densification in biological-based biorefineries: a critical review. Renew Sustain Energy Rev 183:113520. https://doi.org/10.1016/j.rser.2023.113520
8. Fertahi S, Elalami D, Tayibi S et al (2023) The current status and challenges of biomass biorefineries in Africa: a critical review and future perspectives for bioeconomy development. Sci Total Environ 870:162001. https://doi.org/10.1016/j.scitotenv.2023.162001
9. Chen Z, Chen L, Khoo KS et al (2023) Exploitation of lignocellulosic-based biomass biorefinery: a critical review of renewable bioresource, sustainability and economic views. Biotechnol Adv 69:108265. https://doi.org/10.1016/j.biotechadv.2023.108265
10. Poveda-Giraldo JA, Cardona Alzate CA (2024) Valorization of lignocellulosic platform products based on biorefinery schemes through sequential pretreatments. Biomass Convers Biorefin. https://doi.org/10.1007/s13399-023-05140-6
11. Narayanan M (2024) Promising biorefinery products from marine macro and microalgal biomass: a review. Renew Sustain Energy Rev 190:114081. https://doi.org/10.1016/j.rser.2023.114081
12. Barboza-Rodríguez R, Rodríguez-Jasso RM, Rosero-Chasoy G et al (2024) Photobioreactor configurations in cultivating microalgae biomass for biorefinery. Bioresour Technol 394:130208. https://doi.org/10.1016/j.biortech.2023.130208
13. Wei H, Wang Q, Zhang R et al (2023) Efficient biodiesel production from waste cooking oil by fast co-immobilization of lipases from *Aspergillus oryzae* and *Rhizomucor miehei* in magnetic chitosan microcapsules. Process Biochem 125:171–180. https://doi.org/10.1016/j.procbio.2022.12.025

14. Rajesh Banu J, Preethi KS et al (2021) Lignocellulosic biomass based biorefinery: a successful platform towards circular bioeconomy. Fuel 302:121086. https://doi.org/10.1016/j.fuel.2021.121086
15. Ubando AT, Felix CB, Chen W-H (2020) Biorefineries in circular bioeconomy: a comprehensive review. Bioresour Technol 299:122585. https://doi.org/10.1016/j.biortech.2019.122585
16. Gupta VK, Pandey A, Koffas M et al (2022) Biobased biorefineries: Sustainable bioprocesses and bioproducts from biomass/bioresources special issue. Renew Sustain Energy Rev 167:112683. https://doi.org/10.1016/j.rser.2022.112683
17. Culaba AB, Mayol AP, San Juan JLG et al (2022) Smart sustainable biorefineries for lignocellulosic biomass. Bioresour Technol 344:126215. https://doi.org/10.1016/j.biortech.2021.126215
18. Kumar B, Bhardwaj N, Agrawal K et al (2020) Current perspective on pretreatment technologies using lignocellulosic biomass: an emerging biorefinery concept. Fuel Process Technol 199:106244. https://doi.org/10.1016/j.fuproc.2019.106244
19. Yoo HY, Kim SW (2021) The next-generation biomass for biorefining. Bioresources 16:2188–2191. https://doi.org/10.15376/biores.16.2.2188-2191
20. Chen X, Song S, Li H et al (2021) Expanding the boundary of biorefinery: Organonitrogen chemicals from biomass. Acc Chem Res 54:1711–1722. https://doi.org/10.1021/acs.accounts.0c00842
21. Rodionova MV, Bozieva AM, Zharmukhamedov SK et al (2022) A comprehensive review on lignocellulosic biomass biorefinery for sustainable biofuel production. Int J Hydrogen Energy 47:1481–1498. https://doi.org/10.1016/j.ijhydene.2021.10.122
22. Kumar B, Verma P (2021) Biomass-based biorefineries: an important architype towards a circular economy. Fuel 288:119622. https://doi.org/10.1016/j.fuel.2020.119622
23. Liao H, Feng B, Ying W et al (2024) Novel approach for corn straw biorefineries: production of xylooligosaccharides, lignin and ethanol by nicotinic acid hydrolysis and pentanol pretreatment. Bioresour Technol 395:130352. https://doi.org/10.1016/j.biortech.2024.130352
24. Xiong M, Huang J, He X et al (2023) Evaluation of bio-oil/biodiesel production from co-pyrolysis of corn straw and natural hair: a new insight towards energy recovery and waste biorefinery. Fuel 331:125710. https://doi.org/10.1016/j.fuel.2022.125710
25. Yu X, Li X, Jin T et al (2021) Integrated biorefinery process for production of sophorolipids from corn straw. Biochem Eng J 167:107908. https://doi.org/10.1016/j.bej.2020.107908
26. Liu Z, Khurshid K, Saldaña MDA (2023) Biorefinery of barley straw using pressurized fluids: biocompounds and biopolymers production. J Supercrit Fluids 203:106078. https://doi.org/10.1016/j.supflu.2023.106078
27. Puke M, Godina D, Brazdausks P (2024) Catalyzed hydrothermal pretreatment of oat husks for integrated production of furfural and lignocellulosic residue. Polymers (Basel) 16:707. https://doi.org/10.3390/polym16050707
28. Singh S, Kaur G, Singh DP et al (2024) Exploring rice straw's potential from a sustainable biorefinery standpoint: towards valorization and diverse product production. Process Saf Environ Prot 184:314–331. https://doi.org/10.1016/j.psep.2024.01.105
29. Saini S, Kuhad RC, Sharma KK (2023) Valorization of rice straw biomass for co-production of bioethanol, biopesticide and biofertilizer following an eco-friendly biorefinery process. Process Saf Environ Prot 173:823–836. https://doi.org/10.1016/j.psep.2023.03.044
30. Yadav P, Anu KT, S et al (2024) Sugarcane bagasse: an important lignocellulosic substrate for production of enzymes and biofuels. Biomass Convers Biorefin 14:6111–6142. https://doi.org/10.1007/s13399-022-02791-9
31. Eixenberger D, Carballo-Arce A-F, Vega-Baudrit J-R et al (2024) Tropical agroindustrial biowaste revalorization through integrative biorefineries—review part II: pineapple, sugarcane and banana by-products in Costa Rica. Biomass Convers Biorefin 14:4391–4418. https://doi.org/10.1007/s13399-022-02721-9
32. Liu Z, Zhang M, Hou Q et al (2024) Sulfomethylation reactivity enhanced the Fenton oxidation pretreatment of bamboo residues for enzymatic digestibility and ethanol production. Front Bioeng Biotechnol. https://doi.org/10.3389/fbioe.2024.1344964

33. Deshmukh M, Pathan A (2023) Transformations of bamboo into bioethanol through biorefinery. Environ Sci Pollut Res 31:3343–3360. https://doi.org/10.1007/s11356-023-31510-8
34. Long L, Chen J, Cao Y et al (2024) Valorization of straw biomass into lignin bio-ink, bacterial cellulose and activated nanocarbon through the trade-off alkali-catalyzed glycerol organosolv biorefinery. Chem Eng J 484:149549. https://doi.org/10.1016/j.cej.2024.149549
35. Pallín MÁ, González-Rodríguez S, Eibes G et al (2024) Towards industrial application of fungal pretreatment in 2G biorefinery: scale-up of solid-state fermentation of wheat straw. Biomass Convers Biorefin 14:593–605. https://doi.org/10.1007/s13399-022-02319-1
36. Tschulkow M, Pizzol M, Compernolle T et al (2024) The environmental impacts of the lignin-first biorefineries: a consequential life cycle assessment approach. Resour Conserv Recycl 204:107466. https://doi.org/10.1016/j.resconrec.2024.107466
37. Wan Q, Cheng L, Li L et al (2024) The application of process simulations in lignocellulosic biorefinery: a review. Biofuels Bioprod Biorefin 18:617–638. https://doi.org/10.1002/bbb.2589
38. Kumar V, Verma P (2024) Pulp-paper industry sludge waste biorefinery for sustainable energy and value-added products development: a systematic valorization towards waste management. J Environ Manage 352:120052. https://doi.org/10.1016/j.jenvman.2024.120052
39. Henry GBL, Isenborghs A, Walhain E et al (2024) Enhancing the fermentability of brewer's spent grains to short- and medium-chain carboxylates through dilute alkaline pretreatment. Biomass Convers Biorefin. https://doi.org/10.1007/s13399-024-05508-2
40. Roy VC, Park J-S, Haque AR et al (2024) Bio-refinery of brewery spent grain utilizing natural deep eutectic solvent-induced subcritical water. J Supercrit Fluids 204:106108. https://doi.org/10.1016/j.supflu.2023.106108
41. Outeiriño D, Costa-Trigo I, Ochogavias A et al (2024) Biorefinery of brewery spent grain to obtain bioproducts with high value-added in the market. N Biotechnol 79:111–119. https://doi.org/10.1016/j.nbt.2023.12.010
42. Shafiei-Alavijeh R, Aghbashlo M, Tabatabaei M et al (2024) A critical review of sustainable biorefineries utilizing high-solid processing for industrial crop lignocellulosic wastes valorization. Ind Crops Prod 211:118236. https://doi.org/10.1016/j.indcrop.2024.118236
43. Hoyos-Sebá JJ, Arias NP, Salcedo-Mendoza J, Aristizábal-Marulanda V (2024) Animal manure in the context of renewable energy and value-added products: a review. Chem Eng Process Process Intensification 196:109660. https://doi.org/10.1016/j.cep.2023.109660
44. Jana A, Pati S, Chowdhury R (2024) Comparative technoeconomic analysis of algal biorefineries with and without methanol recycle in Aspen Plus. Sādhanā 49:89. https://doi.org/10.1007/s12046-023-02372-0
45. Dias B, Lopes M, Fernandes H et al (2024) Biomass and microbial lipids production by *Yarrowia lipolytica* W29 from eucalyptus bark hydrolysate. Renew Energy 224:120173. https://doi.org/10.1016/j.renene.2024.120173
46. Şen AU, Simões R, Yücedağ C et al (2024) Bark-based biorefineries: anatomical and chemical characterization of the bark of endemic *Quercus vulcanica* of Turkey. Wood Sci Technol 58:333–355. https://doi.org/10.1007/s00226-023-01518-x
47. Lv Y, Zhang Y, Xu Y (2024) Understanding and technological approach of acid hydrolysis processing for lignocellulose biorefinery: Panorama and perspectives. Biomass Bioenergy 183:107133. https://doi.org/10.1016/j.biombioe.2024.107133
48. de Sousa IG, Mota GF, Cavalcante ALG et al (2023) Renewable processes of synthesis of biolubricants catalyzed by lipases. J Environ Chem Eng 11:109006. https://doi.org/10.1016/j.jece.2022.109006
49. Klein BC, Scheidemantle B, Hanes RJ et al (2024) Economics and global warming potential of a commercial-scale delignifying biorefinery based on co-solvent enhanced lignocellulosic fractionation to produce alcohols, sustainable aviation fuels, and co-products from biomass. Energy Environ Sci 17:1202–1215. https://doi.org/10.1039/D3EE02532B
50. Yousuf A, Pirozzi D, Sannino F (2020) Fundamentals of lignocellulosic biomass. In: Lignocellulosic biomass to liquid biofuels. Elsevier, pp 1–15
51. Thanigaivel S, Vickram S, Dey N et al (2022) The urge of algal biomass-based fuels for environmental sustainability against a steady tide of biofuel conflict analysis: is third-generation algal biorefinery a boon? Fuel 317:123494. https://doi.org/10.1016/j.fuel.2022.123494

52. Yadav VG, Yadav GD, Patankar SC (2020) The production of fuels and chemicals in the new world: critical analysis of the choice between crude oil and biomass vis-à-vis sustainability and the environment. Clean Technol Environ Policy 22:1757–1774. https://doi.org/10.1007/s10098-020-01945-5
53. Asghar A, Sairash S, Hussain N et al (2022) Current challenges of biomass refinery and prospects of emerging technologies for sustainable bioproducts and bioeconomy. Biofuels Bioprod Biorefin 16:1478–1494. https://doi.org/10.1002/bbb.2403
54. Mailaram S, Kumar P, Kunamalla A et al (2021) Biomass, biorefinery, and biofuels. In: Sustainable fuel technologies handbook. Elsevier, pp 51–87
55. Vu HP, Nguyen LN, Vu MT et al (2020) A comprehensive review on the framework to valorise lignocellulosic biomass as biorefinery feedstocks. Sci Total Environ 743:140630. https://doi.org/10.1016/j.scitotenv.2020.140630
56. Goswami L, Kayalvizhi R, Dikshit PK et al (2022) A critical review on prospects of bio-refinery products from second and third generation biomasses. Chem Eng J 448:137677. https://doi.org/10.1016/j.cej.2022.137677
57. Manhongo TT, Chimphango AFA, Thornley P, Röder M (2022) Current status and opportunities for fruit processing waste biorefineries. Renew Sustain Energy Rev 155:111823. https://doi.org/10.1016/j.rser.2021.111823
58. Saratale RG, Cho S-K, Bharagava RN et al (2022) A critical review on biomass-based sustainable biorefineries using nanobiocatalysts: opportunities, challenges, and future perspectives. Bioresour Technol 363:127926. https://doi.org/10.1016/j.biortech.2022.127926
59. Usmani Z, Sharma M, Awasthi AK et al (2021) Lignocellulosic biorefineries: the current state of challenges and strategies for efficient commercialization. Renew Sustain Energy Rev 148:111258. https://doi.org/10.1016/j.rser.2021.111258
60. Malhotra M, Aboudi K, Pisharody L et al (2022) Biorefinery of anaerobic digestate in a circular bioeconomy: opportunities, challenges and perspectives. Renew Sustain Energy Rev 166:112642. https://doi.org/10.1016/j.rser.2022.112642
61. Rathour RK, Devi M, Dahiya P et al (2023) Recent trends, opportunities and challenges in sustainable management of rice straw waste biomass for green biorefinery. Energies (Basel) 16:1429. https://doi.org/10.3390/en16031429
62. Shapiro AJ, O'Dea RM, Li SC et al (2023) Engineering innovations, challenges, and opportunities for lignocellulosic biorefineries: leveraging biobased polymer production. Annu Rev Chem Biomol Eng 14:109–140. https://doi.org/10.1146/annurev-chembioeng-101121-084152
63. Lopes M, Miranda SM, Costa AR et al (2022) *Yarrowia lipolytica* as a biorefinery platform for effluents and solid wastes valorization—challenges and opportunities. Crit Rev Biotechnol 42:163–183. https://doi.org/10.1080/07388551.2021.1931016
64. Singh N, Singhania RR, Nigam PS et al (2022) Global status of lignocellulosic biorefinery: challenges and perspectives. Bioresour Technol 344:126415. https://doi.org/10.1016/j.biortech.2021.126415
65. Shahedi M, Habibi Z, Yousefi M et al (2021) Improvement of biodiesel production from palm oil by co-immobilization of *Thermomyces lanuginosa* lipase and *Candida antarctica* lipase B: optimization using response surface methodology. Int J Biol Macromol 170:490–502. https://doi.org/10.1016/j.ijbiomac.2020.12.181

Chapter 2
Biorefining Raw Materials and Feedstock: Challenges and Future Perspectives

Fozy Binhweel and Mardiana Idayu Ahmad

Abstract Raising global consciousness about the environmental and economic problems resulted from overconsumption of non-renewable resources led to utilization of renewable biomass toward useful forms of biobased products. Such sustainable practice was driven by the extraordinary advancement in the biotechnologies. Wide spectrum of crops, agricultural residue, forest products, algae, and wastes were used as biorefining raw materials for biobased products such as biofuel, biochemicals, compost, and supplementary nutrition. In general, biomass feedstocks were classified into four generations, namely edible, inedible, algae, and genetically modified crops. However, the extensive use of first generation of edible biomass in biorefinery systems led to food security issues, reduced arable soils, and increased consumption of water and fertilizer resources, making the overall footprint of using first generation feedstock less environmentally sustainable. In contrast, the second generation of inedible waste and the third generation of algal biomass are promising alternatives since they offer affordable feedstocks that neither compete with the food resources nor have a negative impact on the environment. In these frame points, this chapter explores the four generations of biorefining feedstocks, discusses the biotechnologies used to utilize biorefining feedstocks into added-value bioproducts and commodities, and delves into techno-economic analysis and future prospects in the pursuit of encouraging the sustainable practices.

Keywords Feedstock · Biomass · Biofuel · Bioproduct · Biorefinery · Biotechnology

---

F. Binhweel · M. I. Ahmad (✉)
Environmental Technology Division, School of Industrial Technology, Universiti Sains Malaysia, Penang, Malaysia
e-mail: mardianaidayu@usm.my

M. I. Ahmad
Renewable Biomass Transformation Cluster, School of Industrial Technology, Universiti Sains Malaysia, Penang, Malaysia

F. Binhweel
Faculty of Environmental Science and Marine Biology, Hadhramout University, Hadhramout, Yemen

M. Aslam et al. (eds.), *Production and Biorefining of Biocrude Oil: Current Status and Future Developments*, Advances in Sustainability Science and Technology,
https://doi.org/10.1007/978-981-96-5198-6_2

1 Introduction

The fast economic development induced by the rapid growth of human population all over the world led to a high demand for food resources and energy supplies. The recent predictionary investigations reported that the demand for food will rise to 70% by 2050 [1], and the consumption of energy will increase 71% by 2030 compared to the consumption patterns in the beginning of the current millennium [2]. In addition, nutrition supplements, healthcare materials, and various resources and products were highly demanded in accordance with the growth of civilized communities [3]. Significant amounts of these demands were, and still are, sourced from crude oil and fossil-based products which resulted in severe consequences related to depletion of natural resources, waste accumulation, environmental impact, and increasing prices of products and services. Such unsustainable practices would bring forth negative social, economic, and environmental impacts. Hence, reducing fossil-based resilience and increasing the dependence on renewable and sustainable alternatives would be a common concern for the governments and civilized communities toward welfare of the current and coming generations [4].

Biomass, the organic matter derived from plant or animal origins, can be processed through biorefineries systems to offer alternatives for a variety of marketable products [5]. It can be sourced from edible and inedible sources. However, sources like agricultural waste, industrial residue, and municipal spent biowaste are preferable since they do not compete with human food resources. According to the surveyed articles by Liu et al. [6], agricultural waste and industrial residue were the most employed biomass in biorefineries with percentages of 32.61% and 29.35%, respectively. Sorting the spent biowaste as a feedstock to produce added-value products would reduce the pressure on landfills and disposals, amplify the economic value, and significantly mitigate the environmental impact related to the potentiality of global warming, photochemical ozone formation, nutrient enrichment, and potentiality of acidification, with percentages of 30%, 29%, 28%, and 18%, respectively [7]. Utilizing raw and spent waste biomass that can be converted into added-value bioproducts with the aid of biotechnologies and biorefining systems would promote green economy and sustainability practices.

Biorefining was defined as the sustainable processing of biomass into a spectrum of marketable bioproducts and bioenergy. Biobased products include food for human beings, feed for animals, chemicals, and usable materials, whereas bioenergy sources refer to biofuel, power, and heat [8]. Biorefinery systems can fully utilize biobased feedstock and increase the values related to the economic, social, and environmental benefits. Hence, they are promising alternatives for classic refineries and products that completely depend on fossil crude oil as a raw material. However, commercialization of biorefineries products and services still cannot be compared with fossil-based products due to the proper technologies and feedstock availability [6]. Therefore, recent research investigated affordable feedstocks like agriculture waste, industrial residue, and municipal spent biowaste which are available everywhere with advanced technologies in order to substitute the fossil-based products with sustainable eco-friendly alternative bioproducts. Furthermore, in depth research study about the life cycle assessment of biorefinery and their bioproducts concluded that applying biorefinery systems with applicable techniques can curb the environmental pollutions, minimize the expenditures, and maximize the economic benefits [9].

Feedstock plays a crucial role in the feasibility and sustainability of biorefineries and biorefining systems. Availability of feedstock in the vicinity of biorefinery would secure sustainable production of biobased products. In addition, affordable feedstock has the advantage of producing commercially competitive bioproducts. Biorefining feedstocks were classified into four generations. The first generation is derived from edible food crops. In contrast, inedible crops and plants, including biowaste, can offer the second generation of biorefining feedstocks. Multiple algal biomass represents the third generation of feedstocks that can be processed in biorefineries to produce added-value bioproducts. Finally, the genetically modified crops were considered the fourth generation of biorefining feedstocks. The key factor for choosing the feedstock is the affordable raw material that can be utilized in biorefining systems without threatening human food security nor causing price increases. Beside affordability, consideration must be given to the environmental impact since the first feedstock generation may have a footprint worse than fossil-based product. Overall, biorefining feedstocks should overcome the economic, social, and environmental consequences [10].

Considering the economic, social, and environmental challenges related to the biorefining feedstocks, the current chapter aims to explore the four generations of biorefining feedstocks and elaborate about the pros and cons of each generation. In addition, the current advancement in biotechnologies and conversion techniques to utilize the biomass raw materials into added-value bioproducts were discussed. Moreover, the economic feasibility of biorefining feedstocks and their biobased products was analyzed. Finally, common challenges and prospects for the future scientific research aimed fostering the sustainable practices were addressed.

2 Generations of Biorefining Feedstocks

Biorefining feedstock is the renewable and sustainable biomass material used in biorefinery system. Despite the huge amounts of biomass generated in a daily basis, the maximum production of feedstock is limited by capacity of biosphere. Feedstock plays a key role in the biorefinery facility since it constitutes the greater proportion of the costs which ranges from 40 to 60% [11]. In addition, biorefinery systems should be self-dependent on biorefining feedstock supply during operational lifespan to attain sustainability. Therefore, the unceasing operation of biorefinery system depends basically on the sustainable supply of biorefining feedstock. Determining the proper feedstock for a particular biorefinery system is usually affected by multiple factors including availability of feedstock, affordability, vicinity, climate conditions, socioeconomic issues, local polices, and governmental laws. Since biorefining systems use biomass for nonedible purposes, the controversial debate was stemmed

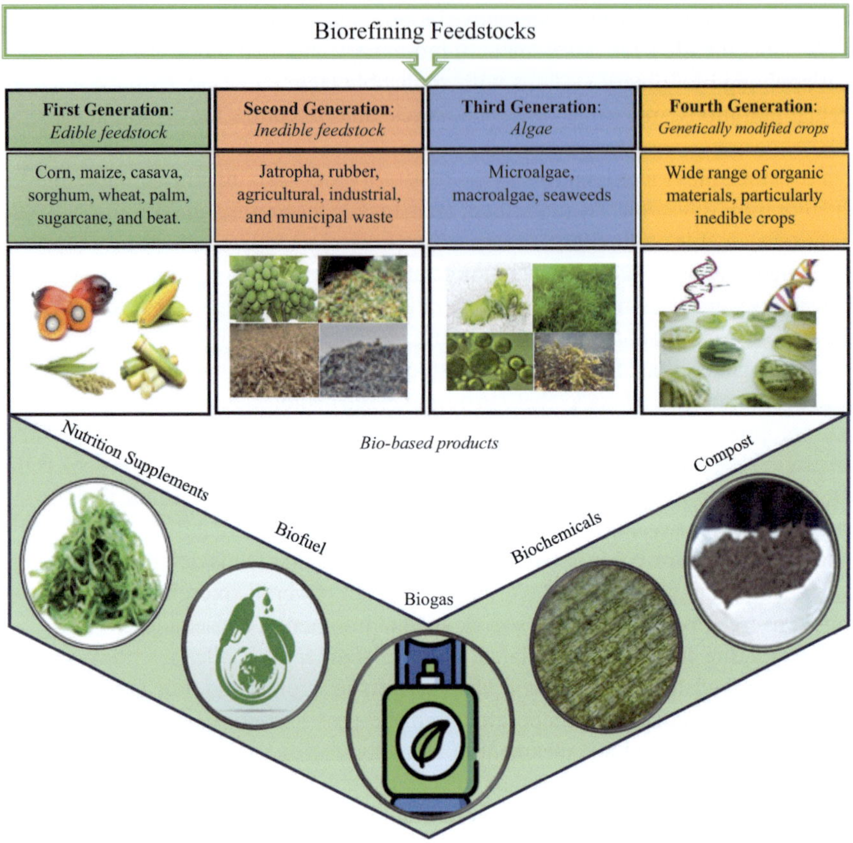

Fig. 1 Four generations of biorefining feedstocks

regarding the ethical considerations of utilizing edible biomass as a refining feedstock. Hence, biorefining feedstocks were categorized into four generations, namely edible, inedible, algae, and genetically modified crops as depicted in Fig. 1.

2.1 *First Generation; Edible Feedstock*

The first generation of biorefining feedstock refers to edible organic materials like corn, wheat, maize, soybeans, sugarcane, and palm. These kinds of crops have high contents of starch, sugar, and lipids which make them the most potential feedstock for biofuel production. Although lipid-based biorefining feedstocks like corn, palm oil, soybeans, and rapeseed were utilized extensively for biodiesel production in early ages of biodiesel industry, it is still the primary source of commercial production of biodiesel fuel. Based on the recent statistics by the International Energy Agency

(IEA), the increment expectation of vegetative oil used for biofuel production is 46% reaching to 54 million tonnes during the period 2022–2027 which shares to cover 23% of biofuel demand [12]. Similarly, the starch and sugar-based biorefining feedstocks like maize, sugar beet, cassava, and wheat were, and still are, widely used for bioethanol production. The demand for the first generation of biorefining feedstock expands with increasing the demand for the biofuel products which raises the concern about the impact of utilizing first generation of feedstock in biofuel production instead of food purposes.

Utilization the edible first generation of biorefining feedstock comes with both benefits and drawbacks. First of all, various agricultural crops of first generation can be used as biorefining feedstocks. It ranges from lipid-based crops like palm oil, sugar-based crops like sugarcane, and starch-based crops like corn. This variety of first generation biorefining feedstocks gives the benefit of alternative options to produce wide range of biobased products. In addition, the first generation of biorefining feedstock is advantaged with the high availability in most world regions under different climate conditions. The areas with temperate climate are the suitable place for wheat cultivation. Cassava and sorghum prefer tropical regions. Sugarcane prefers humid subtropical areas of India and Brazil. In contrast, sugar beet thrives in cold weather. For the barley, it was considered one of the most adaptable crops that can resist the severe drought environments. Beside consistent availability, first generation of biorefining feedstock is featured with easiness and facilitation of conversion processes into biobased products since no purification nor preliminary steps are needed. Furthermore, first generation feedstock is less vulnerable for the environmental contaminations. Despite these benefits, there are conspicuous drawbacks that limit utilizing the first generation as a biorefining feedstock. The extensive use of biorefining edible feedstock increases the competition on food resources. Such behavior of using the edible materials for bioindustry investment instead of feeding starving people raises ethical and moral concerns. Plus, it is substantial contributor for increasing food prices since there is a high demand for food resources as a biorefining feedstock. Finally, the natural resources such as arable, water, and fertilizers were consumed intensively to meet the high demand of biorefining first generation feedstock which questions the total footprint impact of first generation feedstock that may make it worse than fossil fuel resources [13].

2.2 *Second Generation; Inedible Feedstock*

Inedible feedstock is referred to as the second generation of biorefinery feedstock. The ethical concerns of the edible first generation feedstock made the researchers explore nonedible sources as a feedstock for commercial biobased products. This kind of feedstock includes nonfeed vegetations like jatropha, neem, and rubber, and wastes from agricultural, industrial, and municipal residues [14]. In fact, cultivation of nonedible vegetative sources still consumes the natural resources of arable and water. In addition, it competes the edible feedstock sources. Thus, wastes and residues are considered a relief and massive source for the second generation of biorefining feedstock. Agricultural and industrial residues are typical sources to supply such feedstock. Cellulose and lignocellulose biomass is one of the highly abundant and affordable materials sourced from agricultural waste and used for a variety of biobased products [15]. Considering the wide range of vegetative and animal-based wastes, the demand is only 65% of the supplied wastes which makes waste as an affordable source for the second generation of biorefining feedstock. The recent statistics expect increment of waste-based biofuel from 9% in 2021 up to 13% in 2027 due to the implementation of new policies regarding greenhouse gases and feedstock [12].

Despite the availability, there are some pros and cons related to the second generation of biorefining feedstock. It is an eco-friendly feedstock that reduces the accumulated waste generated on a daily basis from agricultural, industrial, and residential activities. According to the fact that the market is dynamic based on supply and demand, the second generation biorefinery feedstock is an affordable source since the supply is higher than demand. Beside that, this feedstock is less competitive on natural resources like arable and water which gives priority access to food resources on the natural resources. On the other side, there are few limitations that constrain the second generation of biorefining feedstock. The vegetative inedible plants usually have low lipid yield like jatropha and karanja which are used for biodiesel production. Moreover, contamination is dominant for the second generation of biorefining feedstock, particularly the waste parts. The contamination factors include chemical hazards and harmful microorganisms that lead to produce unfavorable components in the feedstock [16]. Furthermore, extensive treatments are required for the second generation of biorefining feedstock. For example, the waste-based lipid needs pretreatment to reduce the free fatty acids content in order to get convert it into biodiesel fuel. In addition, the complicated lignocellulosic material existed in agriculture biomass must be pretreated into a reduced form of sugar to be utilized in a biobased products, which is a complicated process compared with other biomass like starch [13, 14].

2.3 Third Generation; Algal Biomass

Agal biomass offers the third generation of biorefining feedstock. Algae are photosynthetic organisms ranging from single-celled microorganism like spirulina and Chlorella to multicellular organisms like seaweed. Algae are usually flourishing in both aquatic medium of saline and freshwater. However, few species can survive in the wet terrestrial environments. There are different classifications of algae. They were classified based on the size as microalgae and macroalgae. The other classification is based on the apparent color which divided the algae into green, red, and brown algae. What makes algae one of the potential sources of biorefining feedstock is the ability to survive and flourish even in harsh environments. With the consideration of growth factors like CO_2, light, ambient temperature, and moderate pH, they grow faster compared to the plant-origin species. The fixation of CO_2 and releasing huge amounts of O_2 during photosynthesis makes algae one of the biological reducers of elevated greenhouses gases in the planet. Waste rich in elements like carbon, nitrogen, phosphate, and sulfur can be utilized for algal cultivation which helps to minimize the waste impact on the environment. In addition, neither many resources nor efforts are required to grow and cultivate the algal biomass in nature. Recently, microalgae have been cultivated indoor with the help of photobioreactors to control the growth factors. In this case, more resources of light, energy, and efforts are needed to cultivate algal biomass. Saccharides, carbohydrates, proteins, and lipids constitute the main components of algal biomass. The absence or low levels of lignin and hemicellulose is a remarkable feature in the algae composition that magnifies the yields of algal hydrolysis and fermentation processes since the woody content needs more treatments for breaking down. Depending on the content, algal biomass can be used to produce nutrition supplements, biochemicals, compost, and biofuel [17, 18].

Even though algae were considered a potential substitute to the first and second generations of biorefining feedstocks, there have been pluses and minuses for such approach. The first advantage is the high productivity of algae compared with the other plant-based organisms. The productivity of algal biomass can be significantly boosted through enhancing the photosynthetic activity [19]. Besides, the indicator of algal growth rate is faster than the other multicellular organisms. Depending on the availability of growth factors, the volume of algal biomass can be doubled between 3 and 10 days [2]. In addition, the cultivation of algal biomass as the third generation of biorefining feedstock consumes fewer natural resources such arable land, water, and fertilizers. Instead of increasing environmental pollution from agricultural and industrial wastes, cultivating algae can alleviate the impact of such wastes through using them as nutritional fertilizers for the algae growth. Given the fact that algae can thrive in harsh environments including wastewaters, it saves efforts and enables the conservation of water for other domestic purposes. Moreover, the footprint of greenhouse effect is low since algae biomass reduces CO_2 from the atmosphere and releases O_2 during photosynthesis. The most favorable aspect is that utilizing the third generation of algae for biorefining feedstock does not compete with the food resources which enhances the ethical and moral values for the human being. In contrast, there are few limitations regarding the utilization of algae as a feedstock for biorefineries systems. The lipid content is quite low compared to first and second biorefining feedstocks. In spite of the fact that algae with 5% of lipid content is considered a potential source for biodiesel production, this quantity may not be feasible economically [20]. Beyond the limited lipid content, extracting lipids from algal biomass is another challenge related to the complexity and efficiency of the extraction method which may be incapable of extracting the entire amount of lipids. Hence, the large production of algae based on industrial scale requires big investments [14].

2.4 Fourth Generation; Genetically Modified Crops

Although the research on the fourth generation of biorefining feedstock is still in the infancy stage, it is considered a progressive and advanced leap in biorefineries systems. The primary objective of the fourth generation of biorefining feedstock is to overcome the limitations of the precedent feedstocks like competitiveness of food crops, competitiveness of natural resources, and low yield. In addition, it enhances the feedstock efficiency and reduces the impact on the environment. Hence, it offers a new horizon toward diverse biomass production and conversion. With the help of synthetic biology, a wide range of organic materials, particularly wastes and inedible crops, can be processed to produce biofuel. Applications like photobiological solar fuels and electro-fuels were introduced as part of this fourth generation. Moreover, a system of photovoltaic, inorganic catalysts, and genetically modified microbes were developed for the purpose of producing and storing biofuel. Such innovative approaches based on natural and human interventional solutions became necessity to meet the energy needs and sustain the existence on the planet as well [14].

Notable perks and pitfalls were associated with the advancements of the fourth generation of biorefining feedstock. The utilization of waste and inedible crops would minimize the pressure of food resources. With the help of genetic engineering, these kinds of inedible crops can thrive in harsh environments which preserves the arable and fertile lands only for food resources. Algae can be incorporated to increase the growth rates. In addition, the rates of CO_2 sequestration would be enhanced. The modifications in the algae and crops culture lead to an increase in the lipid content which results in advancements in biofuel production and energy efficiency. On the other hand, there are some challenges facing the expansion of the fourth generation of biorefining feedstock. Since the research efforts in the fourth generation are still in early stages, it seems that there are complexities related to genetic engineering, cultivating, and cost-effectiveness. Besides, there are concerns about the impact on the ecosystems and surrounding environments. Finally, the production on the industrial scale is still in need for big assets and investments [14, 21]. Table 1 provides examples, along with the pros, and cons of each feedstock generation.

3 Processing Biorefining Raw Materials

3.1 Pretreatment

Biorefining feedstocks usually undergo pretreatment procedures before processing into biobased products. The purpose of pretreatment is to reduce the size of biorefining feedstock, increase the surface area, enhance the conversion process, and reduce unfavorable by-products. Pretreatments were categories into physical, chemical, biological, and electrical pretreatments as illustrated in Fig. 2. Since the biorefining feedstocks are varied in sizes, chemical compositions, and crystalline

Table 1 Pros and cons of the four generations of biorefining feedstock

Biorefining feedstock	Examples	Advantages	Disadvantages
First generation: edible feedstock	Corn, maize, cassava, sorghum, wheat, palm, sugarcane, and beat	– Various crops – Consistent supply – High availability – Processing easiness – Less contaminated	– Food resources competitors – High costs Consuming natural resources of arable, water, and fertilizers
Second generation: inedible feedstock	Jatropha, rubber, karanja, agricultural waste, industrial residue, municipal waste	– Eco-friend feedstock – Affordable sources – Less competitive on natural resources	– Low yield – Contaminated feedstock – Extensive treatments are required
Third generation: algae	Microalgae, macroalgae, seaweeds	– High productivity – Fast growth rate – Uncompetitive with food resources – Less consumption of resources – Less cultivation efforts – Less greenhouse effect	– Large-scale production requires big investments – Low oil content and extraction difficulties
Fourth generation: genetically modified crops	Wide range of organic materials, particularly inedible crops	– Utilization of non-food crops – Fast growth rate – More captured CO_2 – High lipid content – More energy potential	– Early research stage – Impact of genetically modified crops on the ecosystems – Big investments are needed

structures, choosing the appropriate pretreatment method is crucial for favorable outcomes. In some cases, a combination of two pretreatment methods is employed for full use of biorefining feedstock.

3.1.1 Physical Pretreatment

Physical pretreatment usually deals with the physical structure of the biorefining feedstock. Physical pretreatment includes mechanical comminution and thermal pyrolysis [6, 13]. During comminution, the feedstock is chipped, milled, or grinded depending on the nature and structure of the feedstock itself. Mechanical comminution results in substantial reduction in the size of the feedstock, increase in the surface area, and enhancement in the permeability through the smaller particles of the feedstock. The hard woody parts of lignocellulose are usually pretreated mechanically through comminution to reduce the crystallinity structure. For instance, rubber wood was pretreated first mechanically in the form of sawdust with particle sizes

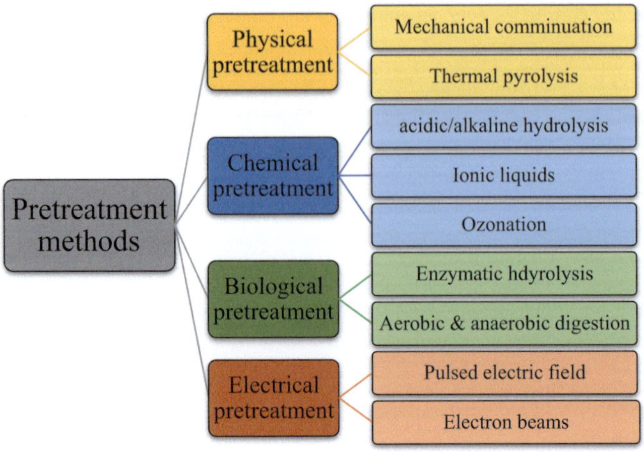

Fig. 2 Pretreatment methods for the biorefining feedstocks

around 686.75 μm to be mycelium-based composite [22]. This pretreatment allows for higher and better reactions with the component materials for further conversion through biorefinery system [6]. The other kind of physical pretreatment is thermal pyrolysis where the biorefining feedstock is heated for specific time under conditions of pressure and temperature below the normal pyrolysis. The parameters used for such thermal pretreatment are temperature 220 °C, pressure of 200–900 kPa, and time 0.25–2 h [13]. This condition is below the parameters of the normal pyrolysis where the temperature reaches 550–900 °C [23]. That is why some researchers preferred to call this process thermal pretreatment instead of pyrolysis. The thermal pretreatment sterilizes the feedstock and increases the degradability characteristics of the resultant substrates. Furthermore, it breaks down the chemical bonds of cellular substances and rigid cell wall which allows to change into aqueous and soluble forms [13].

3.1.2 Chemical Pretreatment

The use of chemical agents to alter the chemical composition of the biorefining feedstock is referred to as the chemical pretreatment. There are three methods have been used for chemical pretreatment which are acidic/alkaline hydrolysis, ionic liquids, and ozonation. The acidic/alkaline hydrolysis is the most commonly used pretreatment, in which concentrated or diluted acid or alkaline is employed on chipped and dried lignocellulose to increase the reaction rate and improve the degradability rate [24]. For better outcomes, organic solvent is added [6], and acetate material is removed during the acidic/alkaline hydrolysis process [13]. This pretreatment method is usually used for biogas production to boost the yield. Despite being widely used, acidic/alkaline hydrolysis pretreatment has some drawbacks related to the high costs of acids and alkali materials and the environmental impact represented in the difficulty of recovering these toxic and corrosive materials [25]. Another chemical pretreatment depends on the stable and volatile ionic liquids which effectively interact with lignocellulose to convert it into simple and tiny materials. Ionic liquids pretreatment method is preferable for bioethanol synthesis. Beside acidic/alkaline hydrolysis and ionic liquids, ozonation is another chemical pretreatment for the biorefining feedstock. It depends on the implementation of ozone to conduct partial oxidation. Ozonation is considered a sterilization method as well, where it reduces the microbes and prevents bad odors. That is why it is commonly used to pretreat the sewage sludge [13].

3.1.3 Biological Pretreatment

Biological pretreatment for the biorefining feedstock is considered economical, safe, and more eco-friendly compared to the physical and chemical procedures. It can be classified into enzymatic hydrolysis and aerobic and anaerobic digestion. The employment of commercial enzymes on the feedstock to break down the complex compounds into simple components is known as the enzymatic hydrolysis. Multiple commercial enzymes such as lipase, dextranase, and protease can be used for biological pretreatment. However, commercial enzymes may increase the costs of the pretreatment process. Thus, microbial enzymes derived from fungi secretions can be applied to reduce the economic costs [13]. The efficiency of enzymatic hydrolysis, based on both commercial and microbial enzymes, depends on employed enzyme and the characteristics of the biorefining feedstock such as surface area, polymerization, and crystallinity [6]. Regardless the economic and environmental benefits offered by enzymatic hydrolysis, the slow rate of enzymatic activity was reported as the major concern for this pretreatment method [26]. The aerobic and anaerobic digestion pretreatment is conducted with the aid of microorganisms. In presence, aerobic digestion, or absent, anaerobic digestion, of oxygen, microorganisms convert lignocellulose into accessible and simple components and produce nitrate, carbon dioxide, and water as by-products. The rate of feedstock biodegradation depends on the microorganism species, structure of the feedstock, and the digestion conditions where optimized pH and temperature are crucial for speeding up the degradation rate. The aerobic and anaerobic pretreatment method is usually employed to produce methane biogas [27, 28].

3.1.4 Electrical Pretreatment

Electrical pretreatment for the biorefining feedstock is seen as an unconventional method in this context. During the electrical pretreatment, a pulsed electric field is used to breakdown the biomass of the feedstock. The external electrical field usually comes with high intensity. It can be used with the solid feedstock or recalcitrant wastewaters. The electrical pretreatment has satisfied results since it speeds up the degradation of the bio-feedstock, increases the surface area, and enhances the permeability inside the biomass [29]. The pulsed electric field was applied on fruits and vegetables wastes for biogas production. The outcomes showed a significant enhancement in biogas yield compared to other pretreatment methods like hydrothermal [30]. In addition, there was enhancement in the porosity of switchgrass and wood chips once they were pretreated with the pulsed electric field [31]. However, lignin feedstock recorded persistence to the pulsed electric field. Hence, electron beams can be used as another approach of electrical pretreatment for biomass with higher lignocellulosic content. The electron beams have the ability to change the chemical bonds within the biomass through ionized energy [29]. This method was used for rich straw feedstock, and notable improvement in the digestibility was recorded through enzymatic hydrolysis [32].

3.2 Biofuel Processing

Biofuel is a pivotal bioproduct synthesized from processing of different biorefining feedstocks. Wide organic materials from the four generations of biorefining feedstock were utilized for biofuel production. However, the raw feedstocks from agriculture wastes, industrial residue, and third generation of algae are preferable since it reduces the competition on food resources. Depending on the applied technology, a wide spectrum of biofuels can be produced through thermochemical and biochemical conversions as illustrated in Fig. 3. The significance of biofuel is that it can partially substitute conventional fossil fuel which helps to mitigate the greenhouse gases effects, secure energy resources, and promote sustainability toward circular economy.

3.2.1 Thermochemical Conversion

The thermochemical conversion processes are the most frequently used techniques for biofuel production due to their straightforward procedures and direct outcomes. The principle in thermochemical processes is utilizing temperature to decompose the raw feedstocks and recover biobased fuel. The employed thermochemical techniques are direct combustion, gasification, liquefaction, pyrolysis, and transesterification. In the direct combustion, the feedstock is burnt directly in oxygenated environment to generate heat and power. Coal fired power plants recorded notable efficiency in the produced energy when organic feedstocks were co-combusted with coal fuel [33]. Gasification is an efficient technique to convert carbonaceous organic matter into useful form of syngas composed of hydrogen, carbon monoxide, carbon dioxide, methane, and nitrogen. The composed syngas can be used as a fuel for transportation automobiles or burnt directly for domestic purposes. The energy content of synthetic gas rich in hydrogen reached 12.48 MJ/Nm^3 from municipal solid waste at 757.65 °C of gasification temperature [34]. Liquefaction is the thermochemical process used to

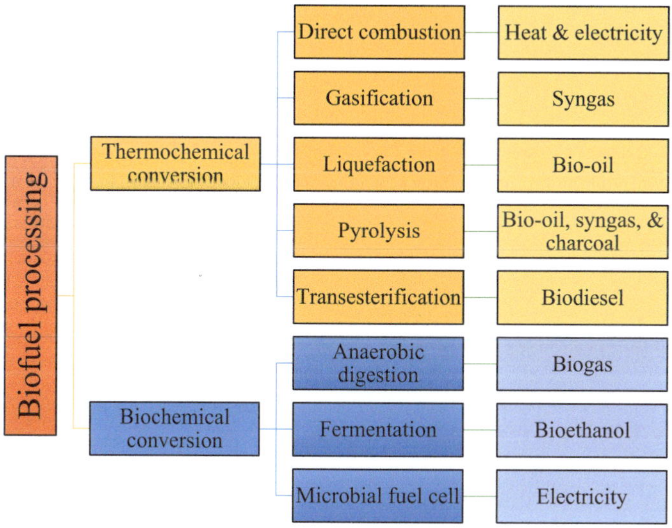

Fig. 3 Conversion processes of biorefining feedstocks into biofuels

convert wet organic materials, usually algae, into liquid and dense bio-oil. The liquefaction process is conducted in the presence of catalyst under moderate temperature range 300–350 °C and high pressure range 5–20 MPa. Several studies reported significant heating values ranged from 34.90 to 45.90 MJ/kg for bio-oils liquefied from algal biomasses [35]. Likewise, the pyrolysis process also decomposes the organic feedstock thermally into liquid fuel. However, there are few differences between pyrolysis and liquefaction in the mechanism and consequently the final product. While liquefaction uses moderate temperature and high pressure with or without oxygen to convert organic matter into liquid fuel, pyrolysis uses high temperatures (550–1300 °C) in oxygen deficiency medium. Based on that, pyrolysis yielded three kinds of bioproducts which are bio-oil, syngas, and charcoal [23]. Based on collective data on algal biomass, the energy contents reported from pyrolyzed bio-oils ranged from 29 to 41 MJ/kg [35]. Lastly, transesterification process is used to synthesize biodiesel fuel. It is a chemical reaction between triglycerides and alcohol in the presence of acid or base catalyst. Foremost, the oil content is extracted from feedstock with the aid of conventional or advanced extraction methods. Then, extracted lipids are subjected to the catalytic transesterification. During the transesterification reaction in Fig. 4, alkyl group of oil is replaced by alkyl group of alcohol resulting in alkyl ester, which is biodiesel fuel, and glycerol as a biproduct that can be used for detergent industry [36]. Reports confirmed that biorefining feedstock is accountable for 70–95% of biodiesel production costs [37].

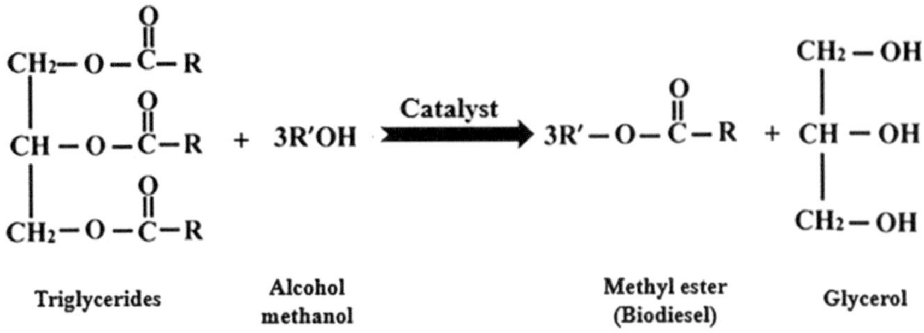

Fig. 4 Transesterification reaction (R: Fatty acids)

3.2.2 Biochemical Conversion

Biochemical conversion techniques utilize the microbial activity to decompose the organic feedstock into useful forms of biofuels [13]. The main techniques used for biochemical conversion of feedstock into biofuel are anaerobic digestion, fermentation, and microbial fuel cell. The anaerobic digestion is defined as the biodegradation of organic matter in the absence of oxygen. Biogas is the main outcome of the anaerobic digestion. Due to the less environmental impact and high energy content of produced biogas, the anaerobic digestion technique was preferred to be employed on organic wastes instead of direct combustion and incineration. Moreover, the elevated energy prices triggered people to search for affordable and applicable energy resources that can substitute the fossil fuel-based gas. Utilizing the organic waste for biogas production through anaerobic digestion can help to meet the part of domestic and transportation energy needs and reduce the impact of accumulated solid waste as well [6]. Biorefining feedstock rich in sugar, starch, and cellulose can be utilized to produce bioethanol fuel through the fermentation process. Such biochemical conversion is called alcoholic fermentation since the final product is alcohol compound. This process is conducted with the aid of the enzymatic activity that can be gain commercially or from microbial activity. Even though some studies used food waste as a raw feedstock for fermentation and reported successful conversion into bioethanol with yields of 13.78% [38] and 16.3 g/L [39]. The yield can be upgraded to 30.3 g/L when glucoamylase enzyme is added to the alcoholic fermentation process [39]. In addition, the pretreatment method for the raw feedstock plays an essential role for boosting bioethanol production [13]. Finally, the organic feedstock is biodegraded under anaerobic conditions using microbial fuel cell, which is an advanced technique to decompose organic waste and generate electricity at the same time. It is considered a promising eco-friendly biotechnology to treat wastewater and solid waste. Studies reported successful installment of microbial fuel cell with voltages of 16 mV from decomposing activated sludge [40]. A little adjustment in the fuel cell membrane led to conspicuous enhancement in the generated voltage till 20 mV [41]. Table 2 explores thermochemical and biochemical processes to produce various biofuels.

Table 2 Literature of biofuel processing through the thermochemical and biochemical conversions

Conversion kind	Conversion technique	Biorefining feedstock	Produced biofuel	Yield	References
Thermochemical conversion	Gasification	Municipal solid waste	Syngas	H_2: 41.36 CO: 28.46 CO_2: 26.05 CH_4: 4.12	[34]
	Gasification	Pine wood	Syngas	CO_2: 37–40% H_2: 27–30% CO: 17–21% CH_4: 10–12%	[42]
	Liquefaction	Human feces	Biocrude oil	34.44%	[43]
	Liquefaction	Domestic sewage	Bio-oil	44.40%	[44]
	Pyrolysis	Beech	Bio-oil	86.1%	[45]
	Pyrolysis	Sewage	Bio-oil	48.7%	[46]
	Transesterification	Fish waste	Biodiesel	87%	[47]
	Transesterification	Azolla macroalgae	Biodiesel	81.90%	[2]
Biochemical conversion	Anaerobic digestion	Municipal solid waste and biodiesel waste	Biogas	CH_4: 52–65% CO_2: 30–40% H_2S: 0.3–0.7%	[48]
	Anaerobic digestion	Food waste	Biogas	9926 ml (feedstock volume: 1.6 L)	[49]
	Fermentation	Food waste	Ethanol	13.78%	[38]
	Fermentation	Food waste	Bioethanol	16.3 g/L	[39]
	Microbial fuel cell	Activated sludge (algae cathode)	Electricity	16 mV	[40]
	Microbial fuel cell	Microalgae	Electricity	20 mV	[41]

3.3 Biochemicals Processing

The four generations of biorefining feedstocks are immense and sustainable sources for biobased products. Beside biofuel, biorefining feedstock can be processed through biorefineries systems to extract a wide spectrum of added-value biochemicals. This approach aligns with the revolution of green chemistry principles that are flourished recently offering pathways toward sustainable and circular economy. Harnessing the organic-based waste for the production of beneficial biochemicals urges industry to come up with innovate eco-friendly alternatives which reduces the accumulated waste on one hand and foster sustainability on the other hand. In this regard, industrial technologies exploited significantly various biorefining feedstocks to come up with added-value biochemicals such as bioactive compounds, platform chemicals, and biomaterials as listed in Table 3. Bioactive compounds are complicated chemicals existed in living organisms tissues. They are well-known with their potential positive effect on human health. In addition, they have applications in pharmaceutical and cosmetic products. The bioactive compounds are sources from natural origins like plants and food waste. Hence, the extraction process is pivotal to obtain the useful form of bioactive compounds. For this reason, conventional solvents are used to extract the targeted biomolecules. Dynamically, the solvent penetrates the feedstock particles and diffuses through the sample allowing the targeted biomolecules to be dissolved in the solvent. Then, the dissolved biomolecules are separated from the solvent. This process usually consumes much time. However, it can be speeded up with the aid of temperature, pressure, microwaves, and ultrasound effect. The other extraction method depends on supercritical fluid, which is more advanced and eco-friendly technology that can avoid the degradation of the bioactive compounds and preserve the biomolecules composition [50]. Bioactive compounds with the sourced biorefining feedstocks were elaborated in Table 3. Beside bioactive compounds, platform chemicals can be processed from the different generations of biorefining feedstocks. Platform chemicals are defined as the prime chemicals used as building blocks to develop a wide spectrum of other chemicals. Thus, it has the role as an intermediate mediocre that can be further processed into final added-value bioproduct. They play an essential role in the industry sector as a starting point for various complicated products. Utilizing organic materials and feedstock to extract platform chemicals promotes sustainable industry. Five groups of platform chemicals, namely hydroxymethyl furfural, levulinic acid, succinic acid, lactic acid, and sorbitol synthesized from cellulosic biorefining feedstocks, were explored in Table 3 [51]. Lastly, biomaterials refer to the chemicals that have applications in fragrances, fabrics, lubricants, and solvents. With the varieties of organic raw materials and feedstocks, biomaterials can replace petrochemical products. However, there are still some challenges related to the isolation, extraction, and synthesis technologies my limit the entire substitution [51]. Table 3 explores some of the added-value products related to biomaterials.

Adapted from [50–52]

3.4 *Compost Processing*

Utilization of biorefining feedstock, particularly the second generation of wastes, as a compost is a sustainable approach to manage waste and nourish the arable lands. Composing was seen as the most suitable method of treating organic matter from

Table 3 Various added-value biochemicals extracted from biorefining feedstocks

Biochemical group	Added-value product	Biorefining feedstock
Bioactive compounds	Starch	Turmeric waste
	Chitosan	Crustaceans, insects, and fungi
	Alginate	Brown seaweeds
	Pectin	Fruit rinds
	Carotenoids	Peels of peach, tomato, apricot, and pumpkin
	Amylose	Pineapple stem
	α-mangostins	Mangosteen peels
	Acid soluble collagen	*Centrolophus niger* skin
	Lycopene	Tomato waste
	Bromelain	Pineapple pulps
	Astaxanthin	Wheat waste
	Ellagic acid	Orange peel
	Oligogalacturonides	Citrus peel
Platform chemicals (hydroxymethyl furfural)	2,5-diformylfuran 2,5-bis (hydroxymethyl) furan 2,5-dimethylfuran 1,6-hexanediol 2,5-furandicarboxylic acid	
Platform chemicals (levulinic acid)	Methyltetrahydrofuran δ-Aminolevulinic acid 1,4-Pentanediol γ-Valerolactone α-Methylene-γ-valerolactone	
Platform chemicals (succinic acid)	1,4-Butanediol γ-Butyrolactone N-Methyl-2-pyrrolidone Tetrahydrofuran Succinic anhydride	
Platform chemicals (lactic acid)	Pyruvic acid Acrylic acid Propanoic acid Acetaldehyde 2,3-Pentanedione	
Platform chemicals (sorbitol)	Isosorbide Ascorbic acid Glycerol Propylene glycol Ethylene glycol	
Biomaterials	Polyethylene Polyethylene glycol Polylactic acid Polytrimethylene terephthalate Polyhydroxybutyrate	

municipal solid waste and agricultural residue since it lowers the greenhouse gases emissions, reduces the disposed waste, and integrates with the economy systems through the agricultural cultivation. In addition, the organic municipal solid waste is featured with the lower micro-pollutants. Such organic feedstock with low content of micro-pollutants results in biobased compost complied with the quality standards that can enhance the soil fertility. However, composting the agricultural waste that contains high quantities of lignocellulose is quite challenging since it is resistant to the fast degradation process. Organic waste with high content of lignin hardly biodegrades aerobically while those rich in hemicellulose shows slow rate of biodegradation. Beside factors like oxygen supply and humidification, the ratios of lignin and cellulosic materials in the organic waste are deemed as an indicator for the rate of aerobic biodegradability and composting process. Based on that, the composting process may range from 90 to 210 days. Hence, saving time, efforts, and resources to come up with a compost having highly valuable quality became an important demand for the researchers. Plus, minimizing greenhouse gases emissions from the composting process was another essential demand to mitigate the environmental impact. Accordingly, several strategies were employed such as amendments in the operational processes, optimization in oxygen flow rate, changing in feedstock pretreatment, supporting the composting process with some strains of microbes, and using supplementary co-composting materials [53]. In addition, integration between the cellulosic materials and green biowastes increases the activity and enhances the productivity of the composition process. Besides that, biomass ash was reported as a noteworthy additive for the composting process to boost the quality of the compost. Moreover, augmented improvement in the microbial degradation and richness with the addition of biochar which catalyzes the microbial activity were confirmed during composting process [54]. Table 4 explores some organic waste and feedstock for composting process.

3.5 Nutrition Supplements Processing

Despite the edible first generation of biorefining feedstock, second and third generations can be utilized for producing nutrition supplements and food additives. The exploitation of algal biomass and organic waste residues for food and nutraceuticals was paced up to recycle the residual feedstock, compensate the shortage in nutrients supply, face the agricultural detriment, and even to provide specific and highly valuable food for space travelers [63]. However, the commercial potential of nutraceuticals from second and third generations of biorefining feedstocks is still untapped. Despite the high nutritional values, there are some restrictions regarding utilization of residual and algal biomass for the supplementary diet of human beings due to the potential toxicity and food safety. Even though, valuable components like proteins, carbohydrates, lipids, vitamins, and pigments were extracted from those biorefining feedstocks as elaborated in Table 5. Proteins are essential components of biorefining feedstocks. They account for 50–70% of the third generation of microalgal

Table 4 Utilization of waste biorefining feedstock for composting

Feedstock	Process	Outcomes	References
Sewage sludge, olive mill waste, and green waste	Two cycles of composting were performed; the first was sewage sludge, and second was sewage sludge and green waste	The higher recorded germination index was 97.36%	[55]
Biomass ash and organic municipal solid waste	The composting process was improved by adding biomass ash	Compost humification was high, and mineral nutrients were observed	[56]
Coal fly ash and organic kitchen waste	Composting system for organic kitchen waste supported by the ash	A combination to decrease the bioavailability of heavy metals and increase the soil productivity	[57]
Fly ash, kitchen biomass, and jaggery	The ash was added in the composting system for boosting the process as an activator	Fly ash has a significant effect on the microbial and enzymatic activity, and consequently enhancing the compost quality	[58]
Plant residues	Plant residue was incubated for long time in the arable soil	The amounts of carbon in the dissolved organic carbon and microbial biomass increased significantly when adding fresh residues	[59]
Aquatic weed *Pistia* and cow dung	Weed *Pistia* was co-composted with cow dung in 1:1 and 3:1 ratio under ambient conditions	Increase in N, P, K, Ca, Zn, Fe, and Cu. Decrease in pH, total organic carbon, and carbon to nitrogen ratio. In addition to higher rate of microbial activity and mineralization	[60]
Horticultural waste	Estimation of the chemical, biological, and microbial characteristics quality of green waste compost and compost products	Higher germination rate was recorded	[61]
Date palm pruning and wastewater sludge	Co-composting palm wastes with sludge is feasible alternative treatment to manage these natural waste streams and produce compost with additional value	One way to manage palm trash is to co-compost it with sewage or agri-food slime. The use of date palm waste increased the amount of salt in the final composts	[62]

biomass. Thus, recovering proteins is helpful to support human and animal nutrition. In this regard, much carefulness is required to avoid the toxic proteins that might be recovered from those valuable feedstocks. Proteins were recovered using different conventional and advanced recovery techniques like centrifugation, filtration, solvent extraction, and supercritical fluid extraction. There are some drawbacks like loss of proteins and the toxicity of the solvents may limit using the conventional extraction methods which gives extra advantage for the supercritical fluid since it is safe and eco-friendly technology. Carbohydrates are stored in the tissues of living things. They are groups of complex compounds such as glucose, starch, cellulose, and polysaccharides. Beside biofuel, polysaccharides have applications in food, therapeutic, and cosmetic additives. The chemical hydrolysis is the chosen extraction method for carbohydrates from tissues of biorefining feedstocks. Lipid is another essential component in the biorefining feedstock. In the third generation of algal biomass, lipid productivity can be enhanced through modification of cultural medium like nitrogen starvation condition. Lipids can be separated from biorefining feedstock using solvent and solvent-free extraction methods. The remaining residue after extraction can be either used to feed animals or pyrolyzed to synthesize biofuel. Depending on the extracted lipids, they can be utilized for bioenergy, nutrition supplements, and health applications. Polyunsaturated fatty acids are part of the lipid composition. They brought interest since they are widely used in human diet to reduce cardiac disorders. They are originally sourced from marine fish. The marine livestock experienced depletion due to the high demand. Hence, advanced technology offered an alternative source for polyunsaturated fatty acids from microalgae culture. In addition, microalgae offered alternative sources of vitamins and trace elements that have direct application in human diet and health products. With particular extraction procedures, B12 and riboflavin vitamins were isolated successfully from third generation of microalgal feedstock. Lastly, natural pigments are obtained from biorefining feedstocks. Carotenoids are a group of essential pigments characterized by their capability to be dissolved in the fat. Astaxanthin is among carotenoids pigments that has benefits like anti-aging and sun proofing agent. Beside carotenoids, chlorophylls are vegetative pigments that are soluble in lipids. It plays an essential role in the photosynthetic activity in plants. Phycobiliproteins are major pigments in algal biomass that exhibit antioxidant, antiviral, and anticancer agents. Isolation of pigments is conducted with standard extraction methods in which the extraction solvent is crucial to determine the targeted pigment. In general, pigments from biorefining feedstock offer food coloring and additives, pharmaceutical agents, and cosmetic products [33].

4 Techno-economic Analysis

Techno-economic analysis is a useful technique for assessing the operations and investment of projects. This analysis is highly required to scale the project up to commercial production. It operates as guidelines to quantify capital and feedstocks, modify technologies, regulate maintenance, estimate costs and profits, and predict future challenges in order to sustain the potential project and develop profitability. There are two major directions in techno-economic analysis which are production technology and economic feasibility. For this purpose, integrated and multidisciplinary tools and indicators are used to come up with alternative pathways in the

Table 5 Supplementary bioproducts from second and third generations of biorefining feedstocks

Feedstock	Extraction/conversion method	Product	References
Fish by-products	Cleaning, drying, milling	Fish power rich in protein and elemental nutrients like iron	[64]
Food waste mixture	Substrates were prepared with assistance of isolated microorganisms	Protein-enriched livestock feed	[65]
Chicken feathers	Feathers were sterilized and fermented with microbial strain	Feed for aquaculture livestock	[66]
S. almeriensis microalgae	Soxhlet	8% of lipids	[67]
P. ellipsoidea microalgae	Microwave extraction	37.50% of lipids	[68]
C. infusionum microalgae	Chemical hydrolysis	89.6% of carbohydrates	[69]
Isochrysis aff. Galbana microalgae	Solvent extraction	5.6% of chlorophyll pigment	[70]
C. saccharophila microalgae	Ultrasonication	37.3% of b-Carotene	[71]
Haematococcus pluvialis microalgae	Cell disruption and solvent extraction	32.5 pg/cell of Astaxanthin	[72]
Piggery wastewater	Co-culture with yeast	Protein additive	[73]

efforts of transferring the production from the lab-scale into industrial viability. While the significant potential in the techno-economic analysis for utilizing biorefining feedstock is the low-cost bioresources, the real challenge is represented in the proper management for the raw materials and innovation for manufacturing added-value biobased products.

Significant studies analyzed the utilization of biorefining feedstock for bioproducts based on techno-economic perspective. Vaskan et al. [74] performed techno-economic assessment for the utilization of abundant residue of palm empty fruit bunches to produce a series of biobased products. The technical procedures included two pathways of pretreatment for the palm empty fruit bunches, dilute acid and hot water, followed by three processes to produce sole bioproduct for each. The first one separates pentose from the pretreated feedstock to come up with animal feed, the second process produces ethanol based on hydrolysis and fermentation of pretreated feedstock, and the last one generates heat and power as a result of the direct combustion of the lignin component within the palm empty fruit bunches. The study concluded that using dilute acid for the pretreatment process proved more efficiency compared to the hot water. However, there is a concern related to the environmental impact due to the huge consumption of chemicals. Another techno-economic analysis was conducted by Hernández et al. [75] for utilizing olive stone by-product for the production of biobased products. The study suggested two tracks of biobased production. The first track produced xylitol, furfural, ethanol, and poly(3-hydroxybutyrate) from raw feedstock of olive stone waste, and the second track burnt the residues from the first track in cogeneration system to generate heat and power. Based on simulation application, the first track was more profitable than the second one (53% vs. 6%) because of high costs required by the cogeneration system. The findings of Mussatte [76] who analyzed the utilization of brewer's spent grains for the production of xylitol, lactic acid, activated carbon, and phenolic acids showed 62.25% of profitable economic margins.

Beside techno-economic analysis, the social and environmental impacts of utilizing biorefining feedstock to produce biobased products are in concern. The projects with the intension to utilize biorefining feedstock, particularly first edible generation, may have interlinked impact on landowners, labor rights, consumers, and surrounding environment. There was a case study for a local company in Kenya aimed to utilize 16 thousand hectares of sugarcane farm to produce biofuels instead of edible crops. Despite the economic benefits gained from this project, it would violate the rights of thousands of farmers who have been working in this area for their families' livelihoods. Another case study was in the USA where there were above 200 plants producing ethanol from corn crops. This behavior would affect the consumers directly because of the competition between food versus energy that leads to increase food prices. Plus, cultivation of corn for biofuel industry may have the same environmental footprint as fossil fuels since it increases the emissions of greenhouse gases [77].

5 Challenges and Future Prospects

With the enormous increase in the global human population, the demand for commodities and goods reached unprecedented level in the current time. The ever-increasing consumers demand created huge pressure on limited resources, particularly those which are non-renewable materials. Hence, exploration of sustainable and alternative resources became imperative matter. Biorefining feedstock and raw materials emerged as an alternative solution for the production of biobased products. With the progress of industrial technologies, the biobased products, synthesized from biomass resources, can help in the avenue of meeting the high demand for commodities in an eco-friendly and responsible manner. However, there are ethical considerations regarding utilization of first generation of edible crops. The potential competitiveness of food security versus manufacturing resources is a pressing global issue. Hereby, prioritizing the second generation of nonedible wastes and third generation of algal biomass as biorefining feedstocks for biobased products is paramount necessity. These renewable feedstocks can be sourced immensely from agricultural residues, industrial residues, forestry by-products, municipal solid wastes, and algae. Utilization of such resources in the bioindustry ensures the alignment with the broader ethical, social, and environmental considerations.

Looking ahead, the future of industrialization of biorefining feedstock is closely interlinked with the progressive advancements in biotechnology. Researchers all over the world are actively investigating innovative approaches to improve the efficiencies and versatilities of biorefinery industry. The optimistic outcomes of the current research projects from genetically modified corps to robust synthesis and degradation processes hold high promise for more sustainable and viable biobased products. Moreover, they paved the ways toward full potential of biorefining feedstocks which makes the seamless and impactful transition into green and circular economy. Hence, the investments in biorefineries systems are urgent necessity to cope with the sustainable practices. Governments and investors are recommended to shift part of resources toward bioindustry development. Expanding investment capital in biobased products would be a strategic move to foster the economic growth and sustain the environment.

6 Conclusion

As the globe grapples with radical issues of resource depletion and environmental degradation, the adoption of biorefineries systems can offer a promising solution. Biorefining feedstocks play a key role as raw materials for biorefineries systems in the pursuit of producing biobased products. Biorefining feedstocks were classified into four generations, namely edible, inedible, algae, and genetically modified crops. Despite being readily available and easily utilized, the use of the first generation of edible biomass for biobased products, like bioenergy, raises ethical and environmental concerns related to the food security and increases the greenhouse gases emissions during cultivation. Hence, the second generation of inedible waste and the third generation of algal biomass are considered promising alternatives. With the progressive advancement in biotechnologies, biotransformation of biomass into added-value bioproducts such as biofuel, biochemicals, compost, and nutrition supplements became seamless. In this view, investments from governmental and private sectors are needed to maximize the economic benefits and minimize the environmental impacts of utilizing biorefining feedstocks toward sustainable biobased products.

Funding The authors would like to thank the International Center for Genetic Engineering and Biotechnology (ICGEB), grant number: 304.PTEKIND.6501382.I151 (R504-LR-GAL008-0006501382-I151) for funding this study.

References

1. Haris M et al (2023) Nanotechnology—a new frontier of nano-farming in agricultural and food production and its development. Sci Total Environ 857:159639. https://doi.org/10.1016/j.scitotenv.2022.159639

2. Ahmad MI, Senusi W, Binhweel F, Alsaadi S (2023) Optimization of base catalytic transesterification toward maximum biodiesel yield from *Azolla filiculoides* macroalgae feedstock. Ind Crops Prod 197:116590. https://doi.org/10.1016/j.indcrop.2023.116590
3. Rusu AV, Trif M, Rocha JM (2023) Microbial secondary metabolites via fermentation approaches for dietary supplementation formulations. Molecules 28(16):6020. https://doi.org/10.3390/molecules28166020
4. Shakir MA et al (2023) Sandwich composite panel from spent mushroom substrate fiber and empty fruit bunch fiber for potential green thermal insulation. Buildings 13(1):224. https://doi.org/10.3390/buildings13010224
5. Shakir MA et al (2023) Effect of fibrillation degree of empty fruit bunch fiber on sandwich composite incorporate with spent mushroom substrate. J Ind Text 53:152808372311536. https://doi.org/10.1177/15280837231153661
6. Liu Y et al (2021) Review of waste biorefinery development towards a circular economy: from the perspective of a life cycle assessment. Renew Sustain Energy Rev 139:110716. https://doi.org/10.1016/j.rser.2021.110716
7. Chi Y, Dong J, Tang Y, Huang Q, Ni M (2015) Life cycle assessment of municipal solid waste source-separated collection and integrated waste management systems in Hangzhou, China. J Mater Cycles Waste Manag 17(4):695–706. https://doi.org/10.1007/s10163-014-0300-8
8. Aristizábal-Marulanda V, Cardona Alzate CA (2019) Methods for designing and assessing biorefineries: review. Biofuels Bioprod Biorefin 13(3):789–808. https://doi.org/10.1002/bbb.1961
9. Ahlgren S et al (2015) Review of methodological choices in LCA of biorefinery systems—key issues and recommendations. Biofuels Bioprod Biorefining 9(5):606–619. https://doi.org/10.1002/bbb.1563
10. Serna-Loaiza S, Carmona-Garcia E, Cardona CA (2018) Potential raw materials for biorefineries to ensure food security: the Cocoyam case. Ind Crops Prod 126:92–102. https://doi.org/10.1016/j.indcrop.2018.10.005
11. Ghatak HR (2011) Biorefineries from the perspective of sustainability: feedstocks, products, and processes. Renew Sustain Energy Rev 15(8):4042–4052. https://doi.org/10.1016/j.rser.2011.07.034
12. IEA (2022) Is the biofuel industry approaching a feedstock crunch? International Energy Agency, 2022. https://www.iea.org/reports/is-the-biofuel-industry-approaching-a-feedstock-crunch. Accessed 14 Nov 14 2023
13. Singh A et al (2023) A comprehensive review of feedstocks as sustainable substrates for next-generation biofuels. BioEnergy Res 16(1):105–122. https://doi.org/10.1007/s12155-022-10440-2
14. Singh D, Sharma D, Soni SL, Sharma S, Kumar Sharma P, Jhalani A (2020) A review on feedstocks, production processes, and yield for different generations of biodiesel. Fuel 262:116553. https://doi.org/10.1016/j.fuel.2019.116553
15. Shakir MA, Yhaya MF, Ahmad MI (2017) Physical effects of pre-hydrolysis and alkali treatment of empty fruit bunch (EFB) fibers. Int J Chem Eng Appl 8(6):387–391. https://doi.org/10.18178/ijcea.2017.8.6.689
16. Ilias MKM, Balakrishnan V, Zuknik MH, Al-Gheethi A, Ghfar AA, Hossain MS (2021) Supercritical CO_2 separation of lipids from chicken by-product waste for biodiesel production: optimization, kinetics, and thermodynamics modeling. Biomass Convers Biorefinery. https://doi.org/10.1007/s13399-021-02092-7
17. Zhang J, Zhang X, Yang M, Singh S, Cheng G (2021) Transforming lignocellulosic biomass into biofuels enabled by ionic liquid pretreatment. Bioresour Technol 322:124522. https://doi.org/10.1016/j.biortech.2020.124522
18. Bolatkhan K, Kossalbayev BD, Zayadan BK, Tomo T, Veziroglu TN, Allakhverdiev SI (2019) Hydrogen production from phototrophic microorganisms: reality and perspectives. Int J Hydrogen Energy 44(12):5799–5811. https://doi.org/10.1016/j.ijhydene.2019.01.092
19. Vecchi V, Barera S, Bassi R, Dall'Osto L (2020) Potential and challenges of improving photosynthesis in algae. Plants 9(1):67. https://doi.org/10.3390/plants9010067

20. Binhweel F, Pyar H, Senusi W, Shaah MA, Hossain MS, Ahmad MI (2023) Utilization of marine *Ulva lactuca* seaweed and freshwater *Azolla filiculoides* macroalgae feedstocks toward biodiesel production: kinetics, thermodynamics, and optimization studies. Renew Energy 205:717–730. https://doi.org/10.1016/j.renene.2023.01.114
21. Yaakob MA, Mohamed RMSR, Al-Gheethi A, Aswathnarayana Gokare R, Ambati RR (2021) Influence of nitrogen and phosphorus on microalgal growth, biomass, lipid, and fatty acid production: an overview. Cells 10(2):393. https://doi.org/10.3390/cells10020393
22. Shakir MA, Ahmad MI, Yusup Y, Rafatullah M (2023) From waste to wealth: converting rubber wood sawdust into green mycelium-based composite. Biomass Convers Biorefinery. https://doi.org/10.1007/s13399-023-05113-9
23. Binhweel F, Hossain MS, Ahmad MI (2023) Recent trends, potentials, and challenges of biodiesel production from discarded animal fats: a comprehensive review. BioEnergy Res 16(2):778–800. https://doi.org/10.1007/s12155-022-10527-w
24. Vu HP, Nguyen LN, Vu MT, Johir MAH, McLaughlan R, Nghiem LD (2020) A comprehensive review on the framework to valorise lignocellulosic biomass as biorefinery feedstocks. Sci Total Environ 743:140630. https://doi.org/10.1016/j.scitotenv.2020.140630
25. Wyman CE, Dale BE, Elander RT, Holtzapple M, Ladisch MR, Lee YY (2005) Coordinated development of leading biomass pretreatment technologies. Bioresour Technol 96(18):1959–1966. https://doi.org/10.1016/j.biortech.2005.01.010
26. Lugani Y et al (2020) Recent advances in bioethanol production from lignocelluloses: a comprehensive review with a focus on enzyme engineering and designer biocatalysts. Biofuel Res J 7(4):1267–1295. https://doi.org/10.18331/BRJ2020.7.4.5
27. Isa MH et al (2020) Improved anaerobic digestion of palm oil mill effluent and biogas production by ultrasonication pretreatment. Sci Total Environ 722:137833. https://doi.org/10.1016/j.scitotenv.2020.137833
28. Abuabdou SMA, Ahmad W, Aun NC, Bashir MJK (2020) A review of anaerobic membrane bioreactors (AnMBR) for the treatment of highly contaminated landfill leachate and biogas production: effectiveness, limitations and future perspectives. J Clean Prod 255:120215. https://doi.org/10.1016/j.jclepro.2020.120215
29. Paudel SR, Banjara SP, Choi OK, Park KY, Kim YM, Lee JW (2017) Pretreatment of agricultural biomass for anaerobic digestion: current state and challenges. Bioresour Technol 245:1194–1205. https://doi.org/10.1016/j.biortech.2017.08.182
30. Safavi SM, Unnthorsson R (2017) Methane yield enhancement via electroporation of organic waste. Waste Manag 66:61–69. https://doi.org/10.1016/j.wasman.2017.02.032
31. Kumar P, Barrett DM, Delwiche MJ, Stroeve P (2011) Pulsed electric field pretreatment of switchgrass and wood chip species for biofuel production. Ind Eng Chem Res 50(19):10996–11001. https://doi.org/10.1021/ie200555u
32. Bak JS, Ko JK, Han YH, Lee BC, Choi I-G, Kim KH (2009) Improved enzymatic hydrolysis yield of rice straw using electron beam irradiation pretreatment. Bioresour Technol 100(3):1285–1290. https://doi.org/10.1016/j.biortech.2008.09.010
33. Chew KW et al (2017) Microalgae biorefinery: high value products perspectives. Bioresour Technol 229:53–62. https://doi.org/10.1016/j.biortech.2017.01.006
34. Chen G et al (2020) Hydrogen-rich syngas production from municipal solid waste gasification through the application of central composite design: an optimization study. Int J Hydrogen Energy 45(58):33260–33273. https://doi.org/10.1016/j.ijhydene.2020.09.118
35. Brennan L, Owende P (2010) Biofuels from microalgae—a review of technologies for production, processing, and extractions of biofuels and co-products. Renew Sustain Energy Rev 14(2):557–577. https://doi.org/10.1016/j.rser.2009.10.009
36. Binhweel F, Bahadi M, Pyar H, Alsaedi A, Hossain S, Ahmad MI (2021) A comparative review of some physicochemical properties of biodiesels synthesized from different generations of vegetative oils. J Phys Conf Ser 1900(1):012009. https://doi.org/10.1088/1742-6596/1900/1/012009
37. Binhweel F (2022) Biofuel in Yemen; an alternative energy resource toward sustainability. Prog Eng Appl Technol 3(2):945–955. https://doi.org/10.30880/peat.2022.03.02.094

38. Thapa B, Patidar SK, Khatiwada NR, KC AK, Ghimire A (2019) Production of ethanol from municipal solid waste of India and Nepal. In: Waste valorisation and recycling. Springer, Singapore, pp 47–58. https://doi.org/10.1007/978-981-13-2784-1_5
39. Prasoulas G, Gentikis A, Konti A, Kalantzi S, Kekos D, Mamma D (2020) Bioethanol production from food waste applying the multienzyme system produced on-site by fusarium oxysporum F3 and mixed microbial cultures. Fermentation 6(2):39. https://doi.org/10.3390/fermentation6020039
40. González del Campo A, Cañizares P, Rodrigo MA, Fernández FJ, Lobato J (2013) Microbial fuel cell with an algae-assisted cathode: a preliminary assessment. J Power Sources 242:638–645. https://doi.org/10.1016/j.jpowsour.2013.05.110
41. Uggetti E, Puigagut J (2016) Photosynthetic membrane-less microbial fuel cells to enhance microalgal biomass concentration. Bioresour Technol 218:1016–1020. https://doi.org/10.1016/j.biortech.2016.07.062
42. Condori O, García-Labiano F, de Diego LF, Izquierdo MT, Abad A, Adánez J (2021) Biomass chemical looping gasification for syngas production using ilmenite as oxygen carrier in a 1.5 kWth unit. Chem Eng J 405:126679. https://doi.org/10.1016/j.cej.2020.126679
43. Lu J et al (2017) Simultaneous production of biocrude oil and recovery of nutrients and metals from human feces via hydrothermal liquefaction. Energy Convers Manag 134:340–346. https://doi.org/10.1016/j.enconman.2016.12.052
44. Couto EA, Pinto F, Varela F, Reis A, Costa P, Calijuri ML (2018) Hydrothermal liquefaction of biomass produced from domestic sewage treatment in high-rate ponds. Renew Energy 118:644–653. https://doi.org/10.1016/j.renene.2017.11.041
45. Boscagli C, Tomasi Morgano M, Raffelt K, Leibold H, Grunwaldt J-D (2018) Influence of feedstock, catalyst, pyrolysis and hydrotreatment temperature on the composition of upgraded oils from intermediate pyrolysis. Biomass Bioenergy 116:236–248. https://doi.org/10.1016/j.biombioe.2018.06.022
46. Cao J-P et al (2013) Nitrogen transformations during fast pyrolysis of sewage sludge. Fuel 104:1–6. https://doi.org/10.1016/j.fuel.2010.08.015
47. Shalfoh E et al (2024) Fish waste oil extraction using supercritical CO_2 extraction for biodiesel production: Mathematical, and kinetic modeling. Renew Energy 220:119659. https://doi.org/10.1016/j.renene.2023.119659
48. Zahedi S, Sales D, García-Morales JL, Solera R (2018) Obtaining green energy from dry-thermophilic anaerobic co-digestion of municipal solid waste and biodiesel waste. Biosyst Eng 170:108–116. https://doi.org/10.1016/j.biosystemseng.2018.04.005
49. Deepanraj B, Sivasubramanian V, Jayaraj S (2017) Effect of substrate pretreatment on biogas production through anaerobic digestion of food waste. Int J Hydrogen Energy 42(42):26522–26528. https://doi.org/10.1016/j.ijhydene.2017.06.178
50. Ng HS, Kee PE, Yim HS, Chen P-T, Wei Y-H, Chi-Wei Lan J (2020) Recent advances on the sustainable approaches for conversion and reutilization of food wastes to valuable bioproducts. Bioresour Technol 302:122889. https://doi.org/10.1016/j.biortech.2020.122889
51. Reshmy R et al (2022) Updates on high value products from cellulosic biorefinery. Fuel 308. https://doi.org/10.1016/j.fuel.2021.122056
52. Binhweel F, Ahmad MI, Zaki SA (2022) Utilization of polymeric materials toward sustainable biodiesel industry: a recent review. Polymers (Basel) 14(19):3950. https://doi.org/10.3390/polym14193950
53. Reyes-Torres M, Oviedo-Ocaña ER, Dominguez I, Komilis D, Sánchez A (2018) A systematic review on the composting of green waste: feedstock quality and optimization strategies. Waste Manag 77:486–499. https://doi.org/10.1016/j.wasman.2018.04.037
54. Awasthi MK et al (2020) Refining biomass residues for sustainable energy and bio-products: an assessment of technology, its importance, and strategic applications in circular bio-economy. Renew Sustain Energy Rev 127:109876. https://doi.org/10.1016/j.rser.2020.109876
55. Asses N, Farhat A, Cherif S, Hamdi M, Bouallagui H (2018) Comparative study of sewage sludge co-composting with olive mill wastes or green residues: process monitoring and agriculture value of the resulting composts. Process Saf Environ Prot 114:25–35. https://doi.org/10.1016/j.psep.2017.12.006

56. Asquer C, Cappai G, De Gioannis G, Muntoni A, Piredda M, Spiga D (2017) Biomass ash reutilisation as an additive in the composting process of organic fraction of municipal solid waste. Waste Manag 69:127–135. https://doi.org/10.1016/j.wasman.2017.08.009
57. Manyapu V, Mandpe A, Kumar S (2018) Synergistic effect of fly ash in in-vessel composting of biomass and kitchen waste. Bioresour Technol 251:114–120. https://doi.org/10.1016/j.biortech.2017.12.039
58. Mandpe A, Paliya S, Kumar S, Kumar R (2019) Fly ash as an additive for enhancing microbial and enzymatic activities in in-vessel composting of organic wastes. Bioresour Technol 293:122047. https://doi.org/10.1016/j.biortech.2019.122047
59. Lerch TZ, Dignac MF, Thevenot M, Mchergui C, Houot S (2019) Chemical changes during composting of plant residues reduce their mineralisation in soil and cancel the priming effect. Soil Biol Biochem 136:107525. https://doi.org/10.1016/j.soilbio.2019.107525
60. Gusain R, Pandey B, Suthar S (2018) Composting as a sustainable option for managing biomass of aquatic weed Pistia: a biological hazard to aquatic system. J Clean Prod 177:803–812. https://doi.org/10.1016/j.jclepro.2017.12.261
61. Milinković M, Lalević B, Jovičić-Petrović J, Golubović-Ćurguz V, Kljujev I, Raičević V (2019) Biopotential of compost and compost products derived from horticultural waste—effect on plant growth and plant pathogens' suppression. Process Saf Environ Prot 121:299–306. https://doi.org/10.1016/j.psep.2018.09.024
62. Vico A et al (2018) Valorization of date palm (*Phoenix dactylifera* L.) pruning biomass by co-composting with urban and agri-food sludge. J Environ Manage 226:408–415. https://doi.org/10.1016/j.jenvman.2018.08.035
63. Javourez U, O'Donohue M, Hamelin L (2021) Waste-to-nutrition: a review of current and emerging conversion pathways. Biotechnol Adv 53:107857. https://doi.org/10.1016/j.biotechadv.2021.107857
64. Abbey L, Glover-Amengor M, Atikpo MO, Atter A, Toppe J (2017) Nutrient content of fish powder from low value fish and fish byproducts. Food Sci Nutr 5(3):374–379. https://doi.org/10.1002/fsn3.402
65. Aggelopoulos T, Katsieris K, Bekatorou A, Pandey A, Banat IM, Koutinas AA (2014) Solid state fermentation of food waste mixtures for single cell protein, aroma volatiles and fat production. Food Chem 145:710–716. https://doi.org/10.1016/j.foodchem.2013.07.105
66. Suharman I (2020) Utilization of feather meal fermented Bacillus subtilis to replace fish meal in the diet of silver pompano, *Trachinotus blochii* (Lacepede, 1801). AACL Bioflux 13(1)
67. Hernández D, Solana M, Riaño B, García-González MC, Bertucco A (2014) Biofuels from microalgae: Lipid extraction and methane production from the residual biomass in a biorefinery approach. Bioresour Technol 170:370–378. https://doi.org/10.1016/j.biortech.2014.07.109
68. Biller P, Friedman C, Ross AB (2013) Hydrothermal microwave processing of microalgae as a pre-treatment and extraction technique for bio-fuels and bio-products. Bioresour Technol 136:188–195. https://doi.org/10.1016/j.biortech.2013.02.088
69. Karemore A, Sen R (2016) Downstream processing of microalgal feedstock for lipid and carbohydrate in a biorefinery concept: a holistic approach for biofuel applications. RSC Adv 6(35):29486–29496. https://doi.org/10.1039/C6RA01477A
70. Bai M-D, Cheng C-H, Wan H-M, Lin Y-H (2011) Microalgal pigments potential as byproducts in lipid production. J Taiwan Inst Chem Eng 42(5):783–786. https://doi.org/10.1016/j.jtice.2011.02.003
71. Singh D, Puri M, Wilkens S, Mathur AS, Tuli DK, Barrow CJ (2013) Characterization of a new zeaxanthin producing strain of *Chlorella saccharophila* isolated from New Zealand marine waters. Bioresour Technol 143:308–314. https://doi.org/10.1016/j.biortech.2013.06.006
72. Kim D-Y et al (2016) Cell-wall disruption and lipid/astaxanthin extraction from microalgae: chlorella and haematococcus. Bioresour Technol 199:300–310. https://doi.org/10.1016/j.biortech.2015.08.107
73. Li H et al (2019) Co-cultivation of *Rhodotorula glutinis* and *Chlorella pyrenoidosa* to improve nutrient removal and protein content by their synergistic relationship. RSC Adv 9(25):14331–14342. https://doi.org/10.1039/C9RA01884K

74. Vaskan P, Pachón ER, Gnansounou E (2018) Techno-economic and life-cycle assessments of biorefineries based on palm empty fruit bunches in Brazil. J Clean Prod 172:3655–3668. https://doi.org/10.1016/j.jclepro.2017.07.218
75. Hernández V, Romero-García JM, Dávila JA, Castro E, Cardona CA (2014) Techno-economic and environmental assessment of an olive stone based biorefinery. Resour Conserv Recycl 92:145–150. https://doi.org/10.1016/j.resconrec.2014.09.008
76. Mussatto, SI (2014) Brewer's spent grain: a valuable feedstock for industrial applications. J Sci Food Agric 94(7):1264–1275. https://doi.org/10.1002/jsfa.6486
77. Jin Q, Yang L, Poe N, Huang H (2018) Integrated processing of plant-derived waste to produce value-added products based on the biorefinery concept. Trends Food Sci Technol 74:119–131. https://doi.org/10.1016/j.tifs.2018.02.014

Chapter 3
Algae as Third-Generation Biorefining Feedstock

Shreya Gupta, Mithilesh Kumar Jha, Mohammad Aslam, and Sanjeev Mishra

Abstract A detailed study of the present global energy landscape reveals the urgent need to shift towards renewable energy sources. Among the various renewable sources, bioenergy, particularly derived from algal biomass, is emphasized as a promising alternative. The chapter discusses the challenges associated with bioenergy production using first- and second-generation feedstocks, highlighting the benefits of algal biomass as a third-generation feedstock with numerous advantages, including rapid growth, CO_2 sequestration, and versatile applications. Commonly used open and closed algal cultivation systems, including open raceways and closed photobioreactors, have been explained in detail. The wastewater treatment efficiency of different algal strains has been discussed in detail. The chapter further explains the biochemical, chemical, and thermochemical conversion of algal biomass into biofuels. The concept of algal biorefinery is introduced, emphasizing its potential to yield a wide range of value-added products, including biofuels, pharmaceuticals, sustainable lubricants, nutraceuticals, and animal feed. The chapter concludes by emphasizing the need for more research and development to optimize algal cultivation and biorefinery processes, thereby unlocking the full potential of this sustainable resource for a greener and more sustainable future.

Keywords Algae · Biorefinery · Biofuels · Biocrude · Conversion pathways

---

S. Gupta · M. K. Jha
Centre for Energy and Environment, Dr B R Ambedkar National Institute of Technology, Jalandhar, India

M. K. Jha
Department of Chemical Engineering, Dr B R Ambedkar National Institute of Technology, Jalandhar, India

M. Aslam
Biofuels Laboratory, Department of Chemistry, National Institute of Technology Srinagar (J&K), Srinagar, India

S. Mishra (✉)
Biochemical Conversion Division, Sardar Swaran Singh National Institute of Bio-Energy, Kapurthala, Jalandhar, India
e-mail: sanjeevmishra@nibe.res.in

M. Aslam et al. (eds.), *Production and Biorefining of Biocrude Oil: Current Status and Future Developments*, Advances in Sustainability Science and Technology,
https://doi.org/10.1007/978-981-96-5198-6_3

1 Introduction

Energy is crucial to modern life, affecting households, commercial spaces, and industries. Currently, 80% of worldwide energy production relies on fossil fuels, contributing significantly to greenhouse gas emissions and global warming [82]. Cutting emissions to nearly half by 2030 and achieving net zero by 2050 is essential to address this crisis. Global programmes to reduce climate change and advance sustainability include the Kyoto Protocol, the Paris Agreement, and the Sustainable Development Goals [75]. In 2023, global primary energy production increased by 1.4%, below the average growth rate from 2010 to 2019. While some countries experienced declines in energy generation, others saw increases. Concurrently, global energy consumption rose by 2.2%, highlighting the need for sustainable energy practices [82]. Bioenergy stands out among renewable energy alternatives due to its ability to reduce greenhouse gas emissions while providing a reliable energy source.

Bioenergy, an environmentally friendly substitute for fossil fuels, is produced from organic resources like plant and algal biomass, offering a cleaner alternative to fossil fuels [6]. The feedstock used to produce bioenergy may be divided into four categories: edible crops (first generation), waste-derived biomass and energy crops (second generation), algae (third generation), and genetically modified organisms (fourth generation) [11]. Algae, as a third-generation biomass, presents multiple benefits, including no arable land requirements, carbon dioxide sequestration, rapid growth rates, wastewater treatment capabilities, and valuable biorefinery products [61]. Algal cultivation systems include closed photobioreactors and open raceways. Closed systems offer controlled environments with fewer contamination risks, while open systems are susceptible to environmental conditions [34]. Bioreactors, such as stirred tank reactors, airlift reactors, and bubble column reactors, are used for microalgal cultivation. Converting algal biomass to biofuels involves biochemical, chemical, and thermochemical processes [15]. Biochemical conversion includes fermentation and anaerobic digestion to produce bioethanol, biogas, and biohydrogen. Chemical conversion involves transesterification to produce biodiesel, while thermochemical processes like pyrolysis, gasification, and liquefaction produce bio-oil, syngas, and biochar [37].

An algae biorefinery produces a range of high-value products such as medicines, biopolymers, nutraceuticals, and biofuels to maximize biomass value [50]. The high protein content of algal biomass can be harnessed to produce pigments, omega-3 fatty acids, and antioxidants for nutraceutical applications. Bioactive components in algal biomass may be used in pharmaceutical research to develop new drugs for antiviral, anti-inflammatory, and anticancer treatments [64]. Applying algal biomass to produce nutrient-rich animal feed for aquaculture and livestock can reduce reliance on traditional feed sources. Industries like Algenol, Euglena, Cellana, Phycal, and Algae Pharma are developing commercial-scale algal biorefineries. Despite advancements, more research is needed to enhance the sustainability and commercial viability of algal biomass [69].

2 World Energy Scenario

Energy is a fundamental component of human existence. In households, businesses, and industries, energy is one of the basic needs utilized both directly and indirectly [75]. About 90% of carbon dioxide and 75% of greenhouse gas emissions worldwide are attributable to the world's 80% reliance on fossil fuels for energy generation [84]. Globally contributing towards intensified global warming and climate change. To address the current energy and environmental problems, emissions must be cut in half by 2030 and eliminated entirely by 2050 [75]. Many goal-oriented policies have been established worldwide to mitigate the worst effects of climate change. The Paris Agreement aims to keep global warming below 2 °C over pre-industrial levels. The Kyoto Protocol and the Sustainable Development Goals (particularly SDG 13) are examples of worldwide efforts to ensure a sustainable future [82].

In 2023, worldwide Global primary energy production recorded an increase of only 1.4%, which is less than the average yearly growth rate of 1.7% between 2010 and 2019 [75]. Saudi Arabia, Russia, Europe, and Australia experienced a decrease in energy generation by 7.8% 3%, 3.2%, and 2.7%, respectively. While the United States, China, India, Indonesia, Brazil, and Iran experienced an increase in energy production by 4.4%, 2.5%, 6.8%, 10%, and 5%, respectively [82]. In contrast to global energy production, global energy consumption increased by 2.2% in 2023, above the historical average. Figure 1 explains the primary energy consumption of the world, OECD, non-OECD, and European (EU) countries from 2012 to 2021 in Gigatonnes of oil equivalent (Gtoe) [82]. This calls for a change in direction towards more sustainable energy practices and greener energy sources. As a renewable energy source, bioenergy can minimize greenhouse gas emissions while providing more sustainable and reliable energy.

3 Advantages of Biomass for Bioenergy Production

The need to switch to renewable energy sources has become urgent due to the world's increasing energy demand and reliance on fossil fuels [6]. Bioenergy is the most reliable and optimum alternative for fossil fuels of all currently accessible renewable energy sources, including hydropower, solar, wind, and bioenergy [43]. This more environmentally friendly energy source is made of organic material, such as plant and algae biomass, and it helps to mitigate climate change and reduce greenhouse gas emissions, promoting sustainable development [46]. Biomass used as feedstock for bioenergy production is categorized into four major groups: first-generation biomass (edible crops like rapeseed, soybean, sunflower, wheat, corn., etc., second-generation feedstock (waste-derived biomass, switch grass, forestry waste, Jatropha crop., etc.; third-generation biomass (algae; and fourth-generation biomass, which involves genetically modified organisms (GMOs) [18].

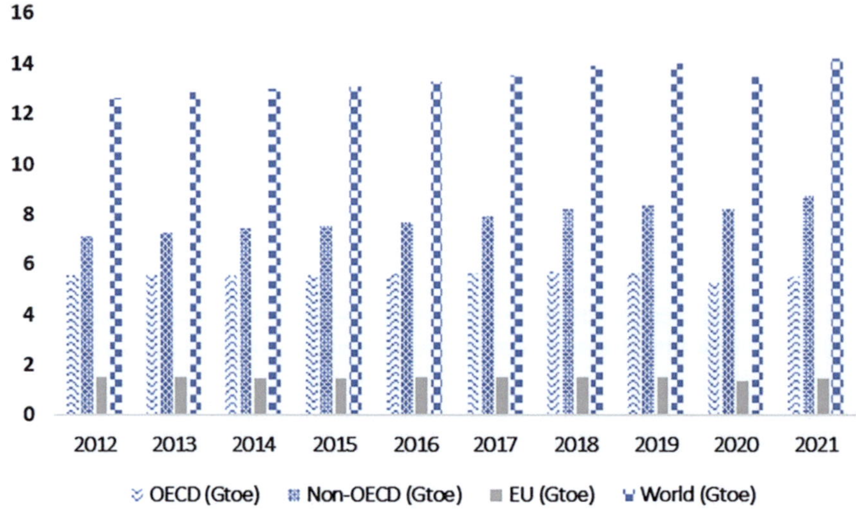

Fig. 1 Primary energy consumption by world, OECD, non-OECD, and European (EU) countries

3.1 Challenges Associated with First- and Second-Generation Feedstock

The first-generation biomass was limited to edible lignocellulosic biomass, which led to a huge food vs fuel debate [7]. This restriction was addressed in second-generation biomass, that is mainly composed of lignocellulosic biomass obtained from waste material or energy crops (e.g., sewage, municipal, forest, and agricultural waste) [86]. Second-generation biomass has several drawbacks, including lack of a continuous supply chain for consistent biomass availability, high lignin content (15–25%) that necessitates energy-intensive pretreatment [20]. An example is Jatropha, an energy crop also known as an oil-seed crop that requires 27 million square kilometres of land to produce enough biomass to meet global energy needs. The land area is comparable to that of the United States and Russia combined. Further, the water requirements for the biomass generation range from 2 to 84 times the amount necessary to generate an equivalent quantity of useable fossil fuel energy [10]. As a result, its ability to solve the global energy issue is limited. Thus, it is essential to promote science and technology for investigation of other more environment-friendly energy generation alternatives. Additionally, advancements in biotechnology and genetic engineering are being explored to improve the efficiency of second-generation biomass production. These innovations aim to reduce the environmental footprint while enhancing the yield of energy crops, thus bridging the gap between sustainable energy production and environmental conservation.

3.2 *Algae as Third-Generation Feedstock*

Algae as the third-generation biomass have emerged as a solution with multiple benefits, including no arable land requirements, CO_2 sequestration ability, rapid growth rate, wastewater treatment efficiency, and commercially valuable biorefinery products leading to a circular economy and zero waste approach [65]. Algae are primarily aquatic, photosynthetic organisms with nuclei that lack stalks, natural roots, specialized multicellular reproductive systems and leaves, as seen in plants. Algae are broadly divided into two major groups based on their size, (i) microalgae (0.004–0.2 μm) and (ii) macroalgae (2–3 m to up to 60 m). The microalgae grow almost 5–10 times faster than any other conventional food crop or lignocellulosic biomass [42]. Algal biomass (dry) is generally composed of carbohydrates (20–30%), lipids (10–20%), and proteins (40–60%) [24]. In addition, the cells also carry more vital components such as polysaccharides, pigments, antioxidants, vitamins, and other cellular biomaterials. These compositions vary based on different factors such as strain type, light intensity, growth conditions, temperature, pH, organic carbon sources, salinity of the culture, stirring and mixing rate, dilution rate, harvesting frequency, media concentration, and cultivation techniques [25].

4 Algal Cultivation Systems

Among various, closed photobioreactors and open raceways are the primary microalgal cultivation systems [28]. The open cultivation system includes unstirred ponds, circular ponds, and raceway ponds. Various factors that affect the open cultivation system include evaporation, climatic conditions, and contaminations [54]. However, these systems are economical and easy to scalable. On the contrary, closed photobioreactors have controlled conditions and fewer chances of contamination [63]. However, these systems are complex and expensive to scale up [22].

Majorly used bioreactors in microalgal cultivation are stirred tank reactors, bubble column reactors, packed bed reactors, airlift reactors, fluidized bed reactors, and flocculated cell reactors [4]. Figure 2 explains various microalgal cultivation systems and resources needed for algal biomass generation. Besides freshwater, wastewater treatment can be integrated with microalgal cultivation as nitrate phosphate is abundantly available in wastewater and an essential component of microalgal growth. Microalgae's minimum requirements for growth make wastewater an ideal substrate for biomass production [34]. Wastewater nutrient removal efficiency of microalgae ranges between 60 and 99% for nitrogen and phosphorus, and 70–95% for COD [4]. A study reported that biomass production (dry cell weight) ranges from 50 to 70 MT ha/year in high-rate algal ponds, whereas in photobioreactors, it is projected to be 150 MT ha/year [1].

A recent report on laboratory-scale study using *Chlorella* sp. *and Spirulina* sp. resulted in total nitrogen removal of 98.4%–99.1% from municipal wastewater in a

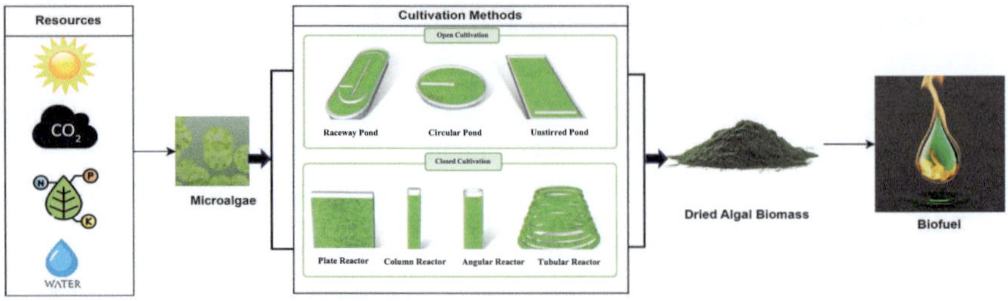

Fig. 2 Microalgal cultivation to biofuel production, adapted from [85]

membrane photobioreactor and generated a biomass yield of 1.58–1.98 g(SS)/L [21]. Another laboratory-scale study reported 97.97 ± 1.12% of total phosphorus removal from second classifier wastewater (WTP) in hydraulic retention time of 8 days with a working volume of 5L using *Scenedesmus rubescens* and *Chlorella vulgari* that resulted in *a* biomass yield of 0.63 ± 0.25 g/L [70]. Besides several other studies are also done using algal strains such as *Pseudanabaena* sp., *Phormidium autumnale*, *Scenedesmus acutus*, *Monoraphidium contortum*, *Chlorella* sp. and *Coelastrella* sp, *Stichococcus* sp. and found sustainability of the process under laboratory-scale studies [3, 45, 51, 71].

Currently, commercial-scale microalgae cultivation is mainly based on raceway ponds, which account for 95% of the total worldwide algal biomass production. Gas exchange requirement and operation cost of thousands of PBRs limit its commercial-scale utilization [34]. Various studies suggested the utilization of *Actinastrum* sp., *Chlorella* sp., *Arthrospira platensis*, *Phaeodactylum tricornotum*, *Scenedesmus* sp., *Micractinium* sp., *Haematococcus pluvialis*, *Anabaena* sp., *Tetraselmis* sp., *Dunaliella salina*, *Nannochloropsis* sp., etc. in raceways ponds [29, 30, 49, 54].

Recently, many advancements to improve the current cultivation system are being implemented keeping in mind the shortcomings of current cultivation techniques. Among those, laboratory and large-scale studies explored biofilm, algal turf scrubber, and immobilized cell adsorption systems. In those systems, algal cultivation studies were focussed on wastewater treatment, river restoration, and specified biomass production targeting commercial use [16]. The above study suggested that the new cultivation techniques that are inexpensive and sustainable provide maximum biomass production under controlled conditions. However, commercial-scale algal biomass production still has many limitations that must be eliminated by developing new and advanced cultivation techniques (Tables 1 and 2).

Table 1 Classification of biomass into first, second, third, and fourth generation

Generation	Feedstock examples	Development stage	Land use	Water use	Food security	GHG reduction	Pros	Cons	Environmental impacts	References
1st	Corn, sugarcane, soybean, rapeseed, etc.	Commercialized	High	High	Poses adverse effect on food security	Low to moderate	Established, high yield, infrastructure in place	Food versus fuel conflict, volatile prices	Fertilizer/ pesticide use, soil erosion	[69]
2nd	Corn stover, wheat straw, sugarcane bagasse), etc	Emerging	Moderate	Moderate	No impact	Moderate to high	Sustainable potential, uses waste products,	Developing technology	Potential land use change if not sourced sustainably	[32]
3rd	Algae	Research and development	Low	Moderate	No impact	High	Tiny powerhouses, very high yield	Developing technology	Nutrient runoff if not managed properly	[25]
4th	Genetically modified organisms (GMOs)	Experimental	Low	Low	No impact	Very high	Future of fuel, very high yield	Genetic engineering concerns	Potential for genetically modified organisms to escape into environment	[37]

Table 2 Laboratory- and large-scale cultivation of microalgal biomass

Scale	Reactor	Strain	Culture medium	T (°C)	pH	Illumination (μmol/s/m^2)	Working volume	HRT (days)	Biomass yield (g/m^2)	References
Lab scale	Glass modular food container	*Chlorella* sp.	Food effluent	25 ± 2	7.2–7.9	150 ± 15	–	14	108	[51]
	Glass modular food container	*Chlorella* sp.	Anaerobic digestate	25	8.03	150 ± 15	–	14	108.64	[52]
	Photo-bioreactors	*C. sorokiniana*	Piggery wastewater	27	7 to 8.5	150	1 L	12 days	–	[26]
	Flow cell (algal biofilms)	*Nitzschia* sp.	Municipal wastewater	22	7	230	400 mL	12	–	[9]
Large scale	Wall-attached microalgal biofilm	*Scenedesmus rubescens* and *Chlorella vulgari*	Second clarifier wastewater (WTP)	25	8.5	7000 lx	5 L	8	0.63 ± 0.25 g/l	[70]
	High-rate algal ponds	*Chlorella vulgaris*	Domestic wastewater	25	–	–	3.8 m^2	21	93.07	[60]
	Integrated algal biofilm	Consortium of *Chlorella* sp., *Scenedesmus* sp.,	Domestic wastewater	–	8.3	200 W LEDs	1.2 MLD	60	–	[31]
	Algal biofilm ponds	*Chlorococcum* sp.	Secondary effluent	20.5 ± 0.5	–	120 μmol photons	48 L	13	15.53	[55]

5 Algal Biomass to Biofuels

There are three forms of algal biomass-to-biofuel conversion: biochemical, chemical, and thermochemical. Based on the composition of feedstock and end product, the required conversion process is selected [74]. Biochemical conversion of biomass mainly includes feedstock with high carbohydrate content. Fermentation of carbohydrate-rich biomass produces bioethanol and biohydrogen [67]. Anaerobic digestion, or the breakdown of organic waste in the absence of oxygen, produces biogas. Chemical conversion is suitable for biomass with high lipid content in biodiesel production through transesterification [65]. Thermochemical conversion includes gasification, hydrothermal liquefaction, pyrolysis, and hydrothermal carbonation of diverse biomass to produce biochar, bio-oil, syngas, and producer gas [35]. A detailed explanation of each process is given in Table 3.

5.1 Biochemical Conversion

Biochemical conversion of any biomass requires the breakdown of organic materials in the presence of microbes. The three main biochemical conversion processes are enzymatic hydrolysis, anaerobic digestion, and fermentation [8]. Microalgal biomass grown in raceways or photobioreactors has been used previously in bioethanol, biogas, and biohydrogen production via biochemical pathways [86]. Carbohydrate-rich biomass has shown significant results, and efforts have been made to produce carbohydrate-rich biomass by varying the C/N ratio, N/P ratio, and nitrogen starvation conditions [17].

Bioethanol is produced by alcoholic fermentation of algal biomass combined with sugars, starch, or cellulose. The biomass is hydrolysed or saccharified to produce sugars, which are then anaerobically broken down by bacteria or yeast cells to form ethanol. Other by-products of alcoholic fermentation include CO_2 and water [20]. It is essential for the ethanol business to purify the diluted product (8–16% ethanol) in order to eliminate impurities such acetaldehyde, 3-methyl butanol, methanol, and butanol [33]. Microalgal biomass and bioethanol production efficiency completely depend on the carbohydrate content of different species. Reactions involved in the Alcoholic fermentation of bioethanol production are expressed in Eqs. 1–4.

$$C_6H_{12}O_6 + 2ADP + 2Pi \rightarrow 2C_3H_4O_3 + 2ATP + 2H_2O \rightarrow \text{(Glucose to pyruvate)} \quad (1)$$

$$2C_3H_4O_3 \rightarrow 2C_2H_4O + 2CO_2 \rightarrow \text{(Pyruvate to acetaldehyde and carbon dioxide)} \quad (2)$$

$$2C_2H_4O + 2NADH \rightarrow 2C_2H_5OH + 2NAD^+ \rightarrow \text{(Acetaldehyde to ethanol)} \quad (3)$$

Table 3 Biomass to biofuel conversion pathways

Conversion pathway	Specific process	Feedstock requirements	End products	Advantages	Disadvantages	References
Biochemical	Fermentation	High carbohydrate content	Bioethanol, other alcohols	Simple, inexpensive, low-energy input	Limited by carbohydrate content, lower energy density products, potential for contamination	[66]
	Anaerobic digestion	High organic matter content	Biogas (methane, carbon dioxide)	Utilizes entire biomass, produces renewable energy and fertilizer	Slow process, lower energy density of biogas, requires upgrading for vehicle use	[33]
Chemical	Transesterification	High lipid content	Biodiesel (fatty acid methyl esters)	Well-established, high-quality biodiesel, lower production costs	Limited by lipid content, requires additional steps for carbohydrate conversion, glycerol by-product	[56]
Thermochemical	Hydrothermal LIQUEFACTION (HTL)	High moisture content, diverse biomass composition	Bio-oil, biogas, biochar	High energy density products, utilize wet biomass, flexible feedstock	High temperature and pressure, expensive reactors, complex upgrading of bio-oil	[13]
	Pyrolysis	Any biomass, including low-quality feedstocks	Bio-oil, syngas, biochar	Versatile feedstock, potential for co-production of chemicals	Requires careful optimization, bio-oil upgrading, char disposal	[72]
	Gasification	Any biomass, including low-quality feedstocks	Syngas (carbon monoxide, hydrogen)	High conversion efficiency, diverse product possibilities, flexible feedstock	High temperature and pressure, expensive reactors, and syngas cleaning required	[23]

$$C_6H_{12}O_6 \rightarrow 2C_2H_5OH + 2CO_2 \rightarrow \text{(Overall Reaction)} \quad (4)$$

A previous study used a chemo-enzymatic hydrolysis technique for lyophilized biomass from different microalgae, followed by the fermentation of hydrolysates with high reducing sugar content [19]. The strategy included acid pretreatment at high pressure and temperature, followed by incubation with Amyloglucosidase and subsequently α-Amylase, reversing the usual enzyme sequence. This method produced maximum yields of at least 34.0 ± 1.0 g total reducing sugar per 100 g dry biomass, increasing the reducing sugar yield by almost one-third. Because of their high reducing sugar content, hydrolysates from *Tetraselmis* sp. (Necton), *Chlorella sorokiniana*, and *Skeletonema* sp. were selected for the generation of bioethanol. With an ethanol productivity of 0.344 ± 0.020 g/L/h. and an ethanol yield of 0.464 $\pm$ 0.013 g/g reducing sugar, *C. sorokiniana* resulted as the best hydrolysate for producing ethanol [19]. Previous research revealed that certain commonly used commercial microalgae might have a carbohydrate content of up to 69.7% [63]. *Chlamydomonas* sp., *Scenedesmus* sp., *Dunaliella* sp., and *Chlorella* sp., are some of the most studied microalgae because of their prominent results in bioethanol, biogas, or biohydrogen production [57].

5.2 *Chemical Conversion*

Chemical conversion of microalgal biomass to biodiesel involves four major steps. Laboratory-scale or large-scale microalgal biomass cultivation followed by dewatering and harvesting processes to get concentrated biomass [65]. Lipid accumulation inside the microalgal cell leads to another step of cell disruption and lipid extraction. Once lipid-extracted biomass is harvested pretreatment followed by transesterification of biomass leads to biodiesel production [47]. Numerous factors, including pH, temperature, aeration, light intensity, salinity, and photoperiod, affect the growth and cultivation of microalgae. The culture technique and mode of cultivation are two additional critical factors that impact biomass productivity overall and growth [56]. A rapid growth rate with considerable lipid content in microalgal cells makes it a sustainable option for biodiesel production.

Many microalgal species like *Chlorella vulgaris* ESP-31, *Scenedesmus abundans, Chlorella pyrenoidosa, Chlorella protothecoides* have shown lipid productivity in the range of 56–687.5 mg/L/d [68, 76, 78, 81]. Lipid transesterification is a crucial procedure that involves extracting and transesterifying lipids from microalgal cells using a catalyst and alcohol. Catalysts play an important role in this process, since they determine the pace of reaction and conversion efficiency [36]. Among acid, base, enzyme, and bio-based catalysts, biocatalysts are gaining popularity since they are inexpensive and enable the sustainable use of waste. Third-generation biodiesel is also gaining increased attention as it supports the availability of sustainable feedstock, which has shown promising results [5].

5.3 Thermochemical Conversion

Thermochemical conversion of biomass includes combustion, pyrolysis, torrefaction, liquefaction, and gasification. These processes differ in temperature and oxygen supply range and produce different end products such as syngas, producer gas, biochar, bio-oil, etc. [79]. Combustion is performed at relatively high temperatures between 900 and 1500 °C with a sufficient oxygen supply [59]. Gasification occurs in the temperature range 600–1300 °C with a limited oxygen supply and results in syngas production [23]. Pyrolysis is performed to obtain bio-oil and biochar without oxygen at 400–800 °C [72]. Liquefaction is mainly performed to extract bio-oil at 200–600 °C with no oxygen supply [13]. Torrefaction is comparable to pyrolysis, although it occurs at a lower temperature, between 200 and 300 °C without oxygen [39].

Previously in a study reported by [38], sorption-enhanced gasification (SEG) of biomass with a CO_2 sorbent was explained, aiming to optimize syngas composition for synthetic fuel production. The effects of the steam-to-carbon (S/C) ratio, CO_2 sorption capacity, and sorbent-to-biomass ratio were analysed in a 30 kWth bubbling fluidized bed gasifier using grape seeds. A high S/C ratio and high sorption capacity reduced higher hydrocarbons to below 1% by volume at 700–740 °C. Adjusting CO_2 separation impacted the H_2, CO, and CO_2 content of syngas, achieving a module $M = (H_2\text{-}CO_2)/(CO + CO_2)$ from 1.2 to nearly 3. In another study by [80], solar-assisted pyrolysis of chicken litter waste was enhanced using CaO and char catalysts to upgrade gas and liquid fuels. In situ pyrolysis with 50% CaO loading yielded the highest CO (63 wt%) and H_2 (15 wt%) at 800 °C, while 50% char catalyst produced 60 wt% CO and 5 wt% H_2. CaO significantly reduced fatty acids in the liquid product to 8% (in situ) and 3% (ex situ). In contrast, char had minimal impact on the deoxygenation of alcohols and fatty acids in bio-oil.

Suriapparao et al. [73] in their study, performed co-pyrolysis of five lignocellulosic biomass residues with two synthetic plastics to improve the bio-oil quality at 450 W. The microwave co-pyrolysis process showed promising results and (63–68%) high efficiency. HHV of the co-pyrolysis bio-oil was found to be (38–42 MJ/kg) with less moisture and lower acidity than pyrolysis bio-oil. Biochar has more energy and mass density, a higher heating value, better hydrophobicity, better powder grindability and flowability, and characteristics similar to coal when compared to biomass [88]. Modern biomass pyrolysis methods like solar- and microwave-assisted pyrolysis have gained significant attention lately, surpassing traditional electricity or heat-assisted pyrolysis technologies [41].

6 Algal Biomass Biorefinery

An algal biorefinery maximizes the value extracted from biomass by utilizing the whole cell to generate various high-value products. [62]. Industries such as Algenol, Euglena, Cellana, Phycal, Algae Pharma have been working on commercial-scale algal biorefinery.

6.1 Algal Biomass to Bio-crude

Algal bio-crude offers a sustainable solution to energy production, utilizing fast-growing algae and minimizing competition with agricultural resources. Its potential to integrate into existing refineries makes it a promising alternative to traditional fossil fuels. Algal biomass has attracted significant interest for its conversion into bio-crude oil, a renewable and sustainable alternative to conventional fossil fuels. The conversion process primarily involves hydrothermal liquefaction (HTL), where wet algal biomass is subjected to high temperatures (around 300–350 °C) and pressures (10–25 MPa) in a water-based environment [44]. This mimics the natural geological processes that form fossil fuels but occurs over a much shorter timeframe. During HTL, the complex organic molecules in the algal biomass break down into smaller hydrocarbons, resulting in a bio-crude that can be further refined into usable fuels like gasoline, diesel, and jet fuel [77]. One of the key advantages of using algae for bio-crude production is the ability to use the whole cell, including non-lipid fractions like proteins and carbohydrates, maximizing the biomass's energy yield. Furthermore, the by-products of HTL, such as aqueous phases rich in nutrients, can be recycled back into the algal cultivation system, enhancing the sustainability of the process [58]. *Nannochloropsis gaditana* and *C. vulgaris* have been commercially used to produce sustainable lubricants for industrial and automotive applications [27]. As research advances, optimizing HTL conditions and integrating bio-crude production into existing biorefinery platforms hold promise for reducing costs and enhancing the commercial viability of algae-based biofuels.

6.2 Nutraceuticals and Functional Foods

The rich protein, vitamin, and mineral content of algae make it an ideal resource for the nutraceutical and functional food industries. Algae such as Spirulina and Chlorella are well-known for their health benefits, offering high levels of essential amino acids, omega-3 fatty acids (such as eicosapentaenoic acid, EPA, and docosahexaenoic acid, DHA), and antioxidants like phycocyanin [87]. These compounds are linked to anti-inflammatory, immune-boosting, and cardiovascular health-promoting effects. Carotenoids, such as beta-carotene and astaxanthin, found in algae like *H.*

pluvialis, have strong antioxidant properties and are used in dietary supplements to support skin health and protect against UV radiation [48]. Algal-based nutraceuticals are growing in demand due to their natural origin, and they are increasingly incorporated into foods, beverages, and supplements marketed for health-conscious consumers. Applying algal biomass to produce sustainable and nutrient-rich animal feed for aquaculture and livestock reduces reliance on traditional feed sources [14]. *Haematococcus pluvialis* and *A platensis* are used to produce Dietary supplements and functional foods [73].

6.3 Pharmaceutical Applications

Algal biomass is a promising source of bioactive compounds for pharmaceutical applications. The diverse metabolic pathways of algae produce a wide range of secondary metabolites with potent biological activities [14]. The bioactive components found in algal biomass may be used in pharmaceutical research to generate new drugs, primarily for antiviral, anti-inflammatory, and anticancer treatments.

Compounds such as sulphated polysaccharides, peptides, and lipids derived from algae exhibit antiviral, anti-inflammatory, and anticancer properties. For example, sulphated polysaccharides from porphyridium have shown antiviral activity against herpes simplex virus (HSV) and HIV [12]. Moreover, fucoidans from brown algae possess anti-inflammatory effects by modulating immune responses, making them candidates for developing new treatments for chronic inflammatory diseases. Algal-derived compounds are also being investigated for their potential to induce apoptosis in cancer cells, providing a natural source of chemotherapeutic agents. Ongoing research aims to isolate and characterize these bioactive molecules for pharmaceutical development.

6.4 Sustainable Animal Feed

Algal biomass is increasingly recognized as a sustainable alternative to traditional animal feeds, such as fishmeal and soybean meal. Algae such as Nannochloropsis, Spirulina, and Chlorella are rich in essential nutrients, including proteins, lipids, vitamins, and minerals. These nutrients support the growth and health of livestock and aquaculture species [2]. For instance, the use of algal-based feed in aquaculture has been shown to enhance fish growth rates, improve immune responses, and increase the nutritional value of fish products, particularly in terms of omega-3 fatty acid content [53]. Additionally, algae can be grown on non-arable land and utilize wastewater, reducing the environmental impact of feed production. The integration of algal biomass into animal feed is seen as a step towards more sustainable and eco-friendly farming practices, contributing to the circular economy.

6.5 *Biopolymers and Biodegradable Products*

The use of algae in producing biopolymers has opened new avenues for sustainable materials. Algae-derived biopolymers, such as alginates, carrageenans, and agar, are used in various applications, including food packaging, pharmaceuticals, and cosmetics [48]. These polymers are valued for their gel-forming properties, biocompatibility, and biodegradability. For example, alginates extracted from brown algae are used in wound dressings due to their ability to form hydrogels that promote healing [78]. Alginates, carrageenans, and agar are examples of algal biorefinery that have paved the way for its commercial scale. *Chlorella vulgaris, Spirulina platensis,* and *Scenedesmus obliquus* are examples of a few commercially used strains in biopolymer industries for the production of Packaging films, disposable utensils, and medical devices. Biodegradable packaging materials made of *Spirulina platensis* and *C. vulgaris* have been a positive effort towards reducing plastic waste [29]. *Chromochloris zofingiensis* is a commercially valuable unicellular microalga and a high-value-added astaxanthin producer [40]. Algae are also being explored to produce biodegradable plastics, replacing petroleum-based plastics and reducing plastic pollution. *Chlorella vulgaris* and *Spirulina platensis* have been studied for their potential to produce bioplastics for packaging materials and disposable utensils, providing eco-friendly alternatives to single-use plastics [78].

7 Conclusion

Using algae biomass as a primary source of biofuel, the book chapter thoroughly examines biomass conversion to biofuels. Key conversion mechanisms, such as chemical, thermochemical, and biological processes, are described. Algal biorefineries, which optimize biomass value by generating a diverse range of co-products in addition to biofuels, are further explained in this chapter. The production of omega-3 fatty acids, antioxidants, and pigments by algae biorefineries is crucial to the nutraceutical business. Algal biomass is utilized to manufacture nutrient-dense animal feed, providing aquaculture and livestock with a sustainable substitute for conventional feed sources. The chapter outlines businesses' initiatives to scale up algal biomass conversion technologies. In order to increase the accessibility of biofuels and other high-value products, these industrial pioneers seek to improve the sustainability and financial feasibility of algae biorefineries. Algal biorefineries provide an encouraging strategy for attaining a sustainable energy future by combining several value chains (Fig. 3).

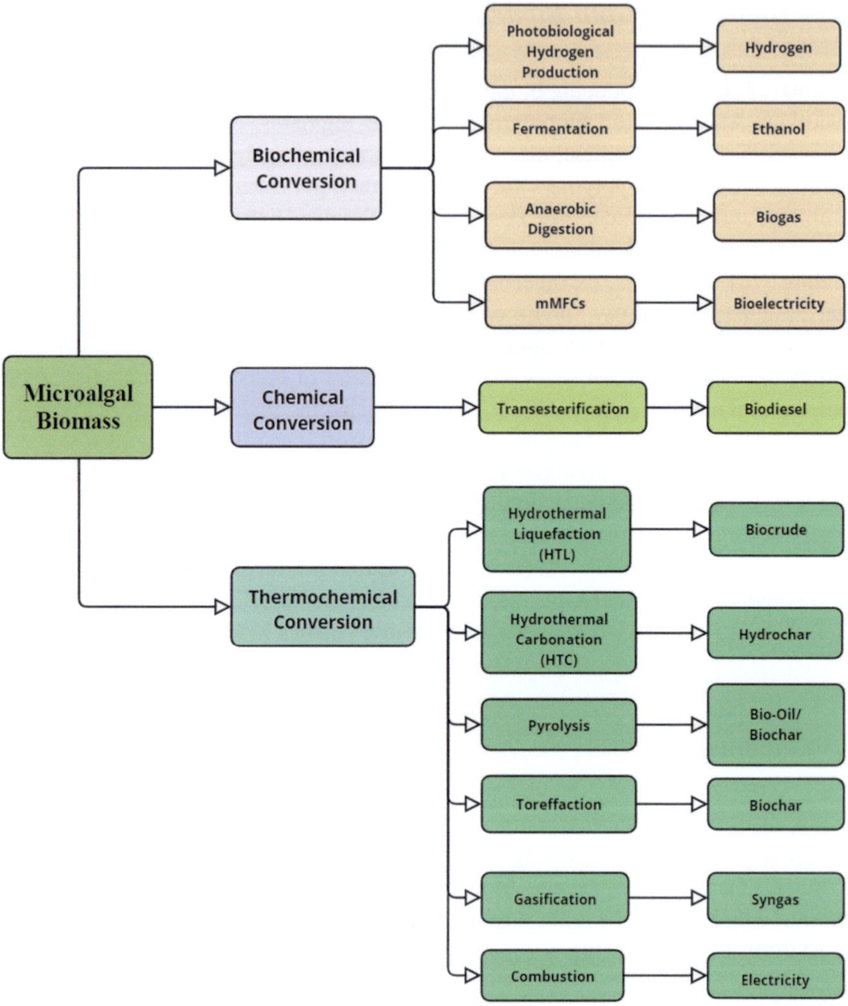

Fig. 3 Microalgal biomass to biofuel conversions, adapted from [83]

References

1. Adesanya VO, Cadena E, Scott SA, Smith AG (2014) Life cycle assessment on microalgal biodiesel production using a hybrid cultivation system. Biores Technol 163:343–355. https://doi.org/10.1016/j.biortech.2014.04.051
2. Ahmad A, Hassan WS, Banat F (2022) An overview of microalgae biomass as a sustainable aquaculture feed ingredient: food security and circular economy. In: Bioengineered, vol 13, Issue 4. Taylor and Francis Ltd., pp 9521–9547. https://doi.org/10.1080/21655979.2022.2061148
3. Akao PK, Singh B, Kaur P, Sor A, Avni A, Dhir A, Verma S, Kapoor S, Phutela UG, Satpute S, Sharma S, Avisar D, Sandha KS, Mamane H (2021) Coupled microalgal–bacterial biofilm

for enhanced wastewater treatment without energy investment. J Water Process Eng. https://doi.org/10.1016/j.jwpe.2021.102029

4. Ali SS, El-Sheekh M, Manni A, Ruiz HA, Elsamahy T, Sun J, Schagerl M (2022) Microalgae-mediated wastewater treatment for biofuels production: a comprehensive review. In: Microbiological research, vol 265. Elsevier GmbH. https://doi.org/10.1016/j.micres.2022.127187
5. Ameen F, Dawoud T, Alabdullatif J, Arif I (2023) CO_2 sequestration and biodiesel production from *Volvox aureus* a newly isolated green microalgal species from industrial wastewater. Environ Res. https://doi.org/10.1016/j.envres.2023.115251
6. Ang TZ, Salem M, Kamarol M, Das HS, Nazari MA, Prabaharan N (2022) A comprehensive study of renewable energy sources: classifications, challenges and suggestions. In: Energy strategy reviews, vol 43. Elsevier Ltd. https://doi.org/10.1016/j.esr.2022.100939
7. Awogbemi O, Kallon DV Von (2022) Valorization of agricultural wastes for biofuel applications. In: Heliyon, vol 8, Issue 10. Elsevier Ltd. https://doi.org/10.1016/j.heliyon.2022.e11117
8. Behera S, Singh R, Arora R, Sharma NK, Shukla M, Kumar S (2015) Scope of algae as third generation biofuels. In: Frontiers in bioengineering and biotechnology, vol 2. Frontiers Media S.A. https://doi.org/10.3389/fbioe.2014.00090
9. Boelee NC, Temmink H, Janssen M, Buisman CJN, Wijffels RH (2011) Nitrogen and phosphorus removal from municipal wastewater effluent using microalgal biofilms. Water Res 45(18):5925–5933. https://doi.org/10.1016/j.watres.2011.08.044
10. Brittaine R, Lutaladio N, & Food and Agriculture Organization of the United Nations (2010) Jatropha: a smallholder bioenergy crop: the potential for pro-poor development. Food and Agriculture Organization of the United Nations
11. Cao L, Yu IKM, Liu Y, Ruan X, Tsang DCW, Hunt AJ, Ok YS, Song H, Zhang S (2018) *Lignin valorization* for the production of renewable chemicals: State-of-the-art review and future prospects. In: Bioresource technology, vol 269, pp 465–475. Elsevier Ltd. https://doi.org/10.1016/j.biortech.2018.08.065
12. Cardozo FTGS, Larsen IV, Carballo EV, Jose G, Stern RA, Brummel RC, Camelini CM, Rossi MJ, Simoes CMO, Brandt CR (2013) In vivo anti-herpes simplex virus activity of a sulfated derivative of *Agaricus brasiliensis* mycelial polysaccharide. Antimicrob Agents Chemother 57(6):2541–2549. https://doi.org/10.1128/AAC.02250-12
13. Carpio RB, Zhang Y, Kuo CT, Chen WT, Schideman LC, de Leon R (2021) Effects of reaction temperature and reaction time on the hydrothermal liquefaction of demineralized wastewater algal biomass. Bioresour Technol Rep. https://doi.org/10.1016/j.biteb.2021.100679
14. Chang H, Feng H, Wang R, Zhang X, Wang J, Li C, Zhang Y, Li L, Ho SH (2023) Enhanced energy recovery from landfill leachate by linking light and dark bio-reactions: underlying synergistic effects of dual microalgal interaction. Water Res. https://doi.org/10.1016/j.watres.2023.119578
15. Chapman RL (2013) Algae: the world's most important "plants"-an introduction. Mitig Adapt Strat Glob Change 18(1):5–12. https://doi.org/10.1007/s11027-010-9255-9
16. Cheenakula D, Hoffstadt K, Krafft S, Reinecke D, Klose H, Kuperjans I, Grömping M (2023) Anaerobic digestion of algal–bacterial biomass of an Algal Turf Scrubber system. Biomass Convers Biorefinery 13(15):13605–13619. https://doi.org/10.1007/s13399-022-03236-z
17. Chen CY, Chang HY, Chang JS (2016) Producing carbohydrate-rich microalgal biomass grown under mixotrophic conditions as feedstock for biohydrogen production. Int J Hydrogen Energy 41(7):4413–4420. https://doi.org/10.1016/j.ijhydene.2015.05.163
18. Chia SR, Nomanbhay SBHM, Chew KW, Munawaroh HSH, Shamsuddin AH, Show PL (2022) Algae as potential feedstock for various bioenergy production. Chemosphere 287(P1):131944. https://doi.org/10.1016/j.chemosphere.2021.131944
19. Constantino A, Rodrigues B, Leon R, Barros R, Raposo S (2021) Alternative chemo-enzymatic hydrolysis strategy applied to different microalgae species for bioethanol production. Algal Res. https://doi.org/10.1016/j.algal.2021.102329

20. Di Donato P, Finore I, Poli A, Nicolaus B, Lama L (2019) The production of second generation bioethanol: the biotechnology potential of thermophilic bacteria. J Clean Prod 233:1410–1417. https://doi.org/10.1016/j.jclepro.2019.06.152
21. Ding M, Wang C, Woo Bae S, Yong Ng H (2022) Enhanced nutrient removal and bioenergy production in microalgal photobioreactor following anaerobic membrane bioreactor for decarbonized wastewater treatment. Biores Technol. https://doi.org/10.1016/j.biortech.2022.128120
22. Fica ZT, Sims RC (2016) Algae-based biofilm productivity utilizing dairy wastewater: effects of temperature and organic carbon concentration. J Biol Eng 10(1). https://doi.org/10.1186/s13036-016-0039-y
23. Giwa SO, Taziwa RT (2024) Adoption of advanced coal gasification: a panacea to carbon footprint reduction and hydrogen economy transition in South Africa. Int J Hydrogen Energy 77:301–323. Elsevier Ltd. https://doi.org/10.1016/j.ijhydene.2024.06.190
24. Goria K, Singh HM, Singh A, Kothari R, Tyagi VV (2024) Insights into biohydrogen production from algal biomass: challenges, recent advancements and future directions. Int J Hydrogen Energy 52:127–151. https://doi.org/10.1016/j.ijhydene.2023.03.174
25. Goswami RK, Mehariya S, Obulisamy PK, Verma P (2021) Advanced microalgae-based renewable biohydrogen production systems: a review. In: Bioresource technology, vol 320. Elsevier Ltd. https://doi.org/10.1016/j.biortech.2020.124301
26. Huang J, Chu R, Chang T, Cheng P, Jiang J, Yao T, Zhou C, Liu T, Ruan R (2021) Modeling and improving arrayed microalgal biofilm attached culture system. Biores Technol. https://doi.org/10.1016/j.biortech.2021.124931
27. Ijaz Malik MA, Kalam MA, Mujtaba MA, Almomani F (2023) A review of recent advances in the synthesis of environmentally friendly, sustainable, and nontoxic bio-lubricants: recommendations for the future implementations. Environ Technol Innov 32. Elsevier B.V. https://doi.org/10.1016/j.eti.2023.103366
28. Janpum C, Pombubpa N, Monshupanee T, Incharoensakdi A, In-na P (2022) Advancement on mixed microalgal-bacterial cultivation systems for nitrogen and phosphorus recoveries from wastewater to promote sustainable bioeconomy. J Biotechnol 360:198–210. Elsevier B.V. https://doi.org/10.1016/j.jbiotec.2022.11.008
29. Jiménez C, Cossío BR, Niell FX (2003) Relationship between physicochemical variables and productivity in open ponds for the production of Spirulina: a predictive model of algal yield. Aquaculture 221(1–4):331–345. https://doi.org/10.1016/S0044-8486(03)00123-6
30. Jorquera O, Kiperstok A, Sales EA, Embiruçu M, Ghirardi ML (2010) Comparative energy life-cycle analyses of microalgal biomass production in open ponds and photobioreactors. Biores Technol 101(4):1406–1413. https://doi.org/10.1016/j.biortech.2009.09.038
31. Katam K, Bhattacharyya D (2021) Improving the performance of activated sludge process with integrated algal biofilm for domestic wastewater treatment: system behavior during the start-up phase. Bioresour Technol Rep. https://doi.org/10.1016/j.biteb.2020.100618
32. Kazemi Shariat Panahi H, Dehhaghi M, Dehhaghi S, Guillemin GJ, Lam SS, Aghbashlo M, Tabatabaei M (2022) Engineered bacteria for valorizing lignocellulosic biomass into bioethanol. Bioresour Technol 344(PA):126212. https://doi.org/10.1016/j.biortech.2021.126212
33. Khan S, Das P, Abdul Quadir M, Thaher MI, Mahata C, Sayadi S, Al-Jabri H (2023) Microalgal feedstock for biofuel production: recent advances, challenges, and future perspective. Fermentation 9(3). MDPI. https://doi.org/10.3390/fermentation9030281
34. Kumar K, Mishra SK, Shrivastav A, Park MS, Yang JW (2015) Recent trends in the mass cultivation of algae in raceway ponds. Renew Sustain Energy Rev 51:875–885. Elsevier Ltd. https://doi.org/10.1016/j.rser.2015.06.033
35. Kumar M, Oyedun AO, Kumar A (2019) A comparative analysis of hydrogen production from the thermochemical conversion of algal biomass. Int J Hydrogen Energy 44(21):10384–10397. https://doi.org/10.1016/j.ijhydene.2019.02.220

36. Kusuma HS, Az-Zahra KD, Saputri RW, Utomo MDP, Jaya DEC, Amenaghawon AN, Darmokoesoemo H (2024) Unlocking the potential of agricultural waste as biochar for sustainable biodiesel production: a comprehensive review. Bioresour Technol Rep 26. Elsevier Ltd. https://doi.org/10.1016/j.biteb.2024.101848
37. Lam MK, Loy ACM, Yusup S, Lee KT (2019) Biohydrogen production from algae. In: Biomass, biofuels, biochemicals: biohydrogen, Second Edition. Elsevier B.V. https://doi.org/10.1016/B978-0-444-64203-5.00009-5
38. Li C, Liu R, Zheng J, Liao L, Zhang Y (2022) Sorption-enhanced steam gasification of woody biomass assisted by Ca/Na/FS (fine slag) composite sorbents. Int J Hydrogen Energy 47(25):12528–12538. https://doi.org/10.1016/j.ijhydene.2022.02.016
39. Lin SL, Zhang H, Chen WH, Song M, Kwon EE (2023) Low-temperature biochar production from torrefaction for wastewater treatment: a review. In: Bioresource technology, vol 387. Elsevier Ltd. https://doi.org/10.1016/j.biortech.2023.129588
40. Liu Y, Tang S, Yan Q, Zhou J, Cai Z (2024) Effectiveness and associated mechanisms of a combination of biofilm attached cultivation and mixotrophy in promoting microalgal biomass. Biores Technol. https://doi.org/10.1016/j.biortech.2023.130077
41. Liu Z, Xie M, Zhou T, Yang J, Yang Y, Liu T, Dai S, Huang Q, Cen Q, Xiao P, Li B, Du H (2024) A review on liquid fuel produced from microwave-assisted pyrolysis of plastic waste. In: Process safety and environmental protection, vol 187, pp 833–844. Institution of Chemical Engineers. https://doi.org/10.1016/j.psep.2024.05.035
42. Loke Show P (2022) Global market and economic analysis of microalgae technology: status and perspectives. Biores Technol. https://doi.org/10.1016/j.biortech.2022.127329
43. Long H, Li X, Wang H, Jia J (2013) Biomass resources and their bioenergy potential estimation: a review. Renew Sustain Energy Rev 26:344–352. Elsevier Ltd. https://doi.org/10.1016/j.rser.2013.05.035
44. Mahata C, Mishra S, Dhar S, Ray S, Mohanty K, Das D (2023) Utilization of dark fermentation effluent for algal cultivation in a modified airlift photobioreactor for biomass and biocrude production. J Environ Manage. https://doi.org/10.1016/j.jenvman.2022.117121
45. Makaroglou G, Marakas H, Fodelianakis S, Axaopoulou VA, Koumi I, Kalogerakis N, Gikas P (2021) Optimization of biomass production from Stichococcous sp. biofilms coupled to wastewater treatment. Biochem Eng J 169. https://doi.org/10.1016/j.bej.2021.107964
46. Manikandan S, Vickram S, Sirohi R, Subbaiya R, Krishnan RY, Karmegam N, Sumathijones C, Rajagopal R, Chang SW, Ravindran B, Awasthi MK (2023) Critical review of biochemical pathways to transformation of waste and biomass into bioenergy. Bioresour Technol 372. Elsevier Ltd. https://doi.org/10.1016/j.biortech.2023.128679
47. Marella TK, Datta A, Patil MD, Dixit S, Tiwari A (2019) Biodiesel production through algal cultivation in urban wastewater using algal floway. Biores Technol 280:222–228. https://doi.org/10.1016/j.biortech.2019.02.031
48. Mehariya S, Goswami RK, Verma P, Lavecchia R, Zuorro A (2021) Integrated approach for wastewater treatment and biofuel production in microalgae biorefineries. In: Energies, vol 14, Issue 8. MDPI AG. https://doi.org/10.3390/en14082282
49. Mendoza JL, Granados MR, de Godos I, Acién FG, Molina E, Heaven S, Banks CJ (2013) Oxygen transfer and evolution in microalgal culture in open raceways. Biores Technol 137:188–195. https://doi.org/10.1016/j.biortech.2013.03.127
50. Mishra S, Roy M, Mohanty K (2019) Microalgal bioenergy production under zero-waste biorefinery approach: recent advances and future perspectives. In: Bioresour Technol 292. Elsevier Ltd. https://doi.org/10.1016/j.biortech.2019.122008
51. Mkpuma VO, Moheimani NR, Ennaceri H (2024) Biofilm and suspension-based cultivation of microalgae to treat anaerobic digestate food effluent (ADFE). Sci Total Environ. https://doi.org/10.1016/j.scitotenv.2024.171320
52. Mkpuma VO, Moheimani NR, Ennaceri H (2024b) Biofilm cultivation of chlorella species. MUR 269 to treat anaerobic digestate food effluent (ADFE): total ammonia nitrogen (TAN) concentrations effect. Chemosphere 354. https://doi.org/10.1016/j.chemosphere.2024.141688

53. Nagappan S, Das P, AbdulQuadir M, Thaher M, Khan S, Mahata C, Al-Jabri H, Vatland AK, Kumar G (2021) Potential of microalgae as a sustainable feed ingredient for aquaculture. J Biotechnol 341:1–20. https://doi.org/10.1016/j.jbiotec.2021.09.003
54. Narala RR, Garg S, Sharma KK, Thomas-Hall SR, Deme M, Li Y, Schenk PM (2016) Comparison of microalgae cultivation in photobioreactor, open raceway pond, and a two-stage hybrid system. Front Energy Res 4(AUG). https://doi.org/10.3389/fenrg.2016.00029
55. Orfanos AG, Manariotis ID (2019) Algal biofilm ponds for polishing secondary effluent and resource recovery. J Appl Phycol 31(3):1765–1772. https://doi.org/10.1007/s10811-018-1731-8
56. Pandey S, Narayanan I, Selvaraj R, Varadavenkatesan T, Vinayagam R (2024) Biodiesel production from microalgae: a comprehensive review on influential factors, transesterification processes, and challenges. In: Fuel, vol 367. Elsevier Ltd. https://doi.org/10.1016/j.fuel.2024.131547
57. Pathy A, Nageshwari K, Ramaraj R, Pragas Maniam G, Govindan N, Balasubramanian P (2022) Biohydrogen production using algae: potentiality, economics and challenges. Bioresour Technol 360. Elsevier Ltd. https://doi.org/10.1016/j.biortech.2022.127514
58. Ravichandran SR, Venkatachalam CD, Sengottian M, Sekar S, Kandasamy S, Ramasamy Subramanian KP, Purushothaman K, Lavanya Chandrasekaran A, Narayanan M (2022) A review on hydrothermal liquefaction of algal biomass on process parameters, purification and applications. Fuel. https://doi.org/10.1016/j.fuel.2021.122679
59. Ren X, Sun R, Meng X, Vorobiev N, Schiemann M, Levendis YA (2017) Carbon, sulfur and nitrogen oxide emissions from combustion of pulverized raw and torrefied biomass. Fuel 188:310–323. https://doi.org/10.1016/j.fuel.2016.10.017
60. Rodrigues de Assis L, Calijuri ML, Assemany PP, Silva TA, Teixeira JS (2020) Innovative hybrid system for wastewater treatment: high-rate algal ponds for effluent treatment and biofilm reactor for biomass production and harvesting. J Environ Manage. https://doi.org/10.1016/j.jenvman.2020.111183
61. Sachin Powar R, Singh Yadav A, Siva Ramakrishna C, Patel S, Mohan M, Sakharwade SG, Choubey M, Kumar Ansu A, Sharma A (2022) Algae: a potential feedstock for third generation biofuel. Mater Today Proceed 63:A27–A33. https://doi.org/10.1016/j.matpr.2022.07.161
62. Safari F, Dincer I (2022) Assessment and multi-objective optimization of a vanadium-chlorine thermochemical cycle integrated with algal biomass gasification for hydrogen and power production. Energy Convers Manage. https://doi.org/10.1016/j.enconman.2021.115132
63. Sánchez-Contreras MI, Morales-Arrieta S, Okoye PU, Guillén-Garcés RA, Sebastian PJ, Arias DM (2021) Recycling industrial wastewater for improved carbohydrate-rich biomass production in a semi-continuous photobioreactor: effect of hydraulic retention time. J Environ Manage. https://doi.org/10.1016/j.jenvman.2021.112065
64. Sarkar O, Katakojwala R, Venkata Mohan S (2021) Low carbon hydrogen production from a waste-based biorefinery system and environmental sustainability assessment. Green Chem 23(1). https://doi.org/10.1039/d0gc03063e
65. Satpati G, Pal R (2018) Microalgae-biomass to biodiesel: a review. In: Journal of algal biomass utilization, vol 2018, Issue 4. https://www.researchgate.net/publication/329424285
66. Selim KA, El-Ghwas DE, Easa SM, Abdelwahab Hassan MI (2018) Bioethanol a microbial biofuel metabolite; new insights of yeasts metabolic engineering. Fermentation 4(1). https://doi.org/10.3390/fermentation4010016
67. Sharma P, Sivaramakrishnaiah M, Deepanraj B, Saravanan R, Reddy MV (2024) A novel optimization approach for biohydrogen production using algal biomass. Int J Hydrogen Energy 52:94–103. https://doi.org/10.1016/j.ijhydene.2022.09.274
68. Shen XF, Liu JJ, Chu FF, Lam PKS, Zeng RJ (2015) Enhancement of FAME productivity of *Scenedesmus obliquus* by combining nitrogen deficiency with sufficient phosphorus supply in heterotrophic cultivation. Appl Energy 158:348–354. https://doi.org/10.1016/j.apenergy.2015.08.057
69. Siddiki SYA, Mofijur M, Kumar PS, Ahmed SF, Inayat A, Kusumo F, Badruddin IA, Khan TMY, Nghiem LD, Ong HC, Mahlia TMI (2022) Microalgae biomass as a sustainable source

for biofuel, biochemical and biobased value-added products: an integrated biorefinery concept. Fuel. https://doi.org/10.1016/j.fuel.2021.121782
70. Su Y, Mennerich A, Urban B (2016) The long-term effects of wall attached microalgal biofilm on algae-based wastewater treatment. Biores Technol 218:1249–1252. https://doi.org/10.1016/j.biortech.2016.06.099
71. Sukačová K, Trtílek M, Rataj T (2015) Phosphorus removal using a microalgal biofilm in a new biofilm photobioreactor for tertiary wastewater treatment. Water Res 71:55–63. https://doi.org/10.1016/j.watres.2014.12.049
72. Suresh Babu KKB, Nataraj M, Tayappa M, Vyas Y, Mishra RK, Acharya B (2024) Production of biochar from waste biomass using slow pyrolysis: studies of the effect of pyrolysis temperature and holding time on biochar yield and properties. Mater Sci Energy Technol 7:318–334. https://doi.org/10.1016/j.mset.2024.05.002
73. Suriapparao DV, Boruah B, Raja D, Vinu R (2018) Microwave assisted co-pyrolysis of biomasses with polypropylene and polystyrene for high quality bio-oil production. Fuel Process Technol 175:64–75. https://doi.org/10.1016/j.fuproc.2018.02.019
74. Tao Q, Gao F, Qian CY, Guo XZ, Zheng Z, Yang ZH (2017) Enhanced biomass/biofuel production and nutrient removal in an algal biofilm airlift photobioreactor. Algal Res 21:9–15. https://doi.org/10.1016/j.algal.2016.11.004
75. Tostes B, Henriques ST, Brockway PE, Heun MK, Domingos T, Sousa T (2024) On the right track? Energy use, carbon emissions, and intensities of world rail transportation, 1840–2020. Appl Energy. https://doi.org/10.1016/j.apenergy.2024.123344
76. Umamaheswari J, Kavitha MS, Shanthakumar S (2020) Outdoor cultivation of Chlorella pyrenoidosa in paddy-soaked wastewater and a feasibility study on biodiesel production from wet algal biomass through in-situ transesterification. Biomass Bioenerg. https://doi.org/10.1016/j.biombioe.2020.105853
77. Usman M, Cheng S, Boonyubol S, Cross JS (2024) From biomass to biocrude: innovations in hydrothermal liquefaction and upgrading. Energy Convers Manag 302. Elsevier Ltd. https://doi.org/10.1016/j.enconman.2024.118093
78. Wang T, Ge H, Liu T, Tian X, Wang Z, Guo M, Chu J, Zhuang Y (2016) Salt stress induced lipid accumulation in heterotrophic culture cells of Chlorella protothecoides: Mechanisms based on the multi-level analysis of oxidative response, key enzyme activity and biochemical alteration. J Biotechnol 228:18–27. https://doi.org/10.1016/j.jbiotec.2016.04.025
79. Wang Y, Wu JJ (2023) Thermochemical conversion of biomass: potential future prospects. Renew Sustain Energy Rev 187. Elsevier Ltd. https://doi.org/10.1016/j.rser.2023.113754
80. Weldekidan H, Strezov V, Kan T, Kumar R, He J, Town G (2019) Solar assisted catalytic pyrolysis of chicken-litter waste with in-situ and ex-situ loading of CaO and char. Fuel 246:408–416. https://doi.org/10.1016/j.fuel.2019.02.135
81. Yeh KL, Chang JS (2012) Effects of cultivation conditions and media composition on cell growth and lipid productivity of indigenous microalga Chlorella vulgaris ESP-31. Biores Technol 105:120–127. https://doi.org/10.1016/j.biortech.2011.11.103
82. Yolcan OO (2023) World energy outlook and state of renewable energy: 10-Year evaluation. Innov Green Develop 2(4). https://doi.org/10.1016/j.igd.2023.100070
83. Zabed HM, Akter S, Yun J, Zhang G, Awad FN, Qi X, Sahu JN (2019) Recent advances in biological pretreatment of microalgae and lignocellulosic biomass for biofuel production. In: Renew Sustain Energy Rev 105:105–128. Elsevier Ltd. https://doi.org/10.1016/j.rser.2019.01.048
84. Zeb K, Ali SM, Khan B, Mehmood CA, Tareen N, Din W, Farid U, Haider A (2017) A survey on waste heat recovery: electric power generation and potential prospects within Pakistan. In: Renew Sustain Energy Rev 75:1142–1155. Elsevier Ltd. https://doi.org/10.1016/j.rser.2016.11.096
85. Zerrouki D, Henni A (2019) Outdoor microalgae cultivation for wastewater treatment. In: Application of microalgae in wastewater treatment, pp 81–99. Springer International Publishing. https://doi.org/10.1007/978-3-030-13913-1_5

86. Zhang Y, Ding Z, Shahadat Hossain M, Maurya R, Yang Y, Singh V, Kumar D, Salama ES, Sun X, Sindhu R, Binod P, Zhang Z, Kumar Awasthi M (2023) Recent advances in lignocellulosic and algal biomass pretreatment and its biorefinery approaches for biochemicals and bioenergy conversion. In: Bioresour Technol 367. Elsevier Ltd. https://doi.org/10.1016/j.biortech.2022.128281
87. Zhao Y, Lu T, Xu G, Luo Y, Zhang X, Wu X, Han X, Tester JW, Wang K (2024) Hydrothermal co-carbonization of rice straw and acid whey for enhanced hydrochar properties and nutrient recovery. Green Energy Resour 2(2). https://doi.org/10.1016/j.gerr.2024.100077
88. Zou R, Qian M, Wang C, Mateo W, Wang Y, Dai L, Lin X, Zhao Y, Huo E, Wang L, Zhang X, Kong X, Ruan R, Lei H (2022) Biochar: from by-products of agro-industrial lignocellulosic waste to tailored carbon-based catalysts for biomass thermochemical conversions. In: Chem Eng J 441. Elsevier B.V. https://doi.org/10.1016/j.cej.2022.135972

Chapter 4
Co-feedstock for Bio-refinery: Status and Future Perspectives

Riya Banik, Kulbhushan Samal, Jitendra Singh Verma, Rashmi Ranjan Sahoo, Chanchal Loha, and Sachin Rameshrao Geed

Abstract Alternative energy sources have drawn a lot of interest for energy security due to rising global energy demand and the depletion of fossil resources. A sustainable, economically viable biorefinery concept has a lot of potential for producing renewable biofuel and value-added byproducts from the variety biomass feedstock. However, a specific feedstock biomass is essential for the biorefinery process to convert biomass into biofuels, but this feedstock is seasonally accessible. The economic sustainability of a biorefinery using a single feedstock is currently not considered feasible, due to the irregular feedstock availability which hider large-scale biofuel production. Therefore, co-feedstock biorefinery would be sustainable with cost-effective processes which can effectively utilize diverse feedstocks. Furthermore, the appropriate co-feedstock combination in the supply chain is necessary for a sustainable and economically viable biorefinery. The current status of the co-feedstock biorefinery for the production of biofuels and value-added byproducts along with insights into future prospective and challenges of co-feedstocks biorefinery is covered in this chapter.

Keywords Bio-refinery · Co-feedstock · Waste utilization · Bio-energy · Biofuel

---

R. Banik · J. S. Verma · S. R. Geed (✉)
CSIR-North East Institute of Science and Technology, Jorhat, India
e-mail: sachingeed@neist.res.in

K. Samal (✉) · R. R. Sahoo · C. Loha
CSIR-Central Mechanical Engineering Research Institute, Durgapur, West Bengal, India
e-mail: k.samal@cmeri.res.in

R. Banik · K. Samal · J. S. Verma · R. R. Sahoo · C. Loha · S. R. Geed
Academy of Scientific and Innovative Research (AcSIR), Ghaziabad, India

M. Aslam et al. (eds.), *Production and Biorefining of Biocrude Oil: Current Status and Future Developments*, Advances in Sustainability Science and Technology,
https://doi.org/10.1007/978-981-96-5198-6_4

1 Introduction

1.1 Alternative Energy Sources

Biofuels have garnered global attention in the recent time. The governments across glob have underscored the importance of attaining energy security for their country, setting a goal to diminish import dependence, specifically in the use of fossil fuels. Concerns regarding the reliance on imported fuels, coupled with environmental pollution issues, have intensified the demand for alternative fuels that not only offer superior environmental benefits but also present economic competitiveness compared to fossil fuels. This underscores the strategic significance of biofuels within the Indian energy landscape. Additionally, this initiative aims to provide better income prospects for farmers, with the objective of doubling their earnings. Simultaneously, it seeks to address environmental concerns arising from the use of fossil fuels and the burning of biomass/waste. The approach aligns with the Swachh Bharat Abhiyan, addressing challenges related to waste management and agri-residue management, while promoting the "Make in India" campaign. Currently, bioethanol is the predominant biofuel utilized in transportation sector, with the industry heavily dependent on high-starch feedstocks, primarily corn, for ethanol production. To meet the biorefineries feed requirements, there is a concerted effort to explore a diverse array of feedstocks. This encompasses alternative crops derived from starch and sugar, an array of crop residues like corn stover, wheat straw, rice straw, and sugarcane straw, dedicated energy crops such as switchgrass, miscanthus, mixed prairie grasses, and short-rotation trees, as well as residues from forests. Moreover, the imperative to address ongoing climate change and the depletion of conventional fuel reserves underscores the urgency for a swift transition from a linear to a circular economy [35].

In response to the gradual depletion of fossil fuels, environmental degradation, and the urgent challenges of climate change, there is a rising focus on sustainable and renewable energy resources. Abundant renewable energy sources, including diverse biomass materials like wood and forestry residues, crops and their byproducts, marine products, and bio-wastes, are being explored for conversion into bio-oil. This utilization of biomass not only addresses energy needs but also provides an optimal solution for waste management. The annual biomass waste generated from the agricultural and forestry industry is estimated to be around 140 gigatons. Conventional methods of handling biomass residues, such as open-air burning, contribute to significant environmental pollution. To combat this issue, these agricultural and forestry wastes can be utilized as cost-effective biomass sources for energy generation, simultaneously addressing challenges related to biomass waste management. Essentially, the conversion of diverse biomass materials into bio-oil offers a dual benefit by serving as a renewable energy source while efficiently managing and repurposing agricultural and forestry waste. This approach aligns with the increasing emphasis on sustainable practices, providing a viable solution to both energy and environmental concerns.

India generates an average of 500 million tonnes (MT) of crop residue annually, according to the Indian Ministry of New and Renewable Energy (MNRE). Out of which 92 MT is burned each year [6]. As of 2023, current data indicates that corn stover ranks as the most extensively produced crop residue in the United States. Despite its adaptability to various deconstruction processes, the actual availability of corn stover is affected by some farmers' hesitancy to authorize its removal from their fields. On the other hand, sugarcane bagasse presents a more readily accessible alternative as it is a by-product of sugarcane juice extraction. Brazil has successfully implemented a commercial-scale conversion of bagasse to ethanol. However, a primary hurdle in converting these widely utilized biomass feedstocks is the rigorous nature of the pretreatment process, often characterized by temperature. The stringent conditions required for high-sugar yields can concurrently lead to the generation of furfural, hydroxymethylfurfural (HMF), acetic acid, and phenolic compounds. On a global scale, the Swedish government has articulated a long-term vision for a sustainable energy supply with the goal of achieving zero greenhouse gas emissions and a non-fossil fuel-powered transportation sector by 2030 [2, 13]. As a response to these movements, current energy systems face heightened examination and the imperative for transformation. Environmental policies are gaining traction and unity, aiming to tackle various societal issues linked to environmental deterioration. These policies underscore considerations for human health, well-being, equity, and the imperative to fundamentally reshape the adverse side effects produced by traditional economic models. Central elements of these environmental policies involve prioritizing resource efficiency and advocating for a circular economy. Examples of such initiatives include the European Green Deal, officially sanctioned in 2019, and the enduring climate strategy endorsed by the European Commission in 2018. This reflects a global shift towards sustainable practices, acknowledging the interconnected nature of environmental, social, and economic considerations.

1.2 The Significance of Feedstock in the Biorefinery Process

The choice of feedstock is crucial in biorefineries, impacting economic viability, environmental sustainability, and overall efficiency. It serves as the raw material for bio-based product production, influencing emissions and resource utilization. Diverse feedstocks offer flexibility, aligning with local practices, and play a vital role in shaping a sustainable bioeconomy. A key consideration is determining the most suitable feedstocks for generating biogas energy while adhering to circular economy and sustainability principles. Life cycle analysis (LCA) provides a theoretical framework, but assessing the practical adoption of sustainable feedstock practices among existing biogas operators is equally essential. Circular bioeconomy, often defined in various ways, consistently emphasizes the circularity of resource use and waste conversion into value-added bio-based products. Researchers highlight the opportunity of circular bioeconomy for implementing circular economy principles in bioresource utilization. Adopting a circular economy approach emphasizes the integration of anaerobic digestion (AD) into waste management practices. This includes energy processing of various organic wastes such as sewage sludge, bio-waste, catering waste, and agricultural by-products like manure and straw. Recycling organic waste through AD not only reduces greenhouse gas emissions associated with landfill disposal but also contributes to energy decentralization, reducing reliance on fossil fuels. The anaerobic phase in the design and operation of anaerobic reactors is markedly influenced by the physical and chemical attributes of feedstocks. These characteristics have a substantial impact on both biogas generation and digestion stability. Key factors, including the C/N ratio, substrate concentration, pH, and temperature, are pivotal in determining the success of biogas production. Nitrogen, essential for amino acid synthesis, influences pH conditions and can lead to nutrient limitation or ammonia toxicity. Anaerobic co-digestion, especially with cattle manure, addresses potential imbalances in pH, alkalinity, and nutrient elements, enhancing biogas production and bioreactor performance [4]. Considering the cost of feedstocks is paramount, as it significantly influences the overall cost of biofuel production. High yield dedicated energy crops are expected to become major feedstocks, necessitating efforts to encourage farmers to transition from grain cultivation to bioenergy feedstocks. Buy-back guarantees can provide assurance to farmers, mitigating reluctance and minimizing the risk associated with the shift in cultivated products. Exploring diverse feedstock prospects is crucial to meeting the demands of a sustainable and efficient bioenergy industry. The overview of feedstock is visually represented in Fig. 1, providing a comprehensive illustration of sources and biomass involved in utilization for biofuel and biochemical production.

1.3 Seasonal Feedstock Availability and Its Impact on Large-Scale Biofuel Production

The challenge of seasonal feedstock availability significantly impacts large-scale biofuel production, causing supply chain fluctuations and efficiency disruptions. Varied growth cycles and harvest seasons create temporal imbalances in biomass availability, necessitating strategic planning, storage solutions, and feedstock source diversification for a continuous, reliable supply. This underscores the need for a resilient and adaptable biofuel production infrastructure to optimize efficiency and mitigate risks associated with unpredictable seasonal feedstock availability. The chemical characteristics of lignocellulosic materials, commonly referred to as biomass, are predominantly regulated by five essential components: cellulose, hemicellulose, lignin, extractives/volatiles, and ash. The composition of these components varies across biomass types, influencing biomass effectiveness, which fluctuates seasonally.

While most studies on biofuel supply chains consider aggregated yearly operations, the often-overlooked feedstock seasonality introduces system interdependence

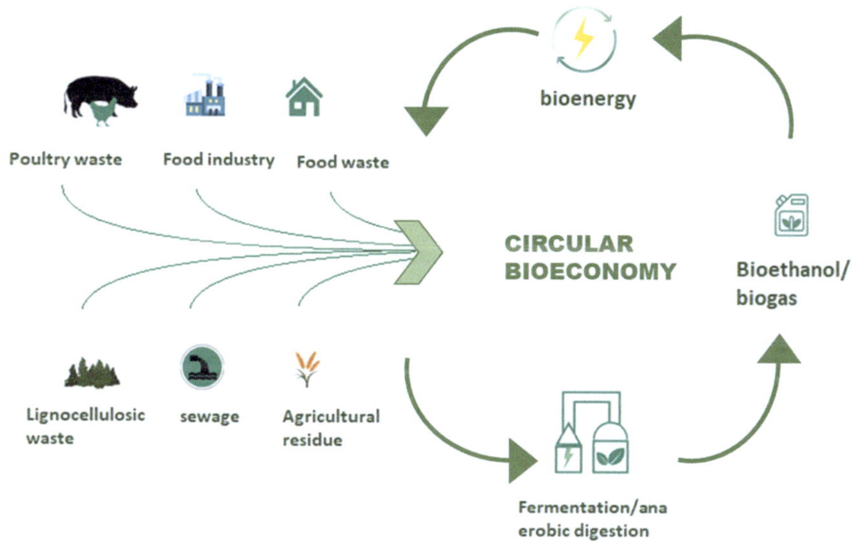

Fig. 1 Overview of co-feedstock biorefineries for addressing energy, resource, and environmental challenges by integrating multiple feedstocks

across temporal dimensions. In biorefinery systems, biomass harvesting limitations, occurring during a brief window, result in processing being restricted to only a few months annually, leaving equipment idle for the remainder of the year. This seasonality also impacts the transport and delivery of feedstock, with varying availability for common feedstocks used in biorefineries, such as sugar cane bagasse, cane trash, forest sawmill residues, agricultural wastes, and algae, each exhibiting distinct production patterns. The seasonality of crops, particularly in fruit and vegetable waste (FVW), affects the anaerobic digestion (AD) process and marketability of feedstock. FVW generation is strongly linked to seasonal variations in quantity and composition, resulting in reported methane production variations of up to 40% across different seasons. Examining the impact of seasonality and particle size reduction on substrate composition and the anaerobic digestion (AD) process for waste from the Mediterranean fruit and vegetable market unveiled significant fluctuations in methane production. This underscores the necessity for adaptive strategies in addressing these variations. The fluctuating availability of feedstocks throughout the year may deter investments in biorefinery plants. However, Pasteur-based biorefinery plants offer a potential solution by operating even with limited feedstock availability. Addressing these challenges requires a holistic understanding of the seasonal dynamics of feedstock availability, enabling the development of resilient and sustainable biofuel production systems.

1.4 The Need for Co-feedstock Biorefineries as a Solution

Co-feedstock biorefineries emerge as a crucial solution to address the challenges of energy security, resource depletion, and environmental protection. Shifting toward the circular economy concept in organic waste management, co-feedstock utilization in co-digestion optimizes existing wastewater digestion capacity in urban environments. While co-feedstocks enhance biogas and nutrient production in farm-scale digesters, challenges arise in sparsely populated remote areas due to the seasonal availability of certain biomasses [12]. Anaerobic digestion of cattle manure as a mono-substrate proves unstable due to low C/N ratios, prompting the exploration of co-digestion to overcome deficiencies and enhance stability. This approach balances the C/N ratio, alleviates inhibitory effects, and promotes stable biogas production [44]. Large-scale livestock operations generate significant challenges to sustainable development through the release of pathogens, excess nutrients, and greenhouse gases from animal waste. Current practices like land spreading contribute to substantial greenhouse gas emissions, particularly during waste storage and application. Co-feedstock utilization offers an alternative by harnessing the untapped bioenergy potential in animal waste, aligning with a circular bioeconomy approach and presenting a more sustainable solution to animal waste management [41, 45]. The integration of various substrates in co-feedstock scenarios demonstrates the potential to overcome individual challenges and exemplifies a circular bioeconomy approach. Co-feedstock biorefineries, as depicted in Fig. 2, are essential for addressing energy, resource, and environmental challenges by integrating multiple feedstocks and providing a sustainable solution to biomass supply variations.

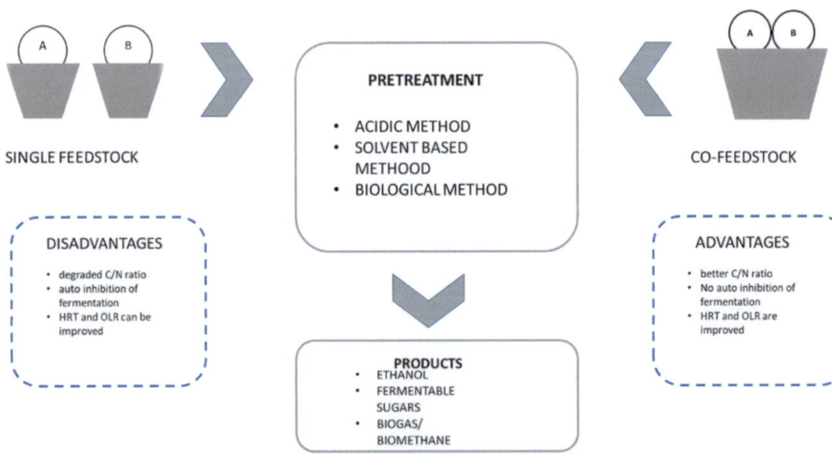

Fig. 2 Comparison of single feedstock and co-feedstock involved in utilization for biofuel and biochemical production

2 Single Feedstock Biorefineries

Biorefineries utilizing single feedstocks represent a focused approach to biofuel production, relying on specific biomass sources for synthesizing biofuels and other valuable products. The simplicity in processing and logistical operations is a distinctive feature of these facilities. However, challenges persist, with seasonal variations in feedstock availability introducing unpredictability and the potential for disrupting production schedules. Moreover, the limited adaptability of single feedstock biorefineries to fluctuating market demands can impact their economic viability. The efficiency of these biorefineries is intricately tied to the characteristics of the selected biomass, including its composition, growth cycle, and geographic factors. These elements influence both the production process and overall biorefinery performance, with transportation costs and supply chain resilience emerging as critical factors determining economic feasibility. Despite these challenges, biorefineries employing single feedstocks play a crucial role in the bioenergy landscape, offering a targeted and streamlined approach, especially when the chosen biomass aligns with local agricultural practices or waste streams. Their contribution to sustainable energy production is noteworthy. To bolster the sustainability and resilience of single feedstock biorefineries, strategic planning becomes imperative. This entails meticulous consideration of feedstock selection, regional factors, and supply chain dynamics. Technological innovations, such as advancements in harvesting and storage techniques, can mitigate the impact of seasonal variations. Additionally, exploring co-feedstock strategies or integrating complementary feedstocks could provide solutions to enhance flexibility and address market dynamics. Despite inherent limitations, these biorefineries remain integral components of the bioenergy landscape, and continuous technological advancements are essential to optimize their efficiency, economic viability, and overall contribution to a sustainable bioeconomy. Table 1 provides a comprehensive overview of various co-feedstock mixtures utilized in biorefinery processes, detailing the associated pretreatment or processing methods and resulting products. The first entry involves mixed hardwood subjected to acid catalyst pretreatment, yielding fermentable sugars [25]. The second entry combines wheat bran and rice straw, utilizing fungal enzyme hydrolysis to produce cellulases with a yield of 60 U g^{-1} for D-xylanases [34]. The third entry employs a combination of aspen and switchgrass, using acid pretreatment to achieve a yield of fermentable sugars ranging from 69 to 93% [7]. Hemlock hog fuel and pin chips, subjected to acid pretreatment, result in a yield of 13.6 g of glucose as fermentable sugars [23]. Sugarcane bagasse combined with straw undergoes acid pretreatment, producing glucose with a yield of 15 g/l [29]. Finally, rice straw and wheat bran, processed through enzyme hydrolysis, yield polysaccharides with a specific activity of 14.71 IU g^{-1}. Table 1 highlight various feedstock mixtures and processing methods in biorefineries, emphasizing the range of products obtained and associated yields. These studies contribute valuable insights into optimizing co-feedstock utilization for enhanced biofuel and bioproduct production.

Table 1 Co-feedstocks and processing methods used in biorefineries

Feedstock mixtures	Pretreatment/ process used	Product	Yield	References
Mixed hardwood	Acid catalyst	Fermentable sugars	–	[25]
Wheat bran + rice straw	Fungal enzyme hydrolysis	Cellulases	60 U g^{-1}	[34]
		D-xylaneses	740 U g^{-1}	
Aspen + switch grass	Acid pretreatment	Fermentable sugars	69–93%	[7]
Hemlock hog fuel + pin chips	Acid pretreatment	Fermentable sugars	13.6 g of glucose	[23]
Sugarcane bagasse + straw	Acid pretreatment	Glucose	15 g l^{-1}	[29]
Rice straw + wheat bran	Enzyme hydrolysis	Polysaccharides	14.71 IU g^{-1}	

2.1 Challenges and Limitations of Single Feedstock Biorefineries

Single feedstock biorefineries face challenges due to their vulnerability to seasonal variations in feedstock availability, potentially disrupting production schedules and affecting economic viability. Limited adaptability to fluctuating market demands makes them less agile in responding to evolving industry needs. The efficiency of these biorefineries depends on the characteristics of the chosen biomass, and transportation costs and supply chain resilience are critical determinants of economic feasibility. The need for specialized equipment and processes tailored to a specific feedstock results in higher capital costs and reduced versatility compared to facilities handling multiple feedstocks. Addressing these challenges requires strategic planning, technological innovations, and potential integration of complementary feedstocks or co-feedstock strategies to enhance their resilience and overall contribution to a sustainable bioeconomy.

Feedstocks often prove unsuitable for mono-digestion due to their unfavorable carbon-to-nitrogen (C/N) ratio [4]. High nitrogen content in pig manure leads to ammonia inhibition during anaerobic digestion, resulting in low bioenergy conversion efficiency [18]. Utilizing animal waste as a sole substrate is deemed impractical and ineffective because of its inherent carbon deficiency, characterized by a low C/N ratio. This deficiency results in a comparatively meager biogas output and, in certain instances, renders the process economically unfeasible. Other mono-substrates, such as dairy effluent and agro-industrial waste, may present challenges as well. These substrates may display low organic loads and elevated nitrogen concentrations, which can impede the activity of methanogenic microorganisms (Archaea). Conversely, dairy effluent and agro-industrial waste might possess low nitrogen content, leading to a slower decomposition process. This, in turn, extends the period of biogas production due to reduced microbial growth. It is important to consider these factors when evaluating the feasibility and efficiency of utilizing different mono-substrates in anaerobic digestion systems. Co-digestion emerges as a viable solution to overcome these challenges, enhancing economic viability through increased biogas production [38].

The anaerobic digestion (AD) of individual substrates (mono-digestion) is associated with specific challenges related to the properties of the substrates involved. For instance, sewage sludge (SS) is characterized by low organic loads, animal manures have low organic loads coupled with high nitrogen concentrations that can inhibit methanogens, and municipal solid waste (OFMSW) contains inappropriate materials along with a relatively high concentration of heavy metals. Crops and agro-industrial wastes, being seasonal substrates, may lack nitrogen, while slaughterhouse wastes (SHW) pose risks due to their high nitrogen concentration and/or the presence of long-chain fatty acids (LCFA), which can potentially inhibit methanogenic activity. To overcome these limitations, anaerobic co-digestion (AcoD) is employed, involving the addition of a co-substrate. This approach effectively addresses the challenges associated with mono-digestion, providing a more versatile and efficient solution to the diverse issues presented by different substrates [27]. Mono-digestion systems, in which anaerobic digestion (AD) plants utilize a single feedstock, exhibit certain drawbacks, including digester instability, limited year-round availability of specific feedstocks, the presence of heavy metals, and a low yield of biogas/methane. The rapid buildup of volatile fatty acids (VFAs) in digesters hampers methanogens in biodegradable organic waste streams, while proteinaceous wastes contribute to the formation of toxic compounds like ammonia (NH_3) and hydrogen sulfide (H_2S). Addressing these issues involves strategies such as pH adjustment, control of retention time, intermittent feeding, micro-aeration, and the implementation of a two-stage AD system. Although the introduction of microbial cultures and biochar can enhance system performance, they also bring about operational complexity and additional costs. The utilization of digestate from mono-digestion raises environmental concerns, including the accumulation of heavy metals, elevated soil salinity, phytotoxicity, and ecotoxicity, issues that co-digestion may potentially mitigate. In the context of food waste mono-digestion, challenges arise, such as a rapid hydrolysis rate leading to a pH drop below 5.5 due to the accumulation of VFAs [21]. Owing to its high biodegradability and a comparatively low C/N ratio, the mono-digestion of food waste may encounter hindrances such as the rapid production of VFAs from starch and the release of free ammonia from protein. Similarly, the mono-digestion of lignocellulosic green waste presents challenges, including inadequate nutrient content, a sluggish start-up process, and extended retention times.

2.2 *Co-feedstock Biorefineries*

Co-feedstock biorefineries pioneer an innovative approach to biofuel production by integrating multiple biomass sources, addressing challenges linked to single feedstock facilities. This integration mitigates the impact of seasonal variations, ensuring a consistent biomass supply and enhancing resource efficiency in biofuel production. Aligned with circular bioeconomy principles, this strategy fosters a more resilient and adaptable system. In urban settings, co-feedstock utilization, particularly in co-digestion processes, optimizes spare digestion capacity in existing wastewater facilities. This not only streamlines organic waste management but also enhances biogas and nutrient production. On farm-scale digesters, co-feedstocks boost self-sufficiency and economic viability, offering a strategic solution to challenges faced in sparsely populated remote areas with highly seasonal biomass availability Moreover, co-feedstock strategies play a pivotal role in mitigating the environmental impact of large-scale livestock operations, which release significant quantities of pollutants. Co-feedstock biorefineries offer a sustainable alternative to conventional waste management practices like land spreading, reducing greenhouse gas emissions and tapping into the bioenergy potential of organic compounds. Co-feedstock biorefineries present a crucial solution to contemporary biofuel production challenges. Their integration of diverse biomass sources improves efficiency, promotes sustainability, and contributes to the advancement of a circular bioeconomy. With ongoing technological advancements and strategic planning, co-feedstock biorefineries are positioned to play a central role in shaping the future of bioenergy production.

2.2.1 Benefits of Co-feedstock Utilization

Co-feedstock utilization in industrial processes brings numerous benefits, enhancing resource efficiency, cutting production costs, and minimizing waste to promote environmental sustainability and circular economy principles. This strategic approach not only fosters adaptability to market changes but also encourages innovation by necessitating advanced technologies for integrating diverse feedstocks, aligning with principles of environmental stewardship and technological advancement. Shifting focus to co-digestion, it offers advantages such as an improved C/N ratio and balanced macro- and micronutrients. In rural settings, anaerobic co-digestion proves effective by incorporating agro-industrial waste and organic municipal solid waste into manure-based digesters, enhancing biogas production. Optimal co-digestion ratios, approximately 70% food waste on a volatile solids basis, benefit sewage sludge and animal manure co-feedstocks, providing essential micronutrients and alkalinity supplementation. Introducing poultry waste as a co-substrate in anaerobic digestion addresses challenges in dilute dairy manure, yielding better methane production performance through synergistic effects. The integration of substrates in co-digestion yields heightened bio-methane production, leveraging potential synergies through the dilution of inhibitory compounds, the addition of valuable nutrients, and microbiome alterations that enhance metabolic processes. Co-digestion's effectiveness is supported by performance indices (CPI) ranging from 0.68 to 1.21, indicating synergistic effects. Contributing to a more robust operation, co-digestion optimally balances the C:N ratio, improves nutrient content, enhances buffering capacity, and dilutes inhibitory compounds. Studies, like Kim et al. [22], emphasize the benefits of anaerobic co-digestion for on-site treatment of household organic waste, demonstrating increased methane production with higher food waste proportions. Additionally, co-feedstock in digestion, as observed by Montoro et al. [28], improves nutrient recovery, enhancing nitrogen and potassium content when sweet potato is added to dairy cattle effluent. Lastly, relevant microbial populations in co-feedstock, highlighted by Montoro et al. [11], enhance product yield by diversifying methanogenic populations during anaerobic co-digestion of pig manure and food waste. Table 2 summarizes successful co-feedstock biorefinery projects, showcasing their locations, feedstock types, sizes, primary products, technologies, and capacities. These include ventures like the INEOS New Planet BioEnergy in Florida, Crescentino Biorefinery in Italy, Abengoa Bioenergy in Kansas, American Process, Inc. in Michigan, Gas Technology Institute in Illinois, ICM, Inc. in Kansas and Missouri, and the Renewable Energy Institute International in Ohio and California. These projects utilize a range of feedstocks, such as vegetative and yard waste, agricultural residues, mixed hardwood, and algae, producing diverse outputs like cellulosic ethanol, green power, ethanol, potassium acetate, renewable gasoline, diesel, and synthetic diesel fuel.

3 Technological Advancements and Innovations in Co-feedstock Processing

Technological advancements in co-feedstock processing mark a pivotal transformation in sustainable industrial practices. The emergence of advanced technologies, such as co-feedstock combinations and the emphasis on selecting the right co-feedstock combinations, represents a crucial stride in optimizing the utilization of diverse raw materials in industrial processes. The significance of carefully choosing co-feedstocks is underscored by the growing awareness of the impact it has on the overall efficiency of the processing system [10]. Integrated bio-refineries and advanced biochemical conversion techniques stand out as key components in this technological evolution. These innovations are expressly designed to enhance the efficiency of co-feedstock utilization, achieving the dual goals of minimizing waste and maximizing resource outputs. As industries increasingly prioritize the selection of suitable co-feedstocks, the integration of these advanced technologies not only supports economic viability but also aligns with sustainable principles by fostering environmental stewardship and resource conservation in industrial operations. The ongoing evolution of co-feedstock processing technologies underscores a promising trajectory toward further improving the feasibility and environmental sustainability of industrial practices.

Table 2 Illustrates successful co-feedstock biorefinery projects, detailing their locations, feedstock types, sizes, primary products, technologies, and capacities

Prime	Location	Feedstock (s)	Feedstock size	Primary products	Technology	Capacity	References
INEOS New Planet BioEnergy	Vero Beach, Florida Capacity	Vegetative and yard waste, MSW	300 dry tons per day	Cellulosic ethanol, renewable power	Commercial	8 million gallons per year and 6 megawatts (gross) of electricity generation	[20]
Crescentino Biorefinery—PROESA™, Italy	Crescentino, Province of Vercelli, Piedmont Region, Italy	Agricultural residues	–	Advanced biofuel (Cellulosic ethanol) Green electricity (from lignin) Biogas	TRL7		[31]
Abengoa Bioenergy U.S. Holdings, LLC	Hugoton, Stevens County, Kansas	Agriculture residues	1200 tons per day	Cellulosic ethanol, green power	Commercial	25 MM gallons ethanol, 20 MW power	[1]
American Process, Inc.	Alpena, Michigan	Mixed northern hardwood and aspen	30 bone-dry tons per day	Ethanol, potassium acetate	Pilot	894,200 gallons per year of cellulosic ethanol and 696,000 gallons per year of aqueous potassium acetate	[3]

(continued)

Table 2 (continued)

Prime	Location	Feedstock (s)	Feedstock size	Primary products	Technology	Capacity	References
Gas Technology Institute	Des Plaines, Illinois (Office and Project Site)	Wood, agricultural byproducts, and algae		Renewable gasoline and diesel	Research and development	Bench scale and 50 kg per day	[14]
ICM, Inc. (Project)	Colwich, Kansas (Headquarters), St. Joseph, Missouri	Corn fiber, switch grass, and energy sorghum	10 tons of cellulosic feedstock per day	Cellulosic ethanol	Pilot	245,000 GY fuel or product	[19]
Renewable Energy Institute International	Toledo, Ohio (IBR site); Sacramento, California, and Maumee, Ohio (IBR Engineering)	25 dry tons per day of agriculture and forest residues		Approximately 54 gallons of drop-in synthetic diesel fuel from one dry ton of biomass	Pilot	–	[32]

3.1 Co-feedstock Combinations

The integration of co-feedstock combinations has become indispensable in modern industrial processes, strategically merging multiple raw materials to optimize resource utilization and enhance efficiency. This innovative approach directly addresses challenges related to resource scarcity and waste management by leveraging synergies among different feedstocks. The selection of appropriate co-feedstock combinations is a critical factor in establishing a balanced and sustainable processing system, considering aspects like feedstock compatibility, nutrient content, and overall environmental impact. Industries are increasingly recognizing the importance of tailored co-feedstock strategies to maximize economic benefits while minimizing ecological footprints. This significance extends across various sectors, particularly in bio-refineries and biochemical conversion processes. Meticulous selection and blending of feedstocks in these applications can lead to improved yields, reduced waste, and enhanced production efficiency. For example, the synergistic combination of agricultural residues with organic municipal waste in biorefineries exemplifies a circular and integrated approach, where the by-products of one process become valuable inputs for another, contributing significantly to overall sustainability. Co-feedstock combinations represent a forward-looking paradigm in industrial processing, offering a holistic and efficient means of utilizing diverse raw materials. As technology continues to advance, ongoing optimization of co-feedstock strategies will play a pivotal role in shaping environmentally conscious and economically viable industrial practices.

Table 3 complements this narrative by providing a comprehensive overview of co-feedstock combinations, detailing primary and secondary substrates, feedstock ratios, reactor types, hydraulic retention times (HRT), resulting products, and production values. The studies highlighted in Table 3, such as those conducted by Cucina et al. [10], Dennehy et al. [11], and Hu et al. [18], underscore the versatility and potential of co-feedstock combinations in enhancing biomethane and biogas production across various substrates and reactor types. These findings offer valuable insights into optimizing organic waste utilization for sustainable energy generation, contributing to the ongoing evolution of environmentally conscious industrial practices.

3.2 Importance of Choosing the Right Co-feedstock Combinations

Choosing appropriate co-feedstock combinations in industrial processes is crucial, influencing process efficiency, economic viability, and environmental sustainability. Optimal combinations enhance efficiency by considering diverse feedstock compositions, maintaining nutrient balance, and minimizing risks through supply chain diversification. Cost-effective co-feedstock choices ensure competitiveness, and environmentally conscious decisions contribute to sustainable practices. Co-feedstock

Table 3 Overview of co-feedstock combinations and factors

Primary substrate	Secondary substrate	Feedstock ratio	Reactor	HRT	Name of product	Production	References
Buckwheat milling residue	Slauterhouse waste	–	–	–	Biomethane	485 ± 38 NL kg/VS	[10]
	Fruit waste					365 ± 30 NL kg/VS	
	Brewery trub					485 ± 38 NL kg/VS	
Pig manure	Food waste	40:60	Batch	21 days	Biomethane	751 L CH_4 m^{-3} days-1(highest recoreded in the work) (VMY)	[11]
Pig manure	Filamentous microalgae (*Tribonema sp.*)	1:1	Batch	–	Biomethane	27.4% increase in yield 580 mL/g VS (BMP)	[18]
Isochrysis galbana	Digested sludge from mesophilic and thermophilic anaerobic reactors	–	Batch	–	biogas	440 ± 25 mLBiogas/gVS	[8]
Selenastrum capricornutum							
Brewery spent grain	Sewage sludge	30G:30L	CSTR	21 days	Biomethane	0.48 ± 0.03 m^3/kg	[35]
Food waste	Chicken manure		Fed Batch (STR)anaerobia digestion	35 days	Biomethane	507.58 ml/g VS	[40]
Pig manure	Food waste	1:1	Batch	–	Biomethane	291.7 NmL/g VS	[41]
Food waste	Green waste	40:60	Batch	27.5 days	Biomethane	272.1 mL/g	[9]
Activated sludge	Algae	–	Batch	45 days	Biogas	262 mL/gVSfed	[39]
Sewage sludge	Algae	–	Fed batch	–	Biogas	2880 mL/day	[37]

combinations often include waste utilization, facilitate regulatory compliance, and enable customization of product properties, underscoring their pivotal role in shaping a resilient industrial landscape. In the realm of co-digestion, plants surpass single-substrate counterparts in total substrate volume, resulting in larger active digester volumes, higher specific methane production (SMP), and volumetric methane production (MP_Vtot) [2]. These benefits span environmental, technological, and economic domains, reducing toxicity, balancing nutrient demand, and enhancing biogas production. Co-digestion feedstocks, such as cattle waste, food waste, and industrial organic waste, contribute to improved biodegradability and volatile solids (VS) in anaerobic reactors, maintaining an optimal carbon-to-nitrogen (C/N) ratio [2, 4].

Crucial micronutrients for anaerobic digestion (AD) include Ni, Co, Mo, Fe, Se, Wo, Zn, Cu, and Mn, provided by manures. Efficient AD requires a C/N ratio around 15, and co-feedstock addition aids in maintaining this balance, preventing ammonia inhibition. The stability of the digestion process depends on substrate compositions, with sudden changes causing disturbances, such as volatile fatty acid accumulation. While protein-based feedstocks may elevate nitrogen load, co-feedstocks often synergize, enhancing biogas production. Specific co-feedstocks improve pH, buffer capacity, and trace element concentrations, enhancing overall AD performance. During lignocellulosic residues treatment, prioritizing easy biodegradable organic matter in co-digestion substrates is crucial. Algae offer a sustainable approach for balancing carbon-to-nitrogen ratios in wastewater treatment, capturing CO_2 for nutrient recovery and increased biogas production [10, 12, 36]. Monodigestion of ammonia-rich PM is prone to ammonia inhibition, causing process instability and low methane production. Co-digestion, particularly with substrates like food waste, emerges as an economically feasible solution [41]. In a separate study, biomass mixtures (aspen, balsam, and switchgrass) underwent dilute acid pretreatment and enzymatic hydrolysis, showing no synergism or antagonism in sugar recovery [7]. Furthermore, sugarcane straw exhibited higher enzymatic digestibility than bagasse, with enzymatic hydrolysis increasing with higher straw content in mixtures [29].

3.3 Factors to Consider in Selecting Co-feedstocks

The selection of co-feedstocks for an industrial process necessitates a comprehensive assessment of various factors to optimize efficiency, cost-effectiveness, and sustainability. Foremost among these considerations is the compatibility of the co-feedstocks with the primary feedstock and the overall production process, ensuring a seamless integration. The role of co-feedstock availability and cost is pivotal, with their affordability significantly influencing the economic viability of the production process. It is essential to account for geographical and seasonal variations in feedstock availability to maintain a consistent supply chain. Environmental factors, including the carbon footprint and ecological impact of co-feedstocks, assume increasing importance in the context of sustainable practices. Furthermore, a thorough analysis of the composition and quality of co-feedstocks is imperative to ensure adherence to desired specifications without compromising the integrity of the final product. By diligently examining these factors, industries can make well-informed decisions that strike a balance between economic, environmental, and operational considerations in the selection of co-feedstocks.

3.3.1 Digest Dewaterability

Digestate dewaterability is crucial in co-feedstock selection for anaerobic digestion. Factors such as polymer demand, organic and inorganic content, and microbial stress conditions influence dewatering efficiency. Optimizing these factors ensures the production of a valuable solid byproduct and facilitates proper management of the liquid fraction. Digestate dewatering, integral to anaerobic digestion, involves removing water to yield a solid, nutrient-rich byproduct [43]. Pre-conditioning, featuring polymer addition and physical separation, is pivotal. Key factors impacting efficiency are polymer demand and digestate dewaterability, influenced by digestate characteristics. Performance relies on the organic and inorganic content, particularly the impact on solid–liquid fraction separation. Microbial stress, generating extracellular polymeric substances, decreases dewaterability and lowers volatile solid removal efficiency. Managing these factors is vital for optimal dewatering, ensuring a valuable solid byproduct and effective liquid fraction disposal. Insights into microbial dynamics underscore the need for a holistic approach, integrating biological and engineering aspects to enhance anaerobic digestion efficiency.

3.3.2 Hydraulic Retention Time (HRT)

Hydraulic retention time (HRT) is a critical factor in selecting co-feedstocks for anaerobic digestion. It directly impacts substrate processing efficiency, influencing the economic feasibility of bioprocesses. Achieving the right balance is crucial for optimal biogas production, emphasizing the need for tailored HRT durations based on specific feedstock characteristics. HRT is a critical parameter closely tied to the substrate processing capacity per unit time, directly impacting the economic viability of a bioprocess. A brief HRT can enhance the rate of hydrogen production, minimizing capital costs by reducing bioreactor size. However, an excessively short HRT may lead to system failure through the washout of crucial, slow-growing bacteria essential for hydrogen production, compromising substrate degradation efficiency. Generally, a short HRT is advantageous for hydrogen-producing bacteria due to their higher growth rates compared to hydrogen-consuming bacteria like methanogens [24]. For co-digestion scenarios involving vegetable oil, pig manure, olive mill wastewater, olive mill solid wastewater, food waste, sewage sludge, and agro-industrial wastes, optimal HRT values have been reported. During the acidogenic phase, a range of 1–5 days is recommended, while the methanogenesis phase is optimal with an HRT between 20 and 25 days [17]. In a separate study by [26], the reported HRT ranged from 15 to 40 days. Notably, methane yields generally increased with extended HRT, with a recommended range of 30–40 days for optimum anaerobic digestion (AD). These findings emphasize the importance of tailoring HRT to specific co-digestion scenarios for maximizing biogas production efficiency. Optimizing hydraulic retention time (HRT) in co-digestion processes is crucial for balancing economic considerations and maximizing methane yields, with specific recommendations varying depending on the substrate composition.

3.3.3 Organic Loading Rate (OLR)

OLR is a crucial factor in co-feedstock selection for anaerobic digestion. It significantly impacts biogas production, requiring a delicate balance to avoid inhibitions and ensure system stability. Careful consideration, along with understanding co-substrate composition, is essential for optimal anaerobic digestion performance. In the study by Grangeiro et al. [16], the organic loading rate (OLR) is highlighted as a critical factor in biogas production during anaerobic digestion (AD). OLR, determined by the amount of organic material per unit reactor volume subjected to AD within a given time period, plays a crucial role in system performance. While increased substrate concentration is linked to higher OLR and improved biogas production, caution is needed, as excessive concentrations may lead to inhibitions and alterations in microbial pathways. During anaerobic co-digestion (AcoD), the risk of organic overload is emphasized, as it can result in the accumulation of inhibitory compounds like VFAs and LCFAs [43]. Co-substrates, categorized into carbohydrate-rich, protein-rich, and lipid-rich materials, exhibit varying impacts during AcoD. High organic loading with carbohydrate-rich substrates can induce VFAs accumulation, causing process instability, while protein-rich substrates lead to the formation of ammonium-N and free ammonia. Lipid-rich substrates, although providing high biogas yields, pose challenges with the accumulation of long-chain fatty acids, suppressing methanogenic microbes and causing system instability. Maintaining a balance and avoiding co-substrate organic overloading is crucial for stable AcoD [43]. Despite digestates being rich in plant macronutrients, challenges arise in their agricultural reuse due to residual phytotoxicity. Factors contributing to phytotoxicity, such as high salinity, Na content, and TVFA concentration, necessitate careful inoculums selection and operational conditions for obtaining agriculturally compatible digestates. This aligns with the findings of Cucina et al. [10]. Studies on co-digestion of manures and organic wastes, exemplified by Dennehy et al. [11], reveal that higher fat and protein content in food wastes enhance biomethane potential and substrate biodegradability, underscoring the significance of feedstock composition. Compared to unicellular microalgae, filamentous microalgae offer easier and cost-effective harvesting methods due to their physical characteristics. Their potential use in bioenergy generation is emphasized by Hu et al. [18], highlighting their large cell size, aggregate-forming ability, and high filterability. Anaerobic co-digestion of organic-rich poultry waste presents challenges, particularly regarding ammonia inhibition. Zhang et al. [45] emphasize the importance of understanding the complex anaerobic food web, with methanogens being susceptible to excessive ammonia. The study by Wang et al. [40] supports the benefits of co-digestion, showing improved results when poultry waste is combined with other biomass. Finally, the total solids (TS) content significantly influences methane production, with higher TS contents impacting methane yield and production rates in various digestion processes [41]. These studies collectively underscore the intricate dynamics involved in co-digestion processes, emphasizing the need for careful consideration of substrate characteristics, process parameters, and microbial interactions to optimize overall system performance.

4 Current Status of Co-feedstock Biorefineries

Co-feedstock biorefineries, driven by a commitment to optimize the utilization of diverse biomass resources while overcoming compatibility and processing challenges. These refineries play a pivotal role in diversifying the bioenergy sector, providing economic and environmental benefits through efficient resource utilization and waste reduction. Anticipated technological advancements, supported by policies and industry collaborations, are poised to elevate co-feedstock biorefineries, playing a crucial role in realizing a greener and more sustainable energy future. The integration of co-digestion and co-feedstock utilization in various biorefineries is currently in the pilot or Technology Readiness Level (TRL) stage. An advantageous aspect of co-digestion lies in its efficient utilization of equipment and cost-sharing, as existing facilities can be modified, or new ones can be constructed to process multiple waste streams in a single unit.

A notable co-digestion case at the Viareggio and Treviso wastewater treatment plants in Italy, as reported by Bolzonella et al. [5], involved treating source-sorted organic fraction of municipal solid waste (OFMSW) with sewage sludge. Another research findings indicated a 50% surge in biogas production when the organic loading of the reactor was raised from 1.0 to 1.2 kg VS/m^3 d. Additionally, the second plant under investigation demonstrated a fivefold increase in monthly biogas production [5]. In the global landscape, the prevalence of biogas systems is noteworthy. China boasts a total of 110,448 biogas systems, with 6972 being large scale, while Europe, with 17,783 plants and 10.5 GW installed capacity, is led by Germany (10,971 plants), followed by Italy, France, Switzerland, and the UK. The USA operates 2200 anaerobic digesters with a 977 MW installed capacity, India has an estimated 300 MW biogas-based installed capacity, and Canada possesses about 180 digesters with 196 MW installed capacity. Collectively, these figures suggest the operation of approximately 132,000 small, medium, or large-scale digesters worldwide, underscoring the global growth of the biogas industry. IRENA statistics reveal a 90% increase in global electricity generation from biogas, surging from 46,108 GWh in 2010 to 87,500 GWh in 2016. The World Biogas Association notes that many of these plants utilize multiple substrates [42].

Furthermore, mixed biomass utilization extends beyond biogas production, finding application in the production of various commodities such as bioethanol, biohydrogen, biogas/methane, polyhydroxyalkanoates, microbial enzymes, and in power, heat, and cooling applications. Mixed biomass has been employed in combination with coal for power generation and in combined heat and power (CHP) systems. Although studies on mixed biomass utilization are comparatively fewer than those focusing on single biomass applications, the potential of adopting mixed biomass utilization (MLB) across these diverse applications holds promise for enhancing the biorefinery concept at large [30]. The current advancements in co-digestion and co-feedstock utilization signal a transformative shift toward sustainable and efficient bioenergy production. This integration of diverse biomass not only optimizes resource usage and cost-sharing but also plays a vital role in minimizing waste and environmental impact. With the global expansion of the biogas industry and the versatility of mixed biomass utilization, it is clear that embracing these technologies is essential for advancing the broader biorefinery concept and promoting a sustainable future for global energy systems.

5 Future Prospects and Challenges

The future prospects for co-feedstocks in biorefineries present a promising trajectory towards advancing sustainable bioenergy production. Integrating a diverse array of organic materials creates opportunities to optimize resource efficiency, elevate biogas quality, and diversify the range of products derived through anaerobic digestion. However, challenges lie in the necessity for robust process optimization to accommodate variations in co-substrate properties, ensuring both stability and efficiency. Addressing concerns such as phytotoxicity, inhibitory compounds, and microbial dynamics is essential for the successful implementation of co-feedstock strategies in biorefineries. Conquering these challenges is pivotal for unlocking the full potential of co-feedstocks, contributing to a resilient and eco-friendly future for bioenergy. While the synergistic digestion process in the absence of oxygen (anaerobic co-digestion) frequently results in improved efficiency, the economic and environmental outcomes may not consistently meet expectations. This discrepancy is attributed to the expenses and environmental effects linked to acquiring and managing co-substrates. The sustained availability of environmentally sustainable co-substrates is imperative to the enduring economic feasibility and environmental efficacy of co-digesting pig manure [18]. The integration of microalgae as an anaerobic digestion feedstock presents an attractive aspect, allowing for a closed loop in microalgal biomass production and anaerobic digestion by-product management [18]. Dewatering digestate is essential for maintaining substrate availability and recovering nutrients, and co-digestion with microalgae has been shown to enhance digestate dewaterability [36].

In wastewater treatment plants (WWTPs), sewage sludge is a significant by-product, posing challenges in terms of disposal and operational costs. Anaerobic digestion of sludge is a common stabilization process in WWTPs, recovering energy through biogas generation. Additionally, WWTPs can serve as a potential nutrient source for microalgae growth, utilizing CO2 and nitrogen/phosphorus present in wastewater [8]. Landfill leachate (LL), known for its toxicity, is a challenging substrate, but pre-treatments like hydrodynamic cavitations (HC) have shown promise in managing bio-refractory compounds and toxic contaminants [35]. The organic fraction of municipal solid waste (OFMSW) presents challenges in terms of regulatory frameworks, pre-treatment requirements, composition variability, and inhibitions caused by factors such as VFA accumulation and toxic substances. Despite these challenges, steam explosion and low-temperature pre-treatments have shown promising results in increasing methane yield. The future implementation of anaerobic co-digestion in agriculture relies on developing economically self-sustainable systems, emphasizing optimal conditions, innovative pre-treatments, and revenue generation through digestate application as fertilizer [27]. Co-feedstocks in biorefineries offer promising prospects for sustainable bioenergy. Overcoming challenges tied to substrate variations, inhibitory compounds, and microbial dynamics is crucial. Integrating diverse organic materials, like pig manure, microalgae, and organic waste, provides opportunities to enhance resource efficiency in anaerobic digestion. Effective management of economic and environmental considerations is essential. The future of co-feedstocks points toward a resilient and eco-friendly bioenergy landscape, emphasizing innovation, efficiency, and environmental stewardship.

6 Conclusion

The exploration of alternative energy sources, prompted by the escalating global energy demand and the depletion of fossil resources, accentuates the promise of a sustainable and economically viable biorefinery concept. Challenges arise from the dependence on a specific feedstock, particularly due to its seasonal availability, rendering the economic sustainability of a single feedstock biorefinery impractical for large-scale biofuel production. The remedy lies in co-feedstock biorefineries, which leverage cost-effective processes to efficiently harness a diverse array of feedstocks. This chapter underscores the critical importance of judiciously selecting co-feedstock combinations within the supply chain, ensuring both sustainability and economic viability. Offering insights into the present state and future outlook of co-feedstock biorefineries, the chapter actively contributes to the ongoing discourse surrounding challenges and potential solutions in this pioneering approach to biofuel and byproduct production.

Acknowledgements The authors express heartfelt gratitude to the Director of CSIR-North East Institute of Science and Technology, Jorhat, and the Director of CSIR-Central Mechanical Engineering Research Institute, Durgapur, for their invaluable support.

References

1. Abengoa Bioenergy Biomass of Kansas (2012) Abengoa bioenergy biomass of Kansas integrated biorefinery for conversion of biomass to ethanol, power, and heat (July 2011), p 2012. https://www1.eere.energy.gov/bioenergy/pdfs/ibr_commercial_abengoa.pdf
2. Ahlberg-eliasson K, Nadeau E, Lev L (2017) Biomass and Bioenergy Production ef fi ciency of Swedish farm-scale biogas plants 97:27–37. https://doi.org/10.1016/j.biombioe.2016.12.002
3. Alpena Biorefinery', alpena biorefinery pilot-scale project (January)DOE (2013), p 8807. Available at: https://www1.eere.energy.gov/bioenergy/pdfs/ibr_arra_api.pdf
4. Avs FTY (2016). Effects of co-substrate on biogas production from cattle manure: a review. https://doi.org/10.1007/s13762-016-1069-1
5. Bolzonella D, Battistoni P, Susini C, Cecchi F (2006) Anaerobic codigestion of waste activated sludge and OFMSW: the experiences of Viareggio and Treviso plants (Italy). Water Sci Technol 53(8):203–211. https://doi.org/10.2166/wst.2006.251
6. Bhuvaneshwari S, Hettiarachchi H, Meegoda JN (2019) Crop residue burning in India: policy challenges and potential solutions. Int J Environ Res Publ Health 16(5):832. https://doi.org/10.3390/ijerph16050832
7. Brodeur-Campbell M, Klinger J, Shonnard D (2012) Feedstock mixture effects on sugar monomer recovery following dilute acid pretreatment and enzymatic hydrolysis. Bioresour Technol 116:320–326. https://doi.org/10.1016/j.biortech.2012.03.090
8. Caporgno MP, Trobajo R, Caiola N, Ibáñez C, Fabregat A, Bengoa C (2015) Biogas production from sewage sludge and microalgae co-digestion under mesophilic and thermophilic conditions. Renew Energy 75:374–380. https://doi.org/10.1016/j.renene.2014.10.019
9. Chen X, Yan W, Sheng K, Sanati M (2014) Comparison of high-solids to liquid anaerobic co-digestion of food waste and green waste. Bioresour Technol 154:215–221. https://doi.org/10.1016/j.biortech.2013.12.054
10. Cucina M, Pezzolla D, Tacconi C, Gigliotti G (2021) Anaerobic co-digestion of a lignocellulosic residue with different organic wastes: relationship between biomethane yield, soluble organic matter and process stability. Biomass Bioenerg 153:106209. https://doi.org/10.1016/j.biombioe.2021.106209
11. Dennehy C, Lawlor PG, McCabe MS, Cormican P, Sheahan J, Jiang Y, Zhan X, Gardiner GE (2018) Anaerobic co-digestion of pig manure and food waste; effects on digestate biosafety, dewaterability, and microbial community dynamics. Waste Manag 71:532–541. https://doi.org/10.1016/j.wasman.2017.10.047
12. Ervasti S, Kostensalo J, Tampio E (2022) Effects of seasonal and local co-feedstocks on the performance of continuous anaerobic digestion of cattle slurry. Bioresour Technol Rep 19:101207. https://doi.org/10.1016/j.biteb.2022.101207
13. Fagerström A, Anderson S (2019) The contribution of Advanced Renewable Transport Fuels to transport decarbonization in Sweden-2030 and beyond
14. Gasoline and Diesel from Wood, Agricultural Waste, and Algae R & D (2012)
15. Gasoline and Diesel from Wood, Agricultural Waste, and Algae R & D' (2012) (December), p 2012. https://www1.eere.energy.gov/bioenergy/pdfs/ibr_arra_gti.pdf
16. Grangeiro LC, de Almeida SGC, de Mello BS, Fuess LT, Sarti A, Dussán KJ (2019) New trends in biogas production and utilization. Sustain Bioenergy Adv Impacts. https://doi.org/10.1016/B978-0-12-817654-2.00007-1
17. Hagos K, Zong J, Li D, Liu C, Lu X (2017) Anaerobic co-digestion process for biogas production: progress, challenges and perspectives. Renew Sustain Energy Rev 76:1485–1496. https://doi.org/10.1016/j.rser.2016.11.184
18. Hu Y, Kumar M, Wang Z, Zhan X, Stengel DB (2021) Filamentous microalgae as an advantageous co-substrate for enhanced methane production and digestate dewaterability in anaerobic co-digestion of pig manure. Waste Manag 119:399–407. https://doi.org/10.1016/j.wasman.2020.10.041
19. Icm, inc. Biorefinery project'Ibr_Arra_Icm' (2012) (December), p 8502. Available at: https://www1.eere.energy.gov/bioenergy/pdfs/ibr_arra_icm.pdf

20. Ineos Demonstration Project Ineos New Planet BioEnergy Commercializes Bioenergy Technology in Florida renewable biofuels and' (2012) 2011(August 2011), p 7165. https://www1.eere.energy.gov/bioenergy/pdfs/ibr_arraprojects_ineos.pdf
21. Karki R, Chuenchart W, Surendra KC, Shrestha S, Raskin L, Sung S, Hashimoto A, Kumar Khanal S (2021) Anaerobic co-digestion: current status and perspectives. Bioresour Technol 330:125001. https://doi.org/10.1016/j.biortech.2021.125001
22. Kim J, Kim J, Lee C (2019) Anaerobic co-digestion of food waste, human feces, and toilet paper: methane potential and synergistic effect. Fuel 248:189–195. https://doi.org/10.1016/j.fuel.2019.03.081
23. Kim KH, Tucker M, Nguyen Q (2005) Conversion of bark-rich biomass mixture into fermentable sugar by two-stage dilute acid-catalyzed hydrolysis. Bioresour Technol 96:1249–1255. https://doi.org/10.1016/j.biortech.2004.10.017
24. Kim MS, Cha J, Kim DH (2013) Fermentative biohydrogen production from solid wastes (1st ed). Biohydrogen. Elsevier B.V. https://doi.org/10.1016/B978-0-444-59555-3.00011-8
25. Lim WS, Lee JW (2013) Effects of pretreatment factors on fermentable sugar production and enzymatic hydrolysis of mixed hardwood. Bioresour Technol 130:97–101. https://doi.org/10.1016/j.biortech.2012.11.122
26. Ma G, Ndegwa P, Harrison JH, Chen Y (2020) Methane yields during anaerobic co-digestion of animal manure with other feedstocks: a meta-analysis. Sci Total Environ 728:138224. https://doi.org/10.1016/j.scitotenv.2020.138224
27. Mata-Alvarez J, Dosta J, Romero-Güiza MS, Fonoll X, Peces M, Astals S (2014) A critical review on anaerobic co-digestion achievements between 2010 and 2013. Renew Sustain Energy Rev 36:412–427. https://doi.org/10.1016/j.rser.2014.04.039
28. Montoro SB, Lucas J, Santos DFL, Costa MSSM (2019) Anaerobic co-digestion of sweet potato and dairy cattle manure: a technical and economic evaluation for energy and biofertilizer production. J Clean Prod 226:1082–1091. https://doi.org/10.1016/j.jclepro.2019.04.148
29. Moutta RDO, Ferreira-Leitão VS, Bon EPDS (2014) Enzymatic hydrolysis of sugarcane bagasse and straw mixtures pretreated with diluted acid. Biocatal Biotransform 32:93–100. https://doi.org/10.3109/10242422.2013.873795
30. Oke MA, Annuar MSM, Simarani K (2016) Mixed feedstock approach to lignocellulosic ethanol production—prospects and limitations. Bioenergy Res 9(4):1189–1203. Available at: https://doi.org/10.1007/s12155-016-9765-8
31. Piero O (2018) Crescentino biorefinery—PROESATM, Italy. Technology Demo/Industrial Scale Implementation. IEA Bioenergy
32. Renewable Energy Institute International Demonstration of a Pilot Integrated Biorefinery for the Economical Conversion of Biomass to Diesel Fuel (2012)
33. Renewable energy institute international Demonstration of a Pilot Integrated Biorefinery for the Economical Conversion of Biomass to Diesel Fuel' (2012), (December), p 2012. Available at: https://www1.eere.energy.gov/bioenergy/pdfs/ibr_arra_reii.pdf
34. Shamala TR, Sreekantiah KR (1986) Production of cellulases and d-xylanase by some selected fungal isolates. Enzyme Microb Technol 8:178–182. https://doi.org/10.1016/0141-0229(86)90109-2
35. Szaja A, Montusiewicz A, Lebiocka M, Bis M (2021) A combined anaerobic digestion system for energetic brewery spent grain application in co-digestion with a sewage sludge. Waste Manag 135:448–456. https://doi.org/10.1016/j.wasman.2021.09.034
36. Thorin E, Olsson J, Schwede S, Nehrenheim E (2018) Co-digestion of sewage sludge and microalgae—biogas production investigations. Appl Energy 227:64–72. https://doi.org/10.1016/j.apenergy.2017.08.085
37. Varol A, Ugurlu A (2016) Biogas production from microalgae (*Spirulina platensis*) in a two stage anaerobic system. Waste Biomass Valorization 7:193–200. https://doi.org/10.1007/s12649-015-9442-8
38. Velásquez Piñas JA, Venturini OJ, Silva Lora EE, Calle Roalcaba OD (2018) Technical assessment of mono-digestion and co-digestion systems for the production of biogas from anaerobic digestion in Brazil. Renew Energy 117:447–458. https://doi.org/10.1016/j.renene.2017.10.085

39. Wang M, Sahu AK, Rusten B, Park C (2013) Anaerobic co-digestion of microalgae Chlorella sp. and waste activated sludge. Bioresour Technol 142:585–590. https://doi.org/10.1016/j.biortech.2013.05.096
40. Wang M, Sun X, Li P, Yin L, Liu D, Zhang Y, Li W, Zheng G (2014) A novel alternate feeding mode for semi-continuous anaerobic co-digestion of food waste with chicken manure. Bioresour Technol 164:309–314. https://doi.org/10.1016/j.biortech.2014.04.077
41. Wang Z, Jiang Y, Wang S, Zhang Y, Hu Y, Hu Z, hu, Wu, G., Zhan, X., (2020) Impact of total solids content on anaerobic co-digestion of pig manure and food waste: insights into shifting of the methanogenic pathway. Waste Manag 114:96–106. https://doi.org/10.1016/j.wasman.2020.06.048
42. World Biogas Association (2019) 'Global potential of biogas', World Biogas Association, p 56. Available at: https://www.worldbiogasassociation.org/global-potential-of-biogas/
43. Xie S, Higgins MJ, Bustamante H, Galway B, Nghiem LD (2018) Current status and perspectives on anaerobic co-digestion and associated downstream processes. Environ Sci Water Res Technol 4:1759–1770. https://doi.org/10.1039/c8ew00356d
44. Zhang C, Xiao G, Peng L, Su H, Tan T (2013) The anaerobic co-digestion of food waste and cattle manure. Bioresour Technol 129:170–176. https://doi.org/10.1016/j.biortech.2012.10.138
45. Zhang Y, Zamudio Cañas EM, Zhu Z, Linville JL, Chen S, He Q (2011) Robustness of archaeal populations in anaerobic co-digestion of dairy and poultry wastes. Bioresour Technol 102:779–785. https://doi.org/10.1016/j.biortech.2010.08.104

Chapter 5
Hydrothermal Liquefaction of Solid Residues and Organic Waste to Bio-crude Oil

Chhavi Shukla, Darpan Lakhmani, and B. Rajasekhar Reddy

Abstract Dwindling fossil fuel stocks, rising demand for petroleum fuels from developing nations, and the burgeoning cost of conventional fossil fuels raise the danger of the global energy supply. Among various renewable energy resources, biomass usage as a sustainable biofuel feedstock represents a viable alternative to fossil fuel. Hydrothermal liquefaction (HTL) is a novel thermochemical process that aims to produce bio-crude at mild operating temperatures, reducing energy consumption and eliminating the need for preliminary drying by using a variety of high-moisture feedstocks. HTL crude bio-oil is often more stable than rapid pyrolysis oil, with lower oxygen and moisture levels, energy density, and calorific value. Solid organic wastes like municipal, industrial and agricultural waste, whose management and applications require extensive research, can be effectively utilized as feedstock for bio-crude production. Plastic as a co-reactant improves bio-oil production and decreases the environmental concerns of waste plastic landfilling, which might result in effective waste management. This chapter reviews various characterization methods of solid wastes, HTL studies on organic and solid wastes, the implication of reaction strategies, and product quantity and quality parameters. This chapter provides an extensive overview of the HTL process, including the fundamental mechanisms, the effects of operating conditions such as temperature, pressure, and residence time, and the challenges associated with feedstock variability and product characterization. Finally, the limitations and future scope for commercial HTL technology are discussed.

Keywords Solid waste · Hydrothermal liquefaction · Biomass · Plastics · Bio-crude

---

Chhavi Shukla, Darpan Lakhmani are equally contributed

C. Shukla · D. Lakhmani · B. Rajasekhar Reddy (✉)
Department of Fuel, Minerals and Metallurgical Engineering, Indian Institute of Technology (Indian School of Mines), Dhanbad, Dhanbad, India
e-mail: rajasekhar@iitism.ac.in

M. Aslam et al. (eds.), *Production and Biorefining of Biocrude Oil: Current Status and Future Developments*, Advances in Sustainability Science and Technology,
https://doi.org/10.1007/978-981-96-5198-6_5

1 Introduction

The basic component of life on Earth is energy. The depletion of fossil fuel supplies as a result of fast industrialization and urbanization has raised worries about greenhouse gas emissions and climate change. Because they account for around 85% of the world's energy needs, coal, natural gas, and petroleum meet that demand [1]. The hunt for alternative energy sources has intensified due to the natural energy reserves being depleted more quickly than anticipated. These sources may be crucial in the fight against climate change.

Tidal, solar, wind, geothermal, and biomass energy are now the available substitute feedstocks for fossil fuels. By using these renewable energy sources, the carbon cycle will be closed and the environmental carbon footprint will be decreased. Compared to other renewable energy sources, biomass is a potential option that can meet 14% of global energy needs [2]. Biofuels can be a good substitute for fossil fuels since they are made from renewable resources like biomass. The primary sources of biomass resources are wood and forest residues (tree barks, small trees, etc.), kitchen scraps, algae, manures, and agricultural crops and residues (nut shells, rice/wheat straws, corn stalks, etc.). Biofuels are divided into three groups: first, second, and third generation biofuels, depending on the kind of feedstock as shown in Fig. 1. Through a variety of conversion platforms, valuable compounds including alcohols, esters, carboxylic acids, phenolics, aldehydes, ketones, and sugars can be extracted from biomass [3]. Therefore, the generation of renewable energy, fuels, and chemicals, as well as the sustainability of energy, depend on biomass, the only renewable source of carbon.

Wastes from homes, streets, industries, plastics, and sanitation leftovers are all considered solid residue (SR). These feedstocks show great promise for producing biofuels due to their abundant availability and generation. Approximately 55 million tonnes of solid residue including organic wastes are produced annually in India, and this number is expected to rise by 5% in the future. The predicted annual increase in

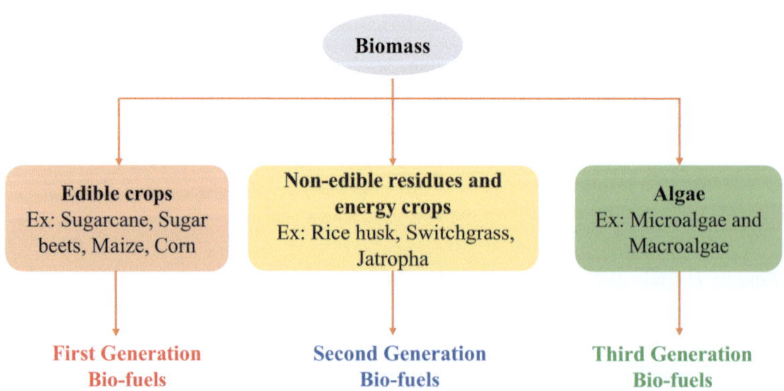

Fig. 1 Classification of bio-fuels obtained from various biomasses

SR generation per capita is 1.33%. Nowadays, the most prevalent thermochemical processes used in SR disposal are combustion and incineration. However, this causes toxic chemicals to be released into the atmosphere, increasing air pollution and the average world temperature. Even though landfilling has been used to keep waste away from the environment, doing it improperly simply serves to turn landfills into dumpsites. Ground water quality is impacted by landfills and dumps because of the leaching of organic materials and harmful metals. Therefore, the carbon cycle can be stabilized by using an environmentally sustainable method to convert SR into usable goods, especially liquid biofuels. Given its accessibility and low cost of production, SR has the potential to be a fourth-generation feedstock for the synthesis of biofuels [4].

Biochemical and thermochemical conversion are the two main methods for converting biomass into biofuels and other valuable products [5]. The fermentation of carbohydrates into biogas, which is normally an anaerobic process that lasts for extended periods of time (10–20 days), is referred to as a biochemical conversion process. Another important stage in the synthesis of bio-ethanol is the biochemical conversion of carbohydrates to sugars. Using heat energy to change the chemical and physical characteristics of biomass is known as thermochemical conversion. The key procedures are gasification, direct combustion, liquefaction, torrefaction, and pyrolysis. Together with the production of energy, the process of gasification and combustion produces a gaseous product known as syn gas or flue gas [6]. One potential alternative fuel to traditional crude oil is bio-crude or bio-oil. Hydrothermal liquefaction (HTL) and pyrolysis are the two main processes that can be used to produce liquid fuels. The latter approach requires drying the feedstock to 10–15 weight percent moisture in order to make the process economically feasible. It also entails high temperatures (600–800 °C) and moderate pressure (1–5 bar) under inert ambiance. On the other hand, the former works at high pressures (200–300 bar) and moderate temperatures (280–370 °C) and does not require feedstock drying. Nicknamed "hydrous pyrolysis," HTL is a process that mimics how Mother Earth produces fossil fuels. When fuel became scarce during World War II, the Pittsburgh Energy Research Centre was the first to liquefy biomass [7]. Any type of organic feedstock can be liquefied using water as a solvent by HTL, and the technique is more cost-effective because there is no need to dewater the feedstock beforehand. HTL offers the benefit of de-oxygenating biomass by using hydrogen found in water. In contrast to the bio-oil derived via pyrolysis, the bio-crude generated via heat treatment yields greater stability at elevated temperatures, a low water content (~5%), a high calorific value (30–35 MJ kg^{-1}), and a high H/C ratio [8].

HTL generates solid residue (bio-char), gases, and water-soluble fraction in addition to bio-crude. Deoxygenation of biomass produces CO, CO, CH, and other low molecular weight oxygenates. Because of the partial conversion of the biomass, solid residue is either created by recombining unstable fragmented components of the biomass [9]. The residue that is solid has a high calorific value of around 20–22 MJ kg^{-1} and burns like coal. The HTL process has grown in popularity as a biofuel production method because of its capacity to convert feedstock with a high moisture content and its mild operating conditions. Techniques including steam reforming, catalytic cracking, and hydrodeoxygenation can raise the quality of bio-crude.

Hydrothermal liquefaction (HTL) is an alternative to classic conversion methods like as pyrolysis and gasification, which often require drying of the feedstock prior to conversion. HTL works in the presence of water, which may act as a solvent, a reactant, and a heat transfer medium, enabling it to convert a wide range of wet biomass feedstocks [10]. The biomass feedstock is supplied into a reactor vessel during the HTL process, where it is heated to a high temperature (typically between 250 and 400 °C) and high pressure (normally between 5 and 25 MPa) in the presence of water. This results in a supercritical water environment, in which water acts as both a solvent and a reactant. A sequence of chemical reactions, such as depolymerization, hydrolysis, and dehydration, quickly break down the biomass into a liquid bio-oil product in this environment.

HTL has the advantage of converting a wide range of biomass feedstocks, including algae, agricultural and forestry wastes, and sewage sludge. As a consequence, it is a versatile technology that may use a variety of waste streams while decreasing reliance on fossil fuels [11]. Moreover, since HTL operates at high pressure and moderate temperature, it may produce beneficial byproducts such as char, gas, and nutrient-rich aqueous phase that may be employed in energy and agricultural applications. Several studies have shown the advantages of HTL over other biomass conversion technologies [12]. For example, compared the conversion of rice straw into bio-crude by HTL to other conversion processes such as pyrolysis and gasification. According to the study, HTL produced the highest yield of bio-crude (about 34%) compared to pyrolysis (about 29%) and gasification (about 5%). In addition, the researchers observed that HTL utilized less energy and produced a bio-crude with a higher heating value than the other techniques. Mahesh et al. [13] studied the feasibility of using municipal solid waste (MSW) as a feedstock for the manufacture of HTL-based bio-crude. According to the study, HTL produced a bio-crude with a high yield (about 60%) and a high energy content. Experts believe that HTL is a viable technology for converting waste into a lucrative fuel.

2 Mechanism of Hydrothermal Liquefaction

Hydrothermal liquefaction is a complex process involving several chemical reactions, some of which are currently unknown. Yet, scientists have made enormous progress in recent years in understanding the physics of this process. Depolymerization is a critical step in hydrothermal liquefaction because it breaks down the large, complex molecules in biomass into smaller, simpler molecules. This reaction is thought to be enhanced by the high temperatures and pressures of the process, as well as the presence of water as a solvent. According to study, the degree of depolymerization increases as temperature, pressure, and reaction time increase. Another critical phase in hydrothermal liquefaction is dehydration, or the removal of water from biomass molecules. This process is thought to be driven by the reactor's high temperatures and low water activity, which fosters circumstances that favor dehydration over hydrolysis. Decarboxylation, which involves the removal of carboxylic acid groups from biomass molecules, is another important stage in hydrothermal liquefaction. This reaction is thought to occur mostly at high temperatures and low pH, and it is aided by the presence of mineral catalysts in the biomass. As a result, hydrolysis, or the breaking of chemical bonds with the addition of water, is an important stage in hydrothermal liquefaction. The presence of acid catalysts in the biomass, as well as the process's high temperatures and pressures, are suggested to help this reaction. HTL bio-oil is a complex mixture of oxygenated molecules such as acids, alcohols, and esters, as well as hydrocarbons. The quality of the bio-oil is determined by the kind of biomass feedstock used, the processing conditions, and the reactor design.

Water is heated and pressurized under HTL operating conditions, which acts as an acid/base catalyst to cause the chemical disintegration of biomass macrostructures. Reaction conditions must be kept at nearly super-critical water temperatures (374 °C and 221 bar) in order to liquefy a feedstock into a water-insoluble bio-crude. High pressure operation also prevents the solvent in the reactor from going into the vapor phase, which saves energy for the enthalpy of operation. The phase diagram of water and the regime where HTL conditions are normally maintained are depicted in Fig. 2.

Changes in water's physical and chemical characteristics at these sub-critical temperatures, such as density, ionic product, hydrogen bonding, viscosity, thermal conductivity, and dielectric constant, give water a non-polar character and increase its ability to dissolve organic biomass components. Since it is crucial to break the C–C and C–O bonds in the feedstock, these characteristics of water can be changed by adjusting the process's pressure and temperature. The usual characteristics of water at 25 °C and at sub-critical conditions (250–340 °C) are displayed in Table 1. Figure 3 depicts the key reaction mechanisms in the formation of different organic compounds from solid residues such as waste plastics and organic waste such as biomass. Dehydration of the hemicellulose component produces phenols and furan derivatives at low temperatures (150–180 °C), and it also causes the creation of

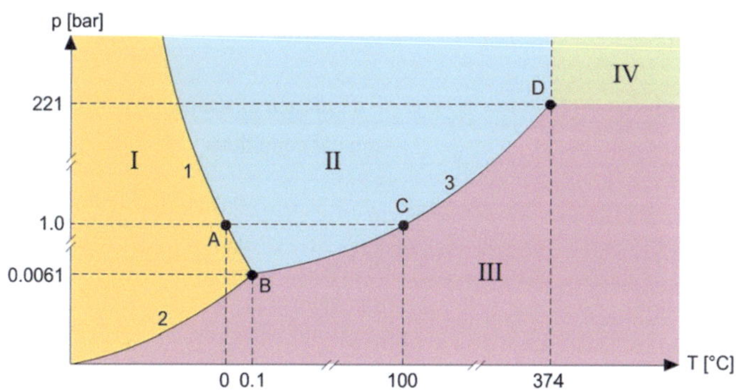

Fig. 2 Phase diagram of water (A—melting point, B—triple point, C—boiling point, D—critical point, 1—melting pressure curve, 2—sublimation curve, 3—vapor pressure curve, I—solid, II—liquid, III—gaseous, IV—supercritical fluid)

gases such as CO and CO_2 by ring scission or the decomposition of acids at higher temperatures. When the cellulosic part breaks down, glucose is produced. Glucose then breaks down further to produce aldehydes and ketones. The primary sources of furan derivatives are cellulose and hemicellulose. Over 300 °C causes the lignin and plastic in SR to depolymerize, resulting in the production of oxygenated chemicals and long-chain aliphatic hydrocarbons. The intermediate compounds of PET, such as acetaldehyde and benzoic acid, are expected to generate linear oxygenates and aromatics, particularly naphthalene derivatives. Cyclo-oxygenated molecules are also produced by chain scission of PET.

The plastic component of SR is seen to disintegrate more quickly at temperatures above 325 °C due to strong cracking reactions. This results in the creation of more aliphatic chemicals, including eicosane, tetradecane, nonadecane, and gaseous molecules in the C2-C4 hydrocarbon range. Polystyrene depolymerizes to styrene at high pressures and temperatures, which yields phenolic chemicals and derivatives of benzene. Lauric acid, decanoic acid, and palmitic acid are among the fatty acids that were also found in the bio-crude; their synthesis is linked to the breakdown of extractives found in SR. The formation of alcohols is mostly from the hydrogenation of glucose and hydration reactions of aliphatic hydrocarbons.

In order to achieve significant energy savings, the reaction pressure should be maintained above the saturation pressure that corresponds to the process temperature around the critical point of water, where the enthalpy of the solvent fluctuates significantly. Another crucial factor to take into account is the solubility of reactants in the solvent, which is reliant on the solvent's dielectric constant. Although water molecules tend to remain polar at super-critical conditions, the polarity of water can be changed on a macroscopic level by adjusting the process parameters. Table 1 makes it clear that water is a polar solvent at room temperature. However, when temperature rises, the dielectric constant drops, which causes the polarity of the liquid to decrease. The ability to dissolve non-polar molecules is improved as a result [13]. A better path for effective heat and mass transmission during HTL is created by a decrease in viscosity, density, and an increase in the ionic product of water, which causes the creation of H_3O^+ and OH^- ions with an increase in temperature.

Table 1 Physico-chemical properties of water as a function of temperature and pressure

Properties	Ambient conditions	Sub-critical water
Temperature (°C)	25	250
Pressure (bar)	1	50
Density (g/cm^3)	0.9	0.8
Dielectric constant (ε)	78.5	27.1
Ionic product	14	11
Viscosity (mpa S)	0.9	0.1
Thermal conductivity (mW m^{-1} K^{-1})	608	620

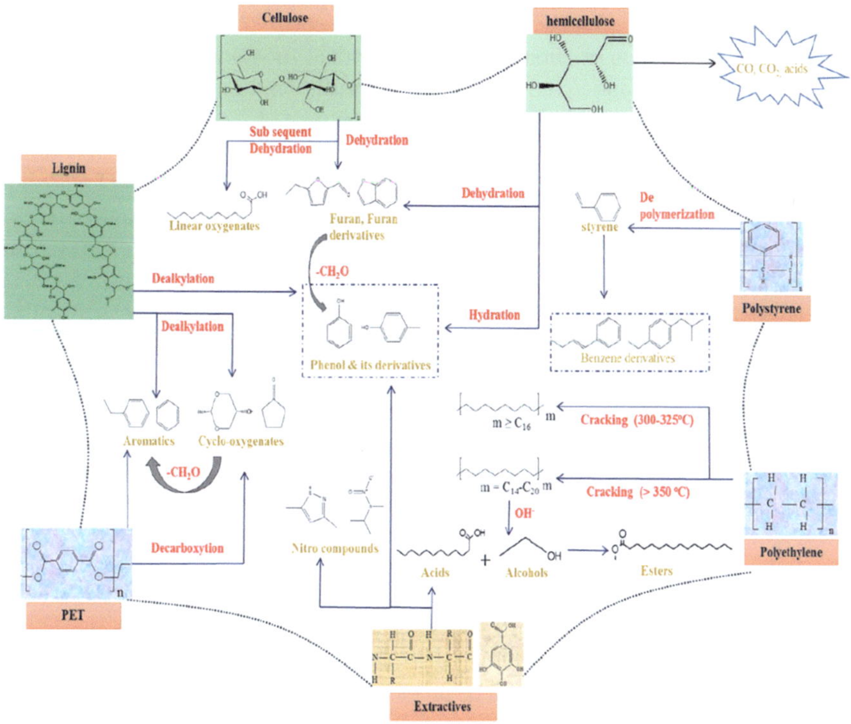

Fig. 3 Plausible reaction pathways for the formation of major organic compounds in bio-crude from various organic wastes and solid residue through HTL

3 Parameters Affecting HTL Process

3.1 *Reaction Temperature*

The temperature is a crucial parameter in HTL because it impacts the pace of chemical reactions as well as the quality of the bio-crude produced. Higher temperatures often result in quicker reaction rates and better bio-crude yields, but they may also cause the bio-crude to deteriorate. Higher temperatures, in general, result in faster reaction rates and larger bio-crude yields. Very high temperatures, on the other hand, might cause the bio-crude to deteriorate, resulting in lesser quality goods. The breakdown process may produce char, water-insoluble solid residues, and other undesirable byproducts. According to research, the ideal temperature for HTL of various feedstocks varies from 250 to 400 °C [13].

Process temperature is a crucial factor in HTL that affects product yields and composition. A rise in temperature accelerates the fragmentation of biomass into smaller parts [14]. Three steps make up the conversion of biomass during HTL: (a) hydrolysis; (b) creation of bio-crude; and (c) breakdown of bio-crude. The reaction temperature needs to be higher than the activation energy of bond scission for the initial depolymerization of biomass to its individual monomers, such as glucose, xylose, and some phenolic compounds, which results in the formation of a water-insoluble product (bio-crude). This depends on the type of feedstock [14].

Elevated temperature initiates the production of free radicals and increases the likelihood of fragmented species repolymerization. Determining the ideal temperature for a hydrothermal process is aided by the competition between fragmentation, repolymerization, and hydrolysis. The bio-crude yield in the majority of HTL tests using lignocellulosic feedstocks diminishes after 350 °C because of condensation/re-polymerization or secondary breakdown reactions that are actively generating gaseous products and solid residue. Because of the inadequate conversion of biomass, the yield of bio-crude is limited at low temperatures (< 280 °C). The breaking down of the ether bonds in biomass is what starts the breakdown of its macromolecules [15]. Because of its heterogeneous structure, the hemicellulosic part of biomass starts to break down in hot compressed water around 180 °C, whereas lignin breaks down at 200 °C and cellulose begins to break down at 250 °C because of its intractable character brought on by its crystalline nature [16]. To maximize liquid yield, HTL operation is typically conducted in the sub-critical water temperature range of 280–370 °C. More gaseous compounds are formed during high temperature (> 380 °C) operation as a result of strong cracking reactions. According to Sasaki et al. [17], HTL operation reduces the residence time required for total cellulose conversion from 1.5 s to 0.01 s when conducted at super-critical conditions (> 374 °C) of water as opposed to the sub-critical area (350 °C). Nevertheless, it was discovered that cellulose changed into a variety of components in both situations, including phenol and furan derivatives.

HTL of sewage sludge, for example, at 330 °C resulted in a bio-crude yield of 38% with a heating value of 34 MJ/kg and a carbon content of 72%, however at 350 °C, the bio-crude yield reduced to 32% and the heating value decreased to 31 MJ/kg [18]. Similarly, HTL of microalgae at 280 °C resulted in a 51% bio-crude yield with a heating value of 41.2 MJ/kg and a nitrogen content of 5.5%, however at 320 °C, the bio-crude yield fell to 43% and the heating value fell to 36.9 MJ/kg [19]. Moreover, several research have looked at the impact of temperature on the content of the bio-crude produced. For example, HTL of sugarcane bagasse at 280 °C produced a bio-crude with a high percentage of aromatic compounds, while HTL at 340 °C produced a bio-crude with a larger proportion of aliphatic compounds [20]. Consequently, it makes sense to consider that polysaccharides would hydrolyze to monomers under both subcritical and supercritical working conditions. Conversely, Carrier et al. [21] observed that chemical distribution changed dramatically with reaction temperature when they examined HTL of lignocellulosic biomass (fern fronds) in water under sub- and supercritical conditions. At sub-critical conditions, cyclopentanones and guaiacols were generated in substantial numbers with comparable amounts of phenolic and benzene compounds; at super-critical conditions, however, no benzene compounds were seen, and cyclopentanones and phenols constituted the majority of the bio-crude. Therefore, in order to produce bio-crude with the necessary yield and quality, it is crucial to set a correct reaction temperature in HTL [22].

3.2 Reaction Pressure

During HTL, pressure is reasonably important in the feedstock's defragmentation process. Ensuring that the entire reaction occurs in a single-phase regime (liquid phase) is the primary goal of high-pressure operation. High enthalpy penalties required for solvent phase change are avoided by HTL at single phase. In relation to the reaction temperature, the pressure needs to be set higher than the saturation pressure. By increasing the rate of hydrolysis and encouraging better defragmentation of biomass, this raises the yield of bio-crude. While the system's temperature can slightly impact the properties of water in its sub-critical state, only pressure changes can affect parameters like the dielectric constant, solvent density, and ionic product in the supercritical area. The solvent can readily permeate the biomass matrix and split it into its component parts because of the increase in solvent density. At extremely critical water conditions, pressure has essentially little effect on product yield [22].

The pressure in HTL affects the solubility of biomass in water and, as a consequence, the reaction rate. Higher pressures often result in higher yields and higher quality bio-crude, but they may also cause bio-crude to degrade. According to research, the optimal pressure for HTL of different feedstocks is between 10 and 30 MPa [20]. HTL of pine sawdust, for example, resulted in a bio-crude yield of 36.5%, a heating value of 33.5 MJ/kg, and a carbon content of 67.6% at 15 MPa, but at 25 MPa, the bio-crude yield climbed to 42.5% and the heating value increased to 35.2 MJ/kg. Similarly, at 20 MPa, HTL of sewage sludge produced a bio-crude yield of 40.9% with a heating value of 32.7 MJ/kg and a carbon content of 69.6%, but at 30 MPa, the bio-crude yield increased to 45.8% with a heating value of 33.8 MJ/kg [9]. Moreover, various research have been conducted to investigate the effect of pressure on the composition of the bio-crude generated [23]. HTL of microalgae, for example, generated a bio-crude rich in fatty acids and phenols at 20 MPa, while at 30 MPa, the bio-crude had more aliphatic hydrocarbons.

3.3 Residence Time

In HTL, residence time aids in determining the feedstock's total conversion, product distribution, and reactor sizing from a process economics standpoint. It is important to provide elementary processes enough time to break up the feedstock's macrostructures, which will ultimately increase the yield of bio-crude. As a result, determining the ideal reaction or residence duration is crucial for the effective conversion of the feedstock's organic components. Biomass has a low thermal conductivity (0.02–0.15 W m^{-1} K^{-1}), hence residence time is adjusted to maximize heat transfer. Lower residence times are said to inhibit secondary or tertiary cracking reactions, such as gasification, thermal cracking, and carbonization, which turn greasy components into solid residue, water-soluble products, and bio-char. Each of these reduces the bio-crude yield [24]. The biochemical makeup of biomass, including the amount of proteins, lipids, carbohydrates, cellulose, hemicellulose, and lignin, affects the recombination reactions of fragmented components.

The residence period refers to the amount of time the biomass spends in the HTL reactor. Longer residence durations often result in better bio-crude yields. Numerous studies have been conducted to evaluate the influence of residence time on the HTL of various feedstocks. For example, HTL of algae with a residence period of 30 min yielded 46.4% bio-crude, whereas a longer residence time of 60 min yielded 53.2% [25]. Similarly, HTL of maize stover with a residence period of 30 min yielded 31.8% bio-crude, whereas a longer residence time of 60 min yielded 43.5%. Nevertheless, too extended residence durations might cause bio-crude deterioration, resulting in inferior grade products. HTL of microalgae, for example, with a residence time of 180 min produced a bio-crude with a lower heating value and a greater oxygen content than HTL with a residence time of 60 min [26]. The influence of residence time on bio-crude composition has also been examined. For example, HTL of sewage sludge with a residence time of 15 min produced a bio-crude rich in fatty acids, while a longer residence time of 60 min produced a bio-crude high in phenols [9].

3.4 *Type of Solvent System*

The kind of solvent used in HTL has a significant impact on how biomass feedstocks are converted. In order to better defragment the feedstock and start the scission of reaction intermediates, the solvent encourages processes including hydration, solvolysis, and hydrolysis [27]. Rubbers, fabric, and plastic materials (polyethylene, polypropylene, polyvinyl chloride, and polystyrene) are examples of compact or dense feedstocks that are difficult to break down in HTL under sub-critical water conditions. Consequently, in order to maximize the amount of oil produced by their effective breakdown, HTL must be operated at high temperatures and pressures. However, because of increased enthalpy inputs and operating concerns under high pressure, this strategy might not always be advantageous. Sub-critical working conditions are thus achieved for the effective degradation of feedstocks by selecting a co-solvent with a density, molecular weight, and critical characteristics greater than those of water. Increasing the solvent mixture's overall polarity and producing more H^+ and OH^- ions are the major goals of utilizing co-solvents, as this improves redox processes. Because a substantial amount of solvent effectively extracts organics, increasing the yields of gas and liquid products, the mass ratio of feedstock to solvent is also a crucial parameter [28]. Hydrogen donor solvents utilized in hydrogenation include tetralin, tetrahydrophenanthrene, and hexahydropyrene [29].

3.5 Feedstock Characteristics

The moisture content of the feedstock might decrease HTL efficiency because too much moisture causes dilution of the products and reduces bio-crude production. Excessively dry feedstock, on the other hand, might result in poor heat transmission and decreased conversion efficiency [30]. Since ash may function as a catalyst and encourage undesired reactions during HTL, the ash level of the feedstock can also impact the production and quality of bio-crude [31]. One of the most critical parameters influencing the output and quality of bio-crude generated by HTL is the chemical composition of the feedstock. The primary components of biomass, such as cellulose, hemicellulose, and lignin, react differently during HTL, affecting the content and quality of the bio-crude. For example, lignin-rich feedstocks may provide greater heating value bio-crude, but cellulose-rich feedstocks yield higher oxygen content bio-crude [32]. Additional feedstock properties, such as particle size and preparation procedures, may also have an impact on the HTL process. Reduced particle size of the feedstock, for example, may increase the surface area accessible for reaction, while pre-treatment procedures like torrefaction or acid hydrolysis can change the chemical composition and reactivity of the feedstock [33].

3.6 Catalyst

Catalysts used in HTL may speed up the breakdown of complex biomass molecules into smaller, more readily liquefiable intermediates, promoting the creation of desired products such as bio-crude. For example, adding a zeolite catalyst to the HTL process has been proven to improve bio-crude production and quality by lowering oxygen concentration and boosting heating value [34]. Similarly, nickel-based catalysts have been demonstrated to boost HTL conversion efficiency by encouraging hydrogenation processes and inhibiting the generation of undesirable products such as char and gas [35]. Moreover, the kind and quantity of catalyst utilized may influence the yield and quality of bio-crude generated by HTL. For instance, the use of larger concentrations of catalyst may lead to better yields of bio-crude, but can also increase the creation of undesirable by-products [36]. Since various catalysts have varied selectivities toward different products and processes, the kind of catalyst utilized will significantly impact the chemical composition of the bio-crude.

3.7 Water-Biomass Ratio

In general, a greater water to biomass or organic waste (W/B) ratio may result in higher bio-crude yields and lower char and gas outputs because more biomass is dissolved in water and transformed into liquid products [37]. A greater W/B ratio, on the other hand, might increase the energy requirements for the operation since more water must be heated to the necessary temperature and pressure. Many studies have been conducted to study the impact of the W/B ratio on the HTL of different kinds of biomass. A research on the HTL of microalgae, for example, discovered that raising the W/B ratio from 5:1 to 10:1 resulted in a better bio-crude output but a lower heating value, while a W/B ratio of 20:1 resulted in a considerable increase in char and gas yields [38]. Another research on maize stover HTL discovered that a W/B ratio of 10:1 resulted in the maximum bio-crude production, but a W/B ratio of 30:1 resulted in a drop in bio-crude yield and an increase in gas yield [39]. Overall, the appropriate W/B ratio for HTL is determined by a number of parameters, including the kind of biomass utilized, the intended product characteristics, and the process's energy needs. Further studies are required to improve the selection and usage of the W/B ratio in HTL in order to increase the process's efficiency and sustainability.

3.8 pH

In general, a lower pH may result in greater bio-crude yields and lower char and gas outputs because more biomass is dissolved in water and transformed into liquid products. A lower pH, on the other hand, might accelerate reactor corrosion and necessitate the use of more costly materials. Since more biomass is converted into gas, a higher pH may result in lower bio-crude yields and greater gas yields. Significant research has been conducted to study the impact of pH on the HTL of different kinds of biomass. A research on the HTL of microalgae, for example, discovered that a pH of 5 resulted in the maximum bio-crude output and quality, but a pH of 9 resulted in a considerable drop in bio-crude yield [40].

3.9 Salinity

In general, increasing salinity may result in lower bio-crude yields and higher gas and char yields because more biomass is dissolved in water and transformed into gas and solids. Yet, increased salinity might exacerbate reactor corrosion and necessitate the use of more costly components. Lower salinity may increase bio-crude yields and product quality, but it may also need extra processing procedures to remove salts from the products. Many research has been conducted to evaluate the impact of salt on the HTL of different kinds of biomass. A research on the HTL of microalgae, for example, found that increasing the salinity from 0 to 30 g/L NaCl resulted in a drop in bio-crude output and an increase in gas and char yields [41].

4 Literature on Bio-crude Production via HTL of Various Biomass

Ye et al. [42] investigated the liquefaction of bamboo shoot shell (BSS) at atmospheric pressure using polyethylene glycol and ethylene glycol (EG) catalyzed by sulfuric acid. The impacts of several parameters on the liquefaction yield of BSS were investigated, including liquid–solid ratio, temperature, time, catalyst, solvents ratio, and material size. Many methodologies were used to examine the properties of the generated products in three fractions. The findings demonstrated that polyols may be successfully produced by liquefying BSS, an agricultural by-product. Cheng et al. [43] investigated the use of alcohol (methanol or ethanol) and water as solvents in the direct liquefaction of eastern white pine sawdust. The most successful solvent, according to the research, was a 50 wt% co-solvent of either methanol–water or ethanol–water, with a bio-oil production of roughly 65 wt.% and a biomass conversion of more than 95% at 300 °C for 15 min. The bio-oil generated was predominantly composed of phenolic compounds, which were followed by aldehydes, ketones, alcohols, esters, organic acids, and ethers. The use of hot-compressed ethanol as a liquefaction solvent aided in the breakdown of lignin into monomeric phenols. The X-ray diffraction patterns revealed that the feedstock's cellulose structure was totally transformed into amorphous carbon at about 300 °C and crystalline carbon at roughly 350 °C.

Yip et al. [44] investigated the liquefaction of bamboo in autoclaves using various solvents and liquid ratios catalyzed by HCl at 180 °C. According to the findings, phenol was the best solvent for bamboo liquefaction, with a yield of up to 99%. Using phenol as a solvent, cellulose, hemicelluloses, and lignin were virtually entirely destroyed. After liquefaction, the high molecular weight of bamboo reduced dramatically to roughly 1800 g mol^{-1}, and the low boiling point products of liquefied bamboo were identical regardless of the solvent utilized. The preparation of bio-oil through direct liquefaction of *Dunaliella tertiolecta* utilizing sub/supercritical ethanol–water as a medium is discussed by Chen et al. [45]. The synergistic effects of ethanol and water were discovered, resulting in a maximum bio-oil output of 64.68% and an ideal *D. tertiolecta* conversion of 98.24%. Several approaches were used to analyses the chemical makeup of the bio-oil, and a feasible reaction mechanism for the primary chemical component in *D. tertiolecta* is provided. Biller et al. [46] looked at the hydrothermal liquefaction (HTL) of microalgae in a continuous flow reactor, which produced 40% bio-crude. Because of its nitrogen and oxygen concentration, the bio-crude had a high heating value of 35 MJ/kg but was unsuitable for direct burning. The use of CoMo and NiMo catalysts lowered nitrogen content by 60% and oxygen content by 85%, yielding 25% gasoline, 50% diesel, and 25% heavy fuel oil fractions. GC–MS, Sim-Dis, and elemental analysis of the oils yielded information on fuel quality and nitrogen fractionation. The research emphasizes HTL's potential for biofuel generation as well as the significance of upgrading to increase fuel quality.

The effects of reaction temperature, rice stalk/ethanol ratio, and residence time on the liquefaction of rice stalk (RS) in ethanol were evaluated by Li et al. [47] under sub- and supercritical conditions in a batch autoclave. An oil output of ~ 55% and an RS conversion of 78.3% were obtained under ideal circumstances. According to GC/MS analysis, phenols and esters, which make up about 50% of the bio-oil product, are the main components of the bio-oil. A high heating value of 28.95 MJ/kg, a kinematic viscosity of 5.63 mm^2/s, and a water content of 2.2% were all characteristics of the produced bio-oil.

Singh et al. [48] examined how the solvents (H_2O, CH_3OH, and C_2H_5OH) affect the distribution of products and their composition after *Ulva fasciata*, a kind of macroalgae, is hydrothermally liquefied. Alcoholic solvents were used to convert acids into methyl and ethyl esters, which considerably enhanced the output of bio-oil. With CH_3OH and C_2H_5OH, the bio-oil production was 44% and 40%, respectively, whereas it was only 11% with water. The bio-aliphatic oil's content increased with the use of alcoholic solvents, according to 1H-NMR data. The bio residue that was produced utilizing alcoholic solvents and H_2O differed according to the findings of FTIR and SEM. According to the research, producing functional hydrocarbons for chemical feedstock by liquefaction with supercritical alcohols is a successful method.

At various temperatures, *Onopordum heteracanthum* stalks were transformed to liquid products in organic solvents with and without a catalyst. The impact of liquefaction parameters were examined by Durak et al. [49]. The maximum yield in acetone was found in the catalytic run with $ZnCl_2$ at 563 K. The liquid products collected were examined and described. The liquid products produced in methanol at 563 K included a total of 106 distinct chemicals. Zhai et al. [50] looked at how sewage sludge-based activated carbons (SSAC) affected the liquefaction of sewage sludge at 350 and 400 °C. At 350 °C, SSAC boosted the bio-production oil's and energy density. SSAC's catalytic factor changed depending on the temperature at which it was produced. At 350 °C and 400 °C, liquefaction with SSAC changed the distribution of heavy metal species in the solid residue, lowering the ecological danger of Cu, Zn, and Pb. Liquefaction at 350 °C with SSAC-550 was favored based on the bio-oil production. The hydrothermal liquefaction of waste furniture sawdust was examined by Jindal et al. [51] under varied reaction circumstances. The best conditions for maximal bio-oil output were found to be 280 °C for 15 min, a water to sawdust ratio of 6, and an initial N_2 pressure of 1.0 MPa. Sawdust conversion and bio-oil output were maximum at 61.6 wt% and 12.7 wt%, respectively. Alcohols, acids, aldehydes, ketones, esters, phenols, and their derivatives were among the organic substances found in the bio-oils (Table 2).

Table 2 Summary of studies on bio-crude production from various biomass types via HTL

Feedstock type	Feed characterization	Operating conditions	Yield	Comments	Ref
Bamboo shoot	C:45% H:6.22% N:1.8% O:47.02% HHV: 15.7 MJ/kg	T:110–150 °C P: normal R.T: 20–110 min	–	The residence duration of 80 min was confirmed to be the most favorable for HTL of bamboo shoot biomass	[42]
White pine sawdust	C:52.5% H:6.32% N:0.10% O:40.6%	T: 300 oC P: 2 bar R.T: 15 min	20–66%	After 15 min residence time, white pine sawdust yielded 66 wt% bio-crude utilizing co-solvents of water + ethanol 50/50 v/v	[43]
Bamboo	C:45% H:6.22%	R.T: 120,360,600,840, 1040 min	80% max	Increased residence time up to 840 min boosted bio-crude production; extending residence time beyond 1080 min resulted in unwanted re-condensation and decreased bio-crude output	[44]
Algae	C:46.19% H:7.38% N:0.41% O: 45.41%	T: 553–616 K R.T: 30 B/S: 10:1	64.6% max	The greatest production of bio-crude was discovered after 30 min of residence time	[45]
Corn stalk	C:56.2% H:6.9% N:7.7% O:28.7% HHV: 19.9 MJ/kg	T: 350 oC P: 206 bar R.T: 30 min	88% max	An ideal residence time of 30 min gave maximum yield	[46]
Rice stalk	C:39.77% H:5.53% N:0.82% O: 53.64% HHV: 17.62 MJ/kg	T: 325 oC P: 30 bar R.T: 60–90 min B/S: 1:10	55% max	Temperature-dependent ideal residence times are 250 °C for 90 min and 325 °C for 60 min	[47]
Algae	Carbohydrate: 46.47% Lipid: 1.83% Protein:14.3% Ash: 25.4% Moisture: 16.05%	T: 300 °C P: 1000–1300 psi R.T: 15 min B/S:	44% max	Depending on the kind of solvents used, the yield of bio-crude rose from 11 to 44 wt% after 15 min (water, $C2H5OH$, $CH3OH$)	[48]

(continued)

Table 2 (continued)

Feedstock type	Feed characterization	Operating conditions	Yield	Comments	Ref
Onopordum heteracanthum Stalks	C:44.93% H:6.61% N:1.67% O:46.79% HHV: 16.29 MJ/kg	T: 523–563 K R.T: 75 min	30.8% max	The maximum yield of bio-crude (30.80 wt%) was obtained after 75 min of residence time in an ethanol solvent	[49]
Sewage sludge	C:28.94% H:4.48% N:4.23% O:13.92%	T:350 and 400 °C P: 35 bar R.T: 60,90,120	15% max	After 30 min of residence time, the yield of bio-crude from activated carbon based on sewage sludge rose	[50]
Waste furniture dust	C:47.42% H:5.67% N:0.2% O:46.71% HHV: 15.79 MJ/kg	T:180–300 °C P:0–2 MPa R.T:0–60 B/S:2–10	12.7% Max	During a residence time of 15 min, a yield of 12.70 wt% of bio-crude was discovered	[51]

5 Literature on Bio-crude Production via Co-HTL of Biomass and Plastics

Mahesh et al. [13] investigated the use of hydrothermal liquefaction with glycerol as a co-solvent to convert municipal solid waste into bio-crude. Results showed that glycerol significantly increased bio-crude yield and quality, with phenolic compounds and cyclooxygenates being the major compounds. The process was found to be energetically feasible, with a maximum energy recovery of 95%. Mukunadan et al. [52] showed an efficient way to convert *Prosopis juliflora* biomass into high-quality bio-oil using polypropylene as a co-reactant and supported metal oxide catalysts. The optimal conditions for bio-oil yield were found to be 420 °C and 25% PP substitution using Nb/Al_2O_3 as the catalyst. The bio-oil can be used for chemicals and transportation biofuels. Nb/Al_2O_3 was the best catalyst with excellent recyclability. Harisankar et al. [53] studied the physical and chemical transformations that occur during the process and how it can be used to transform municipal solid waste into bio-crude. The challenges of scaling up the HTL process with municipal solid waste as a feedstock and the potential to derive value-added products from the process are also discussed.

6 Literature on Bio-crude Production via HTL of Plastics

Liu et al. [54] examined the depolymerization of polyethylene (PE) into valuable products using different liquefaction methods, including hydrothermal liquefaction (HTL), ionic liquids catalyzed HTL, and solvothermal liquefaction (STL). The study finds that solvothermal liquefaction with acetone as a solvent is the most effective method for producing oil products with a high energy recovery rate and comparable heating values to gasoline. The study also suggests that the use of a solvent facilitates the depolymerization reaction by dissolving and dispersing the feedstock and reducing the cracking temperature. Overall, the study demonstrates the feasibility of an effective technique for upcycling polyethylene-based plastic waste.

Lu et al. [55] discussed the use of hydrothermal liquefaction (HTL) to convert ocean-bound plastic waste (OPW), which mainly consists of high-density polyethylene (HDPE) waste, into oil. HTL was found to convert over 78% of OPW into oil at 450–475 °C with a 2–4-h reaction time. The resulting oil and gas products were mainly composed of paraffins, olefins, cyclic hydrocarbons, or aromatics. A reaction pathway for converting HDPE was proposed based on the characterization results. Additionally, post-consumer HDPE waste bottles were converted into oil via HTL, with the oil yield remaining in a similar range to OPW.

Seshasayes et al. [56] examined the effect of batch holding time and temperature on the yield, composition, and quality of oil formed from the hydrothermal treatment of plastic waste such as polypropylene (PP), polystyrene (PS), polycarbonate (PC), and polyethylene terephthalate (PET). The highest oil yields ranged from 16% for PET to 86% for polystyrene, and the depolymerization of plastics to produce oil occurs fastest at supercritical water regime temperatures above 400 °C. The higher heating value of the oil from PP and PS is comparable to gasoline, while the oil from PC and PET has lower heating values that can be improved by further treatment. Increasing the hydrothermal reaction severity reduces oil yields but increases heating values. The oil from PC contained bisphenol-A as a major product, which could provide a revenue stream in addition to oil and improve the process economics.

Chen et al. [57] discussed how polypropylene (PP) waste can be converted into oil using supercritical water, which can help reduce the accumulated plastic waste and associated risks to the environment and human health. The study found that up to 91% of model PP was converted into oil at certain reaction times and temperatures, with the resulting oil consisting of useful components. This conversion process is net-energy positive and potentially has a higher energy efficiency and lower greenhouse gas emissions compared to incineration and mechanical recycling. The oil derived from PP can potentially be used as gasoline blend stocks or feedstocks for other chemicals.

Colnik et al. [58] discussed the study of the degradation of polyethylene (PE) plastics using supercritical water (SCW) in a batch reactor. The study was conducted at temperatures between 380–450 °C and reaction times of 15–240 min, and with the addition of acetic acid. The results showed that during SCW decomposition, oil or wax, aqueous, gas, and solid phases were formed, and the oil phase contained alkanes, alkenes, cycloalkanes, aromatics, and alcohols, while the gas phase mainly contained light hydrocarbons. By increasing the temperature and reaction time, the concentration of long-chain hydrocarbons decreased, while the concentration of gasoline fraction increased. The article also proposes a potential degradation mechanism for PE waste in SCW.

7 Future Scope and Challenges

Although hydrothermal liquefaction (HTL) is a promising method for creating bio-crude from biomass, it also has several limitations and possible concerns. Following are some of the major concerns:

1. Feedstock variability: The unpredictability of the biomass feedstock may have an impact on HTL performance. The composition of biomass varies according to variables such as species, location, and growth circumstances. This may have an effect on the quality and yield of the bio-crude produced.
2. Corrosion and fouling: Since HTL involves high temperatures and pressure, the equipment used in the process may corrode and foul. This may result in higher maintenance expenses and downtime. HTL takes a substantial amount of energy to sustain high temperature and pressure conditions. This may make the procedure both energy-intensive and expensive.
3. Product quality and stability: Factors such as reaction time, temperature, and pressure may all have an impact on the quality and stability of the bio-crude generated by HTL. Managing these factors may be difficult, and the resultant bio-crude may need further processing to fulfil quality standards.
4. Scale-up challenges: Moving HTL from the laboratory to the commercial scale may be difficult since it involves considerable financial investment as well as operational experience. It is also necessary to guarantee that the process is both economically and ecologically sustainable.
5. Waste disposal: HTL generates solid and liquid leftovers that must be appropriately disposed of. These wastes may include dangerous chemicals and heavy metals, which might endanger the environment if not treated properly.

However, HTL process can handle the feedstock of high moisture content comfortably and results in a very high-quality bio-crude with < 10 wt%, making the best quality bio-crude among the other thermochemical methods. With the development of continuous HTL reactor systems, the solid residues and organic waste can be processed effectively thus reducing the waste disposal.

References

1. Hussain Qaisar ÁS, Aquil Ahmad ÁM (2014) Production, consumption and future challenges of coal in India. Int J Curr Eng Technol 34374:3437–3440
2. Mascal M, Dutta S (2020) Synthesis of highly-branched alkanes for renewable gasoline. Fuel Process Technol 197:106192. https://doi.org/10.1016/j.fuproc.2019.106192
3. Terrell E, Dellon LD, Dufour A, Bartolomei E, Broadbelt LJ, Garcia-Perez M (2020) A review on lignin liquefaction: advanced characterization of structure and microkinetic modeling. Ind Eng Chem Res 59:526–555. https://doi.org/10.1021/acs.iecr.9b05744
4. Wang T, Zhai Y, Zhu Y, Li C, Zeng G (2018) A review of the hydrothermal carbonization of biomass waste for hydrochar formation: process conditions, fundamentals, and physicochemical properties. Renew Sustain Energy Rev 90:223–247. https://doi.org/10.1016/j.rser.2018.03.071
5. Li CZ (2007) Some recent advances in the understanding of the pyrolysis and gasification behaviour of Victorian brown coal. Fuel 86:1664–1683. https://doi.org/10.1016/j.fuel.2007.01.008
6. Arpia AA, Chen WH, Lam SS, Rousset P, de Luna MDG (2021) Sustainable biofuel and bioenergy production from biomass waste residues using microwave-assisted heating: a comprehensive review. Chem Eng J 403:126233. https://doi.org/10.1016/j.cej.2020.126233
7. Elliott DC, Biller P, Ross AB, Schmidt AJ, Jones SB (2015) Hydrothermal liquefaction of biomass: developments from batch to continuous process. Bioresour Technol 178:147–156. https://doi.org/10.1016/j.biortech.2014.09.132
8. Qian L, Wang S, Savage PE (2020) Fast and isothermal hydrothermal liquefaction of sludge at different severities: reaction products, pathways, and kinetics. Appl Energy 260:114312. https://doi.org/10.1016/j.apenergy.2019.114312
9. Yang T, Liu X, Li R.D. PR, Li B, Kai X (2018) Hydrothermal liquefaction of sewage sludge to produce bio-oil: effect of co-pretreatment with subcritical water and mixed surfactants. J Supercrit Fluids 144. https://doi.org/10.1016/j.supflu.2018.10.005
10. Zhou Y, Zhao N, Li X, Yu Y, Elshareef H, Qian M et al (2022) Chapter seven—thermal processing of biomass for energy and fuel production. In: Li Y, Zhou YBT (eds) A in B, vol 7. Elsevier, pp 271–341. https://doi.org/10.1016/bs.aibe.2022.05.006
11. Ambaye TG, Vaccari M, Bonilla-Petriciolet A, Prasad S, van Hullebusch ED, Rtimi S (2021) Emerging technologies for biofuel production: A critical review on recent progress, challenges and perspectives. J Environ Manage 290:112627. https://doi.org/10.1016/j.jenvman.2021.112627
12. Venkatachalam CD, Sekar S, Ravichandran SR, Sengottian M, Sukumar KC, Chenniappan D et al (2023) A review on bio-crude production from algal biomass using catalytic hydrothermal liquefaction process. Environ Eng Res 28:220210–220211. https://doi.org/10.4491/eer.2022.211
13. Mahesh D, Ahmad S, Kumar R, Chakravarthy SR, Vinu R (2021) Hydrothermal liquefaction of municipal solid wastes for high quality bio-crude production using glycerol as co-solvent. Bioresour Technol 339:125537. https://doi.org/10.1016/j.biortech.2021.125537
14. Akhtar J, Amin NAS (2011) A review on process conditions for optimum bio-oil yield in hydrothermal liquefaction of biomass. Renew Sustain Energy Rev 15:1615–24. https://doi.org/10.1016/j.rser.2010.11.054
15. Gollakota ARK, Kishore N, Gu S (2018) A review on hydrothermal liquefaction of biomass. Renew Sustain Energy Rev 81:1378–1392. https://doi.org/10.1016/j.rser.2017.05.178
16. Bobleter O (1994) Hydrothermal degradation of polymers derived from plants. Prog Polym Sci 19:797–841
17. Sasaki M, Fang Z, Fukushima Y, Adschiri T, Arai K (2000) Dissolution and hydrolysis of cellulose in subcritical and supercritical water. Ind Eng Chem Res 39:2883–2890
18. Xu D, Lin G, Liu L, Wang Y, Jing Z, Wang S (2018) Comprehensive evaluation on product characteristics of fast hydrothermal liquefaction of sewage sludge at different temperatures. Energy 159:686–95. https://doi.org/10.1016/j.energy.2018.06.191

19. Carpio RB, Zhang Y, Kuo C-T, Chen W-T, Schideman LC, de Leon R (2021) Effects of reaction temperature and reaction time on the hydrothermal liquefaction of demineralized wastewater algal biomass. Bioresour Technol Reports 14:100679. https://doi.org/10.1016/j.biteb.2021.100679
20. Nunes VO, Fraga AC, Silva RVS, Pontes NS, Pinho AR, Sousa-Aguiar EF et al (2021) Chemical characterisation of sugarcane bagasse bio-oils from hydrothermal liquefaction: Effect of reaction conditions on products distribution and composition. J Environ Chem Eng 9:106513. https://doi.org/10.1016/j.jece.2021.106513
21. Carrier M, Loppinet-Serani A, Absalon C, Aymonier C, Mench M (2012) Degradation pathways of holocellulose, lignin and α-cellulose from *Pteris vittata* fronds in sub-and super critical conditions. Biomass Bioenerg 43:65–71
22. Marcilla A, Catalá L, García-Quesada JC, Valdés FJ, Hernández MR (2013) A review of thermochemical conversion of microalgae. Renew Sustain Energy Rev 27:11–19. https://doi.org/10.1016/j.rser.2013.06.032
23. Vlaskin M, Grigorenko A, Chernova N, Kiseleva S, Lipatova I, Popel' O et al (2019) The hydrothermal liquefaction as a promising procedure for microalgae-to-biofuel conversion: a general review and some thermophysical problems to be solved. High Temp Press 48:309–51. https://doi.org/10.32908/hthp.v48.716
24. Yin S, Dolan R, Harris M, Tan Z (2010) Subcritical hydrothermal liquefaction of cattle manure to bio-oil: Effects of conversion parameters on bio-oil yield and characterization of bio-oil. Bioresour Technol 101:3657–3664
25. Chen Y, Dong L, Miao J, Wang J, Zhu C, Xu Y et al (2019) Hydrothermal liquefaction of corn straw with mixed catalysts for the production of bio-oil and aromatic compounds. Bioresour Technol 294:122148. https://doi.org/10.1016/j.biortech.2019.122148
26. Cheng F, Jarvis JM, Yu J, Jena U, Nirmalakhandan N, Schaub TM, et al (2019) Bio-crude oil from hydrothermal liquefaction of wastewater microalgae in a pilot-scale continuous flow reactor. Bioresour Technol 294:122184. https://doi.org/10.1016/j.biortech.2019.122184
27. Jin F, Wang Y, Zeng X, Shen Z, Yao G (2014) Water under high temperature and pressure conditions and its applications to develop green technologies for biomass conversion. Appl Hydrothermal React to Biomass Convers 3–28
28. Sato T, Osada M, Watanabe M, Shirai M, Arai K (2003) Gasification of alkylphenols with supported noble metal catalysts in supercritical water. Ind Eng Chem Res 42:4277–4282
29. Neavel RC (1981) Exxon donor solvent liquefaction process. Philos Trans R Soc London Ser A Math Phys Sci 300:141–156
30. Ramirez JA, Brown RJ, Rainey TJ (2015) A review of hydrothermal liquefaction bio-crude properties and prospects for upgrading to transportation fuels. Energies 8:6765–6794. https://doi.org/10.3390/en8076765
31. Chen W-T, Qian W, Zhang Y, Mazur Z, Kuo C-T, Scheppe K et al (2017) Effect of ash on hydrothermal liquefaction of high-ash content algal biomass. Algal Res 25:297–306. https://doi.org/10.1016/j.algal.2017.05.010
32. Jatoi AS, Shah AA, Ahmed J, Rehman S, Sultan SH, Shah AK, et al (2022) Hydrothermal liquefaction of lignocellulosic and protein-containing biomass: a comprehensive review. Catalysts 12. https://doi.org/10.3390/catal12121621
33. Vidal B, Dien B, Ting K, Singh V (2011) Influence of feedstock particle size on lignocellulose conversion—a review. Appl Biochem Biotechnol 164:1405–1421. https://doi.org/10.1007/s12010-011-9221-3
34. Jazie AA, Haydary J, Abed SA, Al-Dawody MF (2022) Hydrothermal liquefaction of *Fucus vesiculosus* algae catalyzed by Hβ zeolite catalyst for Biocrude oil production. Algal Res 61:102596. https://doi.org/10.1016/j.algal.2021.102596
35. Xiang H, Xin R, Prasongthum N, Natewong P, Sooknoi T, Wang J et al (2022) Catalytic conversion of bioethanol to value-added chemicals and fuels: a review. Resour Chem Mater 1:47–68. https://doi.org/10.1016/j.recm.2021.12.002
36. Shah AA, Sharma K, Haider MS, Toor SS, Rosendahl LA, Pedersen TH et al (2022) The role of catalysts in biomass hydrothermal liquefaction and biocrude upgrading. Processes 10. https://doi.org/10.3390/pr10020207

37. Karagöz S, Bhaskar T, Muto A, Sakata Y (2006) Hydrothermal upgrading of biomass: effect of KCO concentration and biomass/water ratio on products distribution. Bioresour Technol 97:90–98. https://doi.org/10.1016/j.biortech.2005.02.051
38. Guo B, Yang B, Silve A, Akaberi S, Scherer D, Papachristou I et al (2019) Hydrothermal liquefaction of residual microalgae biomass after pulsed electric field-assisted valuables extraction. Algal Res 43:101650. https://doi.org/10.1016/j.algal.2019.101650
39. Liu Q, Xu R, Yan C, Han L, Lei H, Ruan R et al (2021) Fast hydrothermal co-liquefaction of corn stover and cow manure for biocrude and hydrochar production. Bioresour Technol 340:125630. https://doi.org/10.1016/j.biortech.2021.125630
40. Han Y, Hoekman S, Cui Z, Jena U, Das P (2019) Hydrothermal liquefaction of marine microalgae biomass using co-solvents. Algal Res 38. https://doi.org/10.1016/j.algal.2019.101421
41. Eboibi BE, Lewis DM, Ashman PJ, Chinnasamy S (2014) Effect of operating conditions on yield and quality of biocrude during hydrothermal liquefaction of halophytic microalga Tetraselmis sp. Bioresour Technol 170:20–9. https://doi.org/10.1016/j.biortech.2014.07.083
42. Ye L, Zhang J, Zhao J, Tu S (2014) Liquefaction of bamboo shoot shell for the production of polyols. Bioresour Technol 153:147–53. https://doi.org/10.1016/j.biortech.2013.11.070
43. Cheng S, D'cruz I, Wang M, Leitch M, Xu C (Charles) (2010) Highly efficient liquefaction of woody biomass in hot-compressed alcohol−water co-solvents. Energy Fuels 24:4659–67. https://doi.org/10.1021/ef901218w
44. Yip J, Chen M, Szeto YS, Yan S (2009) Comparative study of liquefaction process and liquefied products from bamboo using different organic solvents. Bioresour Technol 100:6674–8. https://doi.org/10.1016/j.biortech.2009.07.045
45. Chen Y, Wu Y, Zhang P, Hua D, Yang M, Li C et al (2012) Direct liquefaction of *Dunaliella tertiolecta* for bio-oil in sub/supercritical ethanol–water. Bioresour Technol 124:190–8. https://doi.org/10.1016/j.biortech.2012.08.013
46. Biller P, Sharma BK, Kunwar B, Ross AB (2015) Hydroprocessing of bio-crude from continuous hydrothermal liquefaction of microalgae. Fuel 159:197–205. https://doi.org/10.1016/j.fuel.2015.06.077
47. Li R, Li B, Yang T, Xie Y (2013) Liquefaction of rice stalk in sub- and supercritical ethanol. J Fuel Chem Technol 41:1459–65. https://doi.org/10.1016/S1872-5813(14)60006-2.
48. Singh R, Bhaskar T, Balagurumurthy B (2015) Effect of solvent on the hydrothermal liquefaction of macro algae *Ulva fasciata*. Process Saf Environ Prot 93:154–60. https://doi.org/10.1016/j.psep.2014.03.002
49. Durak H, Aysu T (2014) Effects of catalysts and solvents on liquefaction of *Onopordum heteracanthum* for production of bio-oils. Bioresour Technol 166:309–17. https://doi.org/10.1016/j.biortech.2014.05.051
50. Zhai Y, Chen H, Xu B, Xiang B, Chen Z, Li C et al (2014) Influence of sewage sludge-based activated carbon and temperature on the liquefaction of sewage sludge: yield and composition of bio-oil, immobilization and risk assessment of heavy metals. Bioresour Technol 159:72–9. https://doi.org/10.1016/j.biortech.2014.02.049
51. Jindal MK, Jha MK (2016) Effect of process parameters on hydrothermal liquefaction of waste furniture sawdust for bio-oil production. RSC Adv 6:41772–41780. https://doi.org/10.1039/C6RA02868C
52. El-Hefnawy ME, Alhayyani S, El-Sherbiny MM, Abomohra AEF, Al-Harbi M (2022) Endogenous bioethanol production by solid-state prefermentation for enhanced crude bio-oil recovery through integrated hydrothermal liquefaction of seaweeds. J Clean Prod 355:131811. https://doi.org/10.1016/j.jclepro.2022.131811
53. Harisankar S, Vinu R (2023) Comprehensive evaluation of municipal solid wastes and mixed feedstocks for commercial hydrothermal liquefaction in bio-refineries. Fuel 339:127236. https://doi.org/10.1016/j.fuel.2022.127236
54. Liu Y, Chandra Akula K, Phani Raj Dandamudi K, Liu Y, Xu M, Sanchez A et al (2022) Effective depolymerization of polyethylene plastic wastes under hydrothermal and solvothermal liquefaction conditions. Chem Eng J 446:137238. https://doi.org/10.1016/j.cej.2022.137238

55. Lu T, Jan K, Chen W-T (2022) Hydrothermal liquefaction of pretreated polyethylene-based ocean-bound plastic waste in supercritical water. J Energy Inst 105:282–92. https://doi.org/10.1016/j.joei.2022.10.003
56. Seshasayee MS, Savage PE (2020) Oil from plastic via hydrothermal liquefaction: production and characterization. Appl Energy 278:115673. https://doi.org/10.1016/j.apenergy.2020.115673
57. Chen W-T, Jin K, Linda Wang N-H (2019) Use of supercritical water for the liquefaction of polypropylene into oil. ACS Sustain Chem Eng 7:3749–3758. https://doi.org/10.1021/acssuschemeng.8b03841
58. Čolnik M, Kotnik P, Knez Ž, Škerget M (2021) Hydrothermal decomposition of polyethylene waste to hydrocarbons rich oil. J Supercrit Fluids 169:105136. https://doi.org/10.1016/j.supflu.2020.105136

Chapter 6
Pyrolysis of Algae/lignocellulosic Biomass/Organic and Inorganic Wastes

Óscar Piña and Jorge Aburto

Abstract The production of fuels and chemical products must focus on a sustainable energy transition where all actors in society must be more responsible for the technologies and processes implemented in a current post-consumer market. There are various emerging technologies such as pyrolysis that focus on the treatment of different types of biomasses and plastic wastes which can be treated independently or by mixtures of two or more products and called as co-pyrolysis. The biofuels and bioproducts production from algae, lignocellulosic biomass and wastes may provide viable options in the value chain with good economic benefits. This review presents the variables that affect the pyrolysis process and the new emerging technologies in the production of biopropane and turquoise hydrogen (biomass hydrogen), as well as a review focused on their life cycle analysis. Some emerging technologies such as biomass intermediate pyrolysis polygeneration (BIPP) or Fluidized Catalytic Cracking to obtain biofuels for the electrical and thermal energy generation, as well as for the production of Bio-LPG, biofuels for ships, renewable propylene, and hydrocarbons biological aromatics (BTX: benzene, toluene, and xylene), are by instance essential for rubber production. Decarbonization helps to reduce emissions of CO_2 and other greenhouse gases (GHG). The energy system takes biomass as a viable option to transform it into energy, based on two technologies: pyrolysis and gasification, where biomass plays a twice important role, that is, it is an energy producer and serves as a raw material to capture the CO_2 emitted towards the atmosphere.

Keywords Pyrolysis · Biocrude oil · Biomass · Plastic waste · Biofuels · Bio-based chemicals · Circular economy

---

Ó. Piña
Postgraduate School, Instituto Mexicano del Petróleo, Mexico City, Mexico

J. Aburto (✉)
Bioenergy consultant, Independent Consultant, Mexico City, Mexico
e-mail: jaaburto@yahoo.com

M. Aslam et al. (eds.), *Production and Biorefining of Biocrude Oil: Current Status and Future Developments*, Advances in Sustainability Science and Technology,
https://doi.org/10.1007/978-981-96-5198-6_6

1 Introduction

The current market of fossil fuels is beginning to orient itself towards sustainable products and processes, that is, those that satisfy current needs without compromising the future resources of new generations. Today, fossil fuels derived from petroleum currently satisfy most of our energy demand, which includes petrochemical building units that are used as raw materials to produce various items, among a large majority of which are plastics that are used daily, but with increasing emissions of GHG to the atmosphere that are provoking the raise of earth's temperature and most likely climate change.

A mitigation strategy to overcome such scenario is through the implementation of low, neutral, or negative carbon emission routes. Hence, biomass and wastes, as an alternative feedstock to produce biofuels and bioproducts, is a recognized strategy to reduce GHG emissions. Such bio-based commodities may be produced through biotechnological and/or thermochemical processes that vary in selectivity, operational conditions, processing time and conversion degree. The former route uses biological entities such as microorganisms, enzymes [1] and/or biomass using traditional or contemporary biotechnological processes. On the other side, biomass, organic and inorganic wastes, and fossil fuels may be treated with thermochemical processes that depend on temperature and different operating conditions such as the flow and type of gases that are supplied to obtain various products (non-condensable gases, condensable gases, and carbon). Here we considered thermochemical processes to the gasification, thermal liquefaction, combustion, and pyrolysis, which are intended to generate energy, power, fuels and/or chemicals [2–4].

Currently, pyrolysis has focused on the treatment of biomass altogether waste plastics to obtain fuels and chemical products through a concomitant thermal decomposition [1–19]. Biomass is a renewable resource and it can be obtained from different industries such as agro-industrial, forestry, agriculture, livestock, aquaculture, and resources such as algae, the organic fraction of municipal solid waste, among others [17]. Biomass is a widely available biological material, and it is mainly composed of macromolecules such as hemicellulose, cellulose, and lignin [2, 3]. Alternative fuels are an important part of the energy transition towards a more sustainable, low-carbon economy, which can help to decrease dependence on fossil fuels. It is essential to guarantee that such fuel production is sustainable, i.e. it does not cause deforestation, does not compete with food feed, and mitigates GHG emissions [20].

In this chapter, we undertake the study of current progress in the pyrolysis of biomass as algae, lignocellulosic residues, and organic wastes. Inorganic wastes like plastics are considered also as an alternative way to valorize such residues and to incorporate them into a circular economy through the making of alternative chemicals and fuels.

2 Pyrolysis

Pyrolysis is an endothermic-exothermic process where matter thermally decomposes in the absence of oxygen [3, 4, 21–23]. The word pyrolysis comes from the Greek (pyro: fire and lysis: breakage). Pyrolysis is a widely used process in coke processing and ethylene production [24] to obtain fuels and petrochemicals with high-energy intensity and CO_2 emissions [24]. But thermal and/or catalytic pyrolysis are nowadays attracting attention to produce low, neutral and/or negative GHG emissions in terms of a circular economy where energy may be supplied by renewable energies [25].

In the specialized literature, different types of pyrolysis and technologies are described, which respond to different needs for obtaining final products and depend on the type and quantity of raw material to be processed. Pyrolysis can be classified into three major processes: slow, fast, and flash which differentiate by the heating speed which in turn defines the residence time (Table 1). It is well known that the parameters affecting the thermal decomposition and the performance of forming certain pyrolytic products are temperature, heating rate, gas flow (inert atmosphere), particle size, reactor design, mixing, and type of raw material, among others [2, 22, 26]. Other approaches may include torrefaction, vacuum, and microwave [1, 2]. Also, pyrolysis can be carried out with or without catalysts to improve conversion, yield, and selectivity of the products [6].

Heating rate influences the pyrolysis performance and product composition. Indeed, a quick heating at moderate temperature (200–500 °C) produces more volatiles and therefore more liquid, while more carbon is produced at slower heating at that temperature [27]. The slow pyrolysis is conducted at heating rates below < 80 °C/min, and can maximize the yield of carbon (35%), bio-oil (30–40%) and non-condensable gas (35%). The yield of each fraction depends on the type and composition of matter, type of reactor, retention time, particle size, and heating rate, among others [1, 3]. Slow heating rates increase retention times variables that affect the different types of pyrolysis, it favors carbon formation while short retention times favor bio-oil formation and decrease secondary chemical recombination [28]. Fast pyrolysis is performed at heating rates of 100–200 °C s^{-1} [2]. Fast pyrolysis is performed at heating rates of 0.5–1.0 °C per seconds, which has the advantage to maximize bio-oil yield (50–60%) in contrast to slow pyrolysis. Fast pyrolysis is

Table 1 Variables that affect the different types of pyrolysis

Type of pyrolysis	*Residence time (s)	Temperature (°C)	Heating rate	Particle size (mm)	Bio-oil yield (wt. %)
Slow	> 1 min	100–800	< 80 °C min^{-1}	5–50	30–40
Fast	0.5–10	600–1000	100–200 °C s^{-1}	< 1	50–60
Flash	< 0.5–2.0	800–1000	> 1000 °C s^{-1}	< 0.2	65–70

**Note* Residence time depends on heating rate

considered a more economical way to produce bio-oil with a low sulfur and nitrogen content, a high-energy density and simple handling [3]. Within fast pyrolysis, 50–60% of the raw material is converted into bio-oil, while the produced charcoal and syngas can be used for internal energy use [5].

However, the heating rate does not define the product quality and quantity. We must also consider the residence time in the reactor. Also, a gradual removal of volatiles during slow heating, allows the occurrence of secondary reactions between the carbon particles and the volatiles, leading to secondary carbon formation [4, 6, 28]. Hence, the heating rate may have a positive or negative effect on the outcome of the pyrolysis process. The choice of a low to moderate heating rate of 5–10 °C min^{-1} can have a positive impact. On the other side, a heating rate of 10–20 °C min^{-1} can have a negative impact on the conversion and process outcome with some exceptions [6].

During the pyrolysis process, multiple reactions are carried out which include deoxygenation, decarboxylation, decarbonylation, aldol condensation, ketonization, dehydration, aromatization, cyclization, oligomerization, alkylation, isomerization, depolymerization and polymerization, among others [2, 6, 29, 30]. Pyrolysis is a complicated process that involves different chemical reactions that depend on temperature with parallel reactions that involve competitive and simultaneous ones.

Hence, particles are heated at a defined rate from ambient temperature to a maximum temperature, known as the pyrolysis temperature. Such temperature influences the pyrolysis process since it defines what kind of reactions may occur by reaching the corresponding activation energy. The pyrolysis process is composed of several thermal decomposition stages which can include the rupture of polymer chains (formation of volatiles and decrease in molecular weight), as well as cross-linking chain saturations and side chain reactions that include side chain elimination [6]. Thermal pyrolysis produces oil containing oxygenated compounds (ketones, aldehydes, carboxylic acids, alcohols, sugars, furans, phenols, among others [4, 9]:

- Drying ($\leq$ 100 °C). Removal of physical water (humidity).
- Initial stage (100–300 °C). It involves dehydration, decarboxylation and decarbonization reactions that form highly volatile compounds, CO, and CO_2 due to the decomposition of low molecular weight compounds, i.e. simple sugars, hemicelluloses.
- In the intermediate stage (200–550 °C) occurs depolymerization reactions along with the formation of primary carbon, condensable gases (vapors and bio-oil precursors) and non-condensable gases.
- Finally, it takes place the breakdown and rearrangement of volatile compounds (secondary reactions) into the formation of secondary carbon, ash, and non-condensable gases at temperatures greater than 550 °C.

Also, the physical structure and composition of the feedstock have an effect on the pyrolysis products. A raw material with a smaller particle size facilitates the escape of condensable gases into the environment before undergoing secondary cracking which increases the liquid formation. Mass transfer limitations and reactant diffusion among the feedstock particles define the conversion, yield, and nature of pyrolysis products. On the contrary, secondary molecule cracking is facilitated in large particles due to the poor diffusivity of the primary pyrolysis product [1, 2, 4, 6]. Then, the composition of the raw material through its hydrogen/carbon (H/C) ratio defines its thermal stability, which has an important influence on the pyrolysis performance [21, 27].

Pyrolysis is a thermal process strongly controlled by reaction kinetics as well as heat and mass transfer [15]. Therefore, the distribution of the product depends on both the raw material used and the reaction conditions, such as temperature, heating rate and residence time. Thus, the chemical reactor and therefore its design are equally important for the obtention of desired products. The types of reactors used in the process can influence parameters such as mass and heat transfer, the mixing efficiency of the reactants and type of catalyst (use or not), the residence time, and the economic efficiency of the process, among others [6]. Some of the most used equipment are fluidized and fixed bed reactors, conical spouted bed reactors (CSBR), and microwave-assisted technologies, among others [31].

Pyrolysis studies have highlighted the importance of thermokinetic studies in the comprehension of the influence of pyrolysis temperature and heating rates on conversion, yield, and product selectivity. Some works present these topics in depth and rigorous way and others of utmost importance are those related to the thermal degradation of lignin and spent coffee as well as the thermokinetic analysis of coffee residues and orange peels, among others [28, 32].

Moreover, new pyrolysis technologies are focusing on obtaining new-generation biofuels such as biopropane, which is obtained along the non-condensable gases in the pyrolysis. Also, hydrotreating upgrading of oil is necessary before it can be co-refined to obtain good quality liquid fuels. The pyrolysis oil can be directly used as fuel, but its fuel properties limit the applications to boilers and heavy fuel burners. The oxygen present in biomass requires a large chemical energy and hydrogen demand to obtain fuels, but pyrolysis may help to coproduce hydrogen that can play a key role in the decarbonization, but also carbon obtained as a byproduct that can be used directly as a fertilizer and help to restore the soil [33]. For instance, a thermal plasma pyrolysis was applied for the decomposition of plastics, reaching temperatures greater than 11,000 K, and the hydrogen-rich gas obtained contains a lower fraction of hydrocarbon compounds (51.2–79.4% by volume of H_2 and 20.6–47.4% by volume of C1–C5 hydrocarbon compounds) depending on the raw material and operating conditions [34]. Combining pyrolysis with biomass gasification where biomass is transformed into a gas mixture containing hydrogen, methane and CO. This process is usually intended for the direct production of synthesis gas to generate energy or synthesize fuels rather than to obtain pure hydrogen [35]. But methane pyrolysis may be an attractive source of hydrogen since it does not cause CO_2 emissions and has a specific energy demand of 37.8 compared to 63.3 for steam reforming or 285.9 kJ/mol H_2 for hydrogen production by electrolysis [33].

In the following sections, we will discuss the fuels and chemicals production using pyrolysis for the valorization of algae, lignocellulosic residues as well as organic and inorganic municipal wastes.

3 Pyrolysis Algae for Biofuels and Chemicals

Algae are considered a promising source for sustainable fuels and chemicals since its worldwide spread, minimal nutrient requirements and growth using ambient CO_2. Algae are divided into unicellular microalgae and multicellular macroalgae [36–38]. Algae are classified as third-generation biomass, very heterogeneous organisms with very different evolutionary paths, and feed on various dissolved nutrients. One of the main nutrients for algae is carbon dioxide (CO_2), which can be captured or sequestered from the atmosphere or from some productive process, leading to a reduction in CO_2 emissions. The oceans absorb more than 1 million tons of anthropogenic CO_2 per hour, causing a drop in the pH of seawater in a process called ocean acidification. Ocean acidification can reduce the calcification of micro- and macroalgae, that is, it can reduce the biogenic silica content in some diatoms, which causes the density of diatoms to decrease. The acidification of the oceans causes a decrease in the competitiveness and fitness of coralline algae caused by a reduction in their calcification, this species of algae corresponds to the reef-forming ones. On the other hand, ocean warming, eutrophication and acidification synergistically enhance reproduction events and shorten the generation period in the "green tide" algae *Ulva rigida*, suggesting that more severe green tides may occur in future oceans [39]. Algae require light to carry out the process of photosynthesis, a process that takes place in chloroplasts where metabolize CO_2 and release oxygen. The classification of algae is diverse since it depends on the type of structures they make up; they can be protists or eukaryotes; some authors classify them as protists. Algae can be classified based on their phylogenetic group, among which diatoms, chlorophytes (green algae), euglenoids, dinoflagellates, chrysophytes (golden algae), phaeophytes (brown or brown algae), rhodophytes (red algae) are identified and cyanobacteria [40].

The chemical composition of algae varies based on the species, its approximate wet content corresponds to proteins (10–27%), lipids (1–5%), ashes (11–35%) and carbohydrates (32–60%) [15, 41, 42]. The biomass from algae offers significant advantages for pyrolytic processes because of its high lipid content, accounting for about 20–50% dried weight and up to 80% under specific conditions. This high lipid content may promote a higher bio-oil yield than traditional lignocellulosic biomass [41] and in addition to a rapid growth rate and low land demand, that make algae biomass a promising third-generation energy crop [43]. The algae pyrolysis oils can contain large quantities of oxygenated and nitrogenous compounds. The algae bio-oil produced by pyrolysis has an oxygen content around 10–30% by weight, compared to the pyrolysis oil obtained from agricultural and forestry waste which is around 45–50% by weight. Now, nitrogen comes from the amino acids algae proteins and can be 5–10%. It is well known that a high oxygen content provokes a low calorific value, then the catalytic deoxygenation and denitrogenation of such pyrolysis oils are necessary to produce fuels [31, 44]. Table 2 shows the comparative data of the pyrolytic bio-oil for different types of materials. The HHV is higher for two types of algae (Blue-green and *C. vulgaris*) and the carbon content is greater than 60%. Algae have recently generated interest in the production of food, bioactive compounds, and biofuels, do not require agricultural land or fresh water for growth and they can also be produced in wastewater, seawater and photoreactors which can be batch, continuous or semi-continuous processes. Algae can also be obtained from the wastewater treatment process, specifically during the coagulation/flocculation and sedimentation stage [45]. Microalgae can be grown with wastewater allowing simultaneously the (1) reduction of the costs of nitrogenous and phosphorus salts as well as the (2) wastewater treatment [44]. Also, the combustion gases coming from industries such as energy generation, cement and steel can be used as an economic source of CO_2 for microalgae growth [44].

Hence, microalgae biomass may serve to produce pyrolytic oils and fuels, but also biodiesel while its carbohydrates can be fermented to produce bioethanol or biogas through anaerobic digestion [44]. Dried microalgae can be combusted to produce heat or electricity, but its low thermal efficiency, the release of harmful gases and high ash content delay its further use. Currently, the production of biofuels is aimed at the valorization of algae, lignocellulosic biomass, waste, plastics, and other materials. Thus, during the pyrolysis process, protein's nitrogen can be transformed into pyrrole and pyridine derivatives through dehydrogenation, deamination, cyclization, dehydration, and dimerization [14]; which makes algae pyrolysis oil attractive as a platform of chemicals. The fraction that is first degraded in the pyrolysis of algae corresponds to carbohydrates, followed by proteins and lipids [14]. Indeed, phenol can be extracted as raw material to obtain antioxidants, resins, surfactant additives, preservatives, biodegradable polymers, adhesives, fertilizer additives, resin precursors, pharmaceutical products, flavoring agents in foods, among others [31].

Table 2 Comparison of properties of pyrolysis oils from diverse sources

Pyrolysis oil	C%	H%	N%	O%	HHV (MJ/kg)	Reference
Blue-green algae blooms	67.58	8.95	7.75	14.48	31.90	[43]
Algal biomass (*C. vulgaris*)	51.40	10.40	12.40	24.80	18.60	[43]
Algae *Cyanobacteria*	69.20	9.10	10.80	10.80	35.00	[43]
Biomass lignocellulosic (woody)	56.00	6.00	0.20	38.00	17.00	[46]
Biomass lignocellulosic (woody)	56.00	6.25	0.20	37.50	17.50	[47]
Biomass lignocellulosic (bagasse)	54.60	6.45	0.73	38.07	22.40	[46]
Biomass lignocellulosic (willow wood). Pyrolysis flash	56.40	6.20	0.20	37.10	23.00	[48]
Biomass waste	57.50	6.45	0.9	37.50	17.50	[49]
Organic municipal waste	49.65	6.12	1.24	41.62	20.78	[50]
Tobacco waste	75.07	9.15	3.80	11.84	30.05	[51]

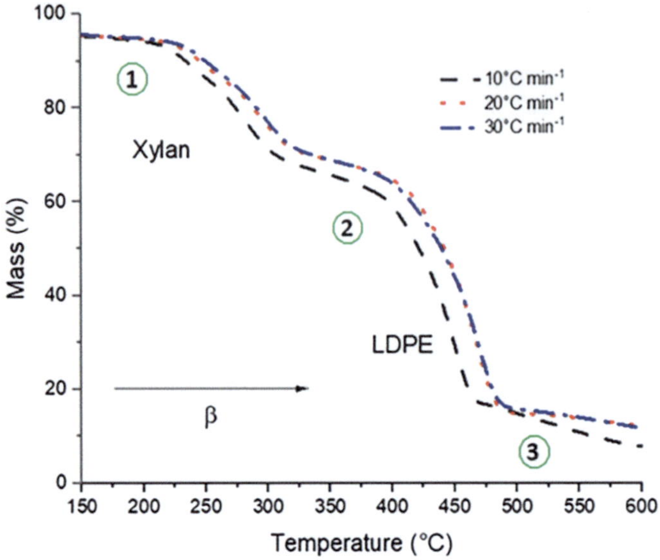

Fig. 1 Thermogram mass (%) versus temperature (°C) of a mixture of xylan and low-density polyethylene (LDPE). Own source

Pyrolysis of xylan/LDPE mixtures was carried out in our group where three different stages were observed: initiation, reaction, and delay (Fig. 1). In the initiation period, the components undergo physical modifications like water evaporation. During the reaction period (second stage, 225–475 °C) there are structural chemical modifications through the breaking of intramolecular bonds that give rise to many reactions generating condensable and non-condensable volatile products. The third stage marks the delay period or end of the reaction in which solids (carbonaceous residues and ashes) are generated.

In a thermogravimetry test and differential thermogravimetry analysis (TG/DTG) [45], the results were presented for a group of cyanobacteria that were collected in Lake Taihu, which were pyrolyzed and their thermal degradation occurred in four stages or temperature intervals (ambient temperature–170, 170–310, 310–550 and 550–800 °C, respectively). The weight losses of each stage were 6.7, 23.7, 32 and 3.4% respectively, with a solid residue of 34.3%. For TG/DTG pyrolysis of blue-green algal blooms, thermal degradation was carried out in three temperature ranges, the decomposition mainly occurs in the temperature range of 160–400 °C. Another work on pyrolysis of algae from lake blooms showed that 500 °C corresponds to the optimal pyrolysis temperature, where the maximum yield of bio-oil was 59% using a horizontal tubular reactor. Here, the yield was similar in the pyrolysis of pine sawdust under the same conditions. The algae bio-oil obtained contains 26.8 times more nitrogen than pine sawdust. Decarboxylation and dehydration reactions increase when temperatures > 700 °C, so the content of polycyclic aromatic hydrocarbons (PAHs) is increased in the bio-oil, such as phenanthrene and pyrene [14].

Moreover, algae charcoal has the potential to be a profitable fertilizer since it has a high content of P, N, and other nutrients with a greater porous structure and a strong ion exchange capacity [52]. The production of algae charcoal may also be implemented as a carbon capture and storage strategy since the algae's high growth rate may facilitate a quicker rate of atmospheric carbon removal and its sequestration. The growth rate of algae ranges from 10 to 27 g/m^2 per day [43]. Algae contain also a large amount of lipids, which represent around 20–50% of the dry matter and up to 80% under specific conditions. This high level of lipids may promote a higher bio-oil yield than traditional lignocellulosic materials [1].

Numerous studies have demonstrated the great techno-economic advantages of co-pyrolysis compared to single pyrolysis [31]. Co-pyrolysis is the process where one or more feedstocks are pyrolyzed simultaneously and the activation energy of thermal decomposition of every feedstock component is affected as well as the yield and composition of obtained products. Co-pyrolysis is a simple and efficient method to produce high-quality bio-oil from two or more materials. Indeed, tires, plastics, and bamboo waste are the materials that improve the performance and quality of pyrolysis oil [31]. Some catalysts, such as CaO and Cu/H-ZSM-5, can improve the quality of the obtained oil by increasing the aromatic content and decreasing oxygenated and nitrogenous compounds [38]. Co-pyrolysis has proven to be an efficient way to solve problems related to high oxygen and nitrogen content in pyrolysis oil by increasing the caloric value and stability, decreasing the high acidity and nitrogen contamination and corrosivity [31, 53]. For instance, the co-pyrolysis of *Enteromorpha clathrata*, a seaweed, and rice husk using ZSM-5 leads to the improvement of pyrolysis oil yield. But, MCM-41 increases the selectivity to a specific component (5-methyl- 2-furancarboxaldehyde), when compared with that obtained solely from *Enteromorpha clathrata* [43].

The costs of cultivating microalgae currently represent two-thirds of the total cost of biofuel production. Within the pyrolysis process, the biofuel with the lowest estimated production cost is bio-oil, since when the cultivation of microalgae is integrated with the treatment of effluents and combustion gases, the estimated production cost is $USD 0.61, and by using microwave-assisted pyrolysis and defatted microalgae biomass, the estimated cost was reduced to US$0.49. The application of charcoal as solid fuel with a selling price of US$50/ton contributes approximately 2.8% of the total revenue generated by the process (2016). By converting charcoal into activated carbon, its selling price can be increased to US$1188/ton, increasing its contribution to 51% of total revenue [16].

4 Pyrolysis of Lignocellulosic Biomass

Lignocellulosic biomass is composed mainly of hemicelluloses, celluloses, and lignin and is worldwide available with an estimated resource of 146 billion tons per year [16]. Currently, exploited biomass represents only 14% of the world's primary energy, representing approximately 56 million TJ/year [53]. Cellulose is one of the major components of plant cells with an elemental formula of $(C_6H_{12}O_6)_n$ composed by anhydro glucose units linked by β-1,4-O-glycosidic bonds with a degree of polymerization (n) of 700–10000. The reactivity of cellulose is directly related to the proportion of crystallinity of the amorphous material [31, 53, 54]. Now, hemicelluloses are heterogeneous polysaccharides that are bound to the surface of cellulose microfibrils. The content of hemicelluloses varies on the type of biomass that constitutes it, thus they are mainly composed of many sugars with five carbon atoms which correspond to ribose, arabinoses, and xylose; while hexoses are galactose, mannose and rhamnose with a molecular mass less than 30,000 units [53–55]. Hemicellulose decomposes thermally between 180 and 350 °C, thereby producing non-condensable gas, carbon and oxygen-rich compounds like ketones, aldehydes, acids, and furans [4, 53]. Lignin is the third main component of lignocellulosic biomass with 25–30% weight and is a copolymer of three basic phenylpropane units (monolignols): p-coumaryl, coniferyl, and sinapyl alcohols corresponding to the p-hydroxyphenyl (H), guaiacyl (G), and syringyl (S) units in lignin, respectively [3, 10, 53]. The complex structure of lignin contains numerous aromatic rings, hydroxyl and methoxy groups, ether bonds and therefore a high O_2 content. Lignin plays a role that provides robustness and resistance to cellulose microfibril systems surrounded by hemicellulose [30, 56]. The cleavage of β-O-4 bonds is the most probable route for solid lignin fragmentation from theoretical and experimental points of view. The thermal degradation of lignin is between 130 and 450 °C [32].

It was found that the pyrolysis of a mixture of bamboo sawdust and linear low-density polyethylene diminishes the activation energies for the thermal decomposition of components. This means that there is a co-pyrolytic process in the mixture that reduced 36% de activation energies when compared to separated feedstocks [57]. Now, the pyrolysis of blends of cellulose and high-density polyethylene (HDPE) showed two thermal decomposition stages: the first corresponds to the decomposition of cellulose between 260 and 410 °C, and the second to HDPE (410–527 °C) and showed higher mass loss of the mixtures which suggests a synergistic effect in the co-pyrolysis process [58].

According to Chen et al. [9], they performed a thermal decomposition of microcrystalline cellulose, lignin, and xylan in a TG-FTIR analyzer, a 15 mg sample was pyrolyzed up to 800 °C at four heating rates (10, 20, 30 and 40 °C min^{-1}), with a nitrogen flow of 70 mL min^{-1}. They demonstrated that hemicellulose presented the highest CO_2 yield, while lignin had the highest methane yield due to the aromatic rings and methoxy groups present in the lignin structure, cellulose demonstrated the highest CO yield at temperatures higher than 550 °C. The main components of cellulose bio-oil included anhydro-sugars and furans. While bio-oils produced from hemicellulose and lignin showed the highest relative content of phenols and acids. Increasing the temperature favored the reactions of deoxygenation, dihydroxylation, decarboxylation, decarbonylation and demethoxylation were promoted. This favored the decrease in the number of functional groups that contain oxygen (such as –OH, –C=O, –COOH, and $–(OCH)_3$). They also favored polycondensation reactions that formed more polycyclic aromatic hydrocarbon units [9].

Besides the aim to produce biofuels or value-added chemical products by pyrolysis, it also may generate electrical and heat power trough combustion of pyrolysis oil and/or non-condensable gases. The released heat present in the combustion gases can be recovered for cooling or heating systems [9].

5 Organic and Inorganic Wastes: A Valorization Alternative

Waste generation is closely related to human development and the urbanization process and is defined as the material or product whose owner or holder considers it unvaluable and discards it. Waste may be further susceptible to recovery, recycling, treatment, or final disposition [59]. The predominance of organic or inorganic waste is associated with the economic condition of the population: the organic composition predominates in lower-income countries, while the waste is mainly inorganic, with a significant number of manufactured products, in higher-income countries [36]. The organic fraction represents the half of total municipal solid waste (MSW) in Mexico (Table 3) with a per capita generation of 1.2 kg [36], which is equivalent to approximately 126 million MSW/day. Such organic fraction is generated from various domestic, industrial, livestock, aquaculture, and forestry activities, among others. The inorganic fraction becomes important because of the materials that can be recovered either to obtain new products or to produce fuels.

Table 3 Composition of MSW in Mexico (SEMARNAT 2020) adapted [36]

Type of waste	Generation (thousands of tons)	Percentage
Organics	23,043	51.6
Nonferrous metals	307	0.7
Ferrous metals	522	1.2
Textiles	642	1.4
Aluminum	798	1.8
Glass	2718	6.1
Plastics	4929	11.0
Cardboard, paper, and their products	6338	14.2
Others	5350	12.0
Overall	44,647	100.0

The plastics, cardboard, paper, and their products are equivalent to 11 and 14.2% respectively per day, which are generated as waste in Mexico (Table 3); but an equivalent percentage of plastic waste is produced in the United States, China and the European Union corresponds to 13, 11 and 9%, respectively [18]. Some of the inorganic waste in Mexico is recovered and incorporated back into recycling processes, that is, it is used to produce other products (plastics, cardboard, paper, and glass) and the metals are taken to foundries for reuse. In 2014 approximately 258 million tons of municipal solid waste (MSW) was generated and over 89 million tons were recycled and composted (34.6 wt%), over 33 million tons were combusted with energy recovery and 136 million tons were placed in landfills in the United States [18].

Plastics are polymers that are primarily derived from fossil fuels and its global production has increased by almost 80% since 2002 (Fig. 2). We observe a trend of gradual increase in plastic production on a global scale [11]. Depending on the degree of recyclability, plastics are classified into seven main groups: polyethylene terephthalate (PET), high-density polyethylene (HDPE), polyvinyl chloride (PVC), low-density polyethylene (LDPE), polypropylene (PP), polystyrene (PS) and miscellaneous plastics. Ethylene is used in the production of several common materials including low- and high-density- polyethylene (LDPE, HDPE), polyethylene terephthalate (PET), polystyrene and polyvinyl chloride (PVC). It is also used in the manufacture of ethanol, acetaldehyde and ethyl acetate. Moreover, propylene is used in the manufacture of a wide range of chemicals and materials including the manufacture of polypropylene, a common light weight and durable thermoplastic used in a variety of applications including packaging, textiles, carpets, furniture, among others. Other propylene derivatives include phenol propylene glycol and acrylic acid, [16].

Four percent of fossil fuels produced globally are used to make plastics and another 4% are used to generate energy for plastics production [60]. The main concern

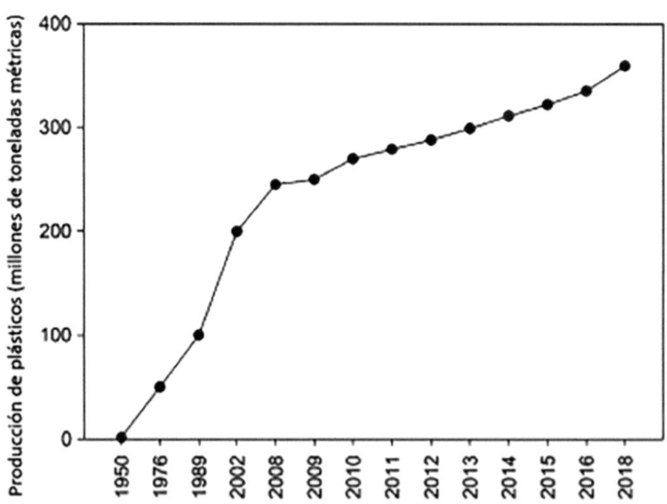

Fig. 2 Plastic production from 1950 to 2020. Adapted from Nanda [11]

about common plastics is that they are resistant to physical, chemical, and biological degradation, fractionate into micro plastics and they stand in landfills, oceans, and coastal seawater for decades to centuries.

Some relevant data on the consumption of plastics and their environmental effects from a life cycle assessment perspective includes the global production of plastics referring to plastic waste as of 2019, around 150–200 million tons are generated each year. Plastic waste generation is expected to double in the next 15 years, and the approximately 8 million tons per year are improperly disposed of in the oceans. Replacing plastics in packaging and consumer products with alternative materials does not necessarily bring environmental benefits, since it could quadruple environmental costs. The total quantities of post-consumer plastic waste generated in the European Union in 2016 correspond to 27.1 million tons of plastic waste in total (post-consumer) of which 31.1% was used for mechanical recycling, the 41.6% was used to recover waste into energy and remaining 27.3% was deposited in landfills. Thus, chemical recycling of plastic waste can be understood as the conversion of monomers or production of new raw materials by changing the chemical structure through pyrolysis, gasification or depolymerization, excluding incineration.

Chemical recycling involves obtaining pyrolysis oil which is used as raw material in the chemical industry which replaces gasoline (hydrocarbon mixtures). The post-treatment of the pyrolysis oil corresponds to the elimination of residual chlorine and oxygen; hydrogenation and separation are the final stages. The pyrolytic oil contains a calorific value of 44 MJ/kg which is produced by a pyrolysis process of three waste plastics (polyethylene, polypropylene, and polystyrene) in which they managed to eliminate 10% of impurities [61]. During the co-pyrolysis process, tires, plastics, and bamboo waste have shown the best synergistic effects in improving the performance and higher heating value (HHV) of bio-oil with biomass. Meanwhile, due to the important advantages of simplicity and efficiency, co-pyrolysis is also considered an ideal technology for waste management. The use of waste, wood sawdust, peanut shells, and sewage sludge as raw materials together with biomass, has also achieved notable results. The use of catalysts can improve the selectivity of the pyrolysis process and the properties of the bio-oil by limiting the distribution of the product, reducing the activation energy, reducing acidity, and reducing oxygenated and nitrogenous compounds [62, 63]. Co-processing of plastic waste with heavy oil waste is another alternative for the manufacturing of transportation fuels and chemical [6]. The performance of thermal pyrolysis oil has found multiple applications as fuel in boilers, turbines, furnaces, and diesel engines.

One short-term solution to the waste plastic problem is to process them by pyrolysis with the generation of primary products such as oil, non-condensable gases, and solids to obtain biofuels and, sometimes, even chemical products, as well as the generation of thermal and electrical energy. Different authors have highlighted the importance and study of different types of biomass and plastics in thermal degradation to obtain biofuels and/or chemicals [1–3, 5, 6, 28]. Studies on polymer pyrolysis have been carried out in a temperature range of 300–900 °C. Nevertheless, the temperature range of 500–550 °C is the most favorable for plastic waste [6]. Thermal and catalytic decomposition of plastic waste through pyrolysis may be one of the best approaches for plastic waste management since the scale of pyrolysis plants is more flexible than that of incineration plants [18] but also because the latter creates pollution unless the CO_2 is captured. The co-processing of waste plastics with waste oil and other used oils has opened a new perspective of recovery and upgrading of pyrolysis oil. Another study shows that LDPE pyrolysis, for 20 min at 500 °C, 10 °C/min using a fixed bed reactor with N_2 as fluidizing gas, yielded 95% by weight of pyrolysis oil with low gases and CO_2. A similar experiment using a batch reactor, at 550 °C and heating rate of 5 °C/min, reported 93.1% w/w pyrolysis oil [6]. Other studies indicate a yield of pyrolysis oil of 75.6% w/w at 430 °C but using a pressurized batch reactor (0.8–4.3 MPa) allowed to increase the yield of the pyrolysis oil greater than 89.5% [6]. Now, adverse effects from melting and agglomeration must be considered in the pyrolysis of plastics. Feed systems may need to be cooled to avoid clogging problems. Although cellulose, hemicellulose, lignin and plastics have different polymeric structures, co-pyrolysis can transform them into similar aromatic and olefinic products.

6 Pyrolysis: A Process Towards Decarbonization of Industry

Decarbonization gained importance with the increase in atmosphere temperature observed in recent years. Thus, the Paris Agreement carried out in 2015 was decisive in which several countries approved limiting the increase in global temperature to 2 °C until 2050 but to continue efforts to reduce it to 1.5 °C. Decarbonization needs alternative ways of living and working that reduce emissions and capture and store carbon in our soil and vegetation. This involves a paradigm in transformative change and can only be effective if supported by behavioral changes, new policies and regulatory frameworks, greener financial flows, international cooperation, and radical collaboration [64]. The decarbonization of a certain sector may not be useful for the decarbonization of the entire energy system, as it could simply shift carbon emissions from one sector to other sector [19].

We could understand that decarbonization is the process of reducing carbon emissions into the atmosphere, mainly carbon dioxide (CO_2), and other greenhouse gases such as methane or gases with global warming potential. Based on the United States Environmental Protection Agency (EPA), the gases that catch heat in the atmosphere are called "greenhouse gases", and within this group of gases belong carbon dioxide (CO_2), methane (CH_4), nitrous oxide (N_2O) and fluorinated gases [65].

The National Inventory of Emissions of Greenhouse Gases and Compounds, presented by the National Institute of Ecology and Climate Change [66], presented the estimates of anthropogenic emissions of greenhouse gases and compounds and absorption by sinks in Mexico (2018). Mexico emitted 683 million tons of carbon dioxide equivalent ($MtCO_2e$) of greenhouse gases (GHG) in 2015. The most relevant gas is carbon dioxide with 71% of emissions followed by methane with 21%. Fossil fuels contribute with 64% of total emissions followed by the livestock production systems with 10%. Other activities contribute with less than 8% of emissions like industrial processes, waste management, fugitive emissions from oil, gas, and mining extraction and agricultural activities. The inventory also recorded 148 $MtCO_2e$ absorbed by vegetation, mainly in forests and jungles. The net balance between emissions and removals was 535 $MtCO_2e$ in 2015 [66]. Biomass is neutral in terms of CO_2 impacts, in an idealistic balance, it emits the same amounts of CO_2 when consumed, which depends on the type of process (thermochemical or biological) as when it is sequestered or absorbed into the atmosphere, therefore, the net effect is zero [18].

Currently, there is a lot of specialized literature that addresses the issue of decarbonization, where alternative technologies, processes, raw materials, and other variables are presented that can help reduce CO_2 emissions and their equivalents. The decarbonization of the energy system takes biomass as a viable option to transform it into energy, based on two technologies: pyrolysis and gasification with Fischer–Tropsch processes for the latter [19]. Biomass plays a twice important role, that is, it is an energy producer and serves as a raw material to capture the CO_2 emitted into the atmosphere as mentioned earlier. Hydrogen is also considered a key resource for decarbonizing the economy. Several technologies are currently being developed to reduce the carbon footprint of hydrogen from methane. Methane pyrolysis (turquoise hydrogen or CO free) is an alternative, carbon capture and storage from methane pyrolysis is a form of carbon byproduct. Decarbonization is a commitment towards a responsible future where all forms of life have the same rights without forgetting that the actions are today, not tomorrow. The concept of circular bioeconomy with biorefinery, which focuses on the field of biomass, takes as a framework of reference biogenic resources which are indicated as carbon neutral from the perspective of life cycle assessment.

Hence, we present various pyrolysis projects to valorize different types of biomasses and under different operating conditions focused on decarbonization. Each project follows a specific objective, which is to obtain certain products of industrial interest to satisfy certain specific demands of the fuel market and the chemical industry (Table 4).

It is well known that pyrolysis oil can be combusted in boilers on a commercial scale, but its fuel properties and storage (oxidation) stability make it an unsuitable fuel. Upgrading of bio-oil to produce direct transportation fuels is under development, however, improving the catalytic properties of the oil through hydrotreatment can also produce biopropane that can be used for plant cogeneration [70]. The Ensyn company has designed and launched two fast pyrolysis plants, one of 3 and another of 10 million gallons pyrolysis oil per year in Ontario and Quebec cities, respectively. A joint venture produces 22 million gallon/y of pyrolysis oil from eucalyptus waste in Brazil that can be sent to heating plants and refineries in the US [39]. Also, the company BTG-BTL established its first commercial-scale fast pyrolysis plant in the Netherlands in 2015, which produces 20 million liters per year of FPBO (fast pyrolysis bio-oil) to generate heat and power. In addition, BTG-BTL has signed a contract for the construction of a new pyrolysis plant with Green Fuel Nordic in

Table 4 Pyrolysis projects implemented by different companies worldwide [7, 8, 67–69, 72]

Country	Project	Technology	Operation date	Feedstocks	Primary product	Capacity	Capital investment
Brazil	RPR Petrobras Braskem, Ultrapar	Fluidized catalytic cracking	May 2023	Soy oil	Bio-LPG, biofuels for ships, renewable propylene, and aromatic hydrocarbons (BTX: benzene, toluene, and xylene), essential to produce synthetic rubber, nylon, and PVC	2000 tons of soybean oil	USD 9.15 M
Brazil	Aracruz	Fast pyrolysis-circulation fluidized bed reactor	Unknown	Eucalyptus forest residues	Pyrolysis bio-oil	83 ML y^{-1} pyrolysis bio-oil production	UDS 77 M
Canada	Ontario	Fast pyrolysis-circulation fluidized bed reactor	Unknown	Mill and forest Woody residues	Pyrolysis oil	Fuel oil 11 ML y^{-1}	CAD 4 M
Finland	Fortum	Fast pyrolysis—VTT fluid bed riser reactor	2013	Forest residues, sawdust	Pyrolysis oil	50 kt y^{-1}	EUR 38 M
Finland	Green fuel	Fast pyrolysis rotating cone reactor	2020	Sawdust and Wood residue	Pyrolysis oil	20 ML y^{-1}	EUR 25 M
Netherlands	Empyro	Fast pyrolysis rotating cone reactor	2015	Clean Woody biomass	Pyrolysis oil	20 ML y^{-1}	EUR 19 M
Sweden	Pyrocell AB	Fast pyrolysis rotating cone reactor	2021	Sawdust	Pyrolysis oil, oil refinery co-feed	25,000 t y^{-1}	EUR 12 M

(continued)

Table 4 (continued)

Country	Project	Technology	Operation date	Feedstocks	Primary product	Capacity	Capital investment
USA	Georgia	Fast pyrolysis—circulation fluidized bed reactor	Unknown	Mill Wood residues, forest residues	Pyrolysis oil	76 ML y^{-1}	USD 70 M
USA	Red Arrow	Fast pyrolysis—circulation fluidized bed reactor	1995	Wood residues	Specialty chemicals	30–40 t/d 9 ML y^{-1}	Unknown
Canada	Abri-tech	Fats pyrolysis auger reactor	Unknown	Woody residues	Pyrolysis bio-oil	6 ML y^{-1}	CDA 7 M
China	Dali Conty	Fast pyrolysis downdraft circulating fluidized bed reactor	2016	Rice husks	Pyrolysis oil	2–6 ML bio-oil y^{-1}	Unknown
China	Hefei facility	Fast pyrolysis fluidized bed	2010	Wood and agricultural waste	Pyrolysis oil	2 ML bio-oil y^{-1}	Unknown
India	Mash energy	Pyrolysis	Unknown	Waste materials, nuts, sewage sludge	Pyrolysis oil	Unknown	Unknown
USA	Biogas energy project	Fast pyrolysis-ablative reactor	2020	Demolition Wood, bark beetle infested trees, forest, and agriculture residues	Pyrolysis oil	1.3 bio-oil ML y^{-1}	Unknown

Finland. Other examples are the Setra consortium that announced the development of a fast pyrolysis plant in Sweden, to produce 25,000 tons of pyrolysis oil per year; and CRI/Shell is developing a catalytic pyrolysis process at the Shell Technology Center in Bangalore, India. This is a process where forest residues are transformed into pyrolysis oil, which is upgraded to produce direct hydrocarbons with a production plant capacity of 5 metric tons per day [39].

One integral approach considers the development of a next-generation biomass intermediate pyrolysis (BIPP) polygeneration system as a short-term alternative in China [71]. The term "polygeneration" refers to multiple outputs of an integrated technological system, which refers to a pyrolysis reactor coupled to a heat recovery system, a microgenerator to combust 80% of the biofuels to generate electrical energy, as well as a storage and distribution system to supply the remaining 20% of biofuels for domestic use. Based on this approach, the authors indicate that BIPP systems could contribute to GHG reductions from 198 to 3156 Mt CO_2-eq before 2030 [71]. By controlling pyrolysis temperature variables and production capacity, the project shows evidence that positive financial returns from biofuel production can be achieved along with significant climate mitigation benefits as the carbon-rich biochar is applied to land. Likewise, they observed that the net energy production of biomass intermediate pyrolysis (BIPP) increases with increasing temperature, the net electrical efficiency shows a large increase between 250 and 350 °C and BIPP systems operating at higher temperatures above 550 °C generate higher returns. The results of the economic analysis indicate that the proposed novel BIPP system with biochar sequestration can be profitable in a mature carbon market without subsidies, the environmental effects are potentially positive in the reduction of carbon levels emitted into the atmosphere, the BIPP plants with a biochar sequestration can lead to GHG emission reductions of 136.45 g CO_2 eq MJ^{-1} [71].

7 Conclusions

Thermal valorization of algae, waste and lignocellulosic biomass through pyrolysis presents a great opportunity in the energy transition, where decarbonization may be achieved by the production of fuels, energy and chemical products towards sustainable processes. Pyrolysis offers the advantage over other technologies that its CO_2 emissions may be neutral. Co-processing waste plastics with biomass can help increase the yield and improve the composition of pyrolytic products. Indeed, the conversion and yield of co-pyrolytic processes may lead to different compositions and concentrations of biochar, pyrolysis oil, and non-condensable gases. During the study of pyrolysis, it must be clear that the variables that affect this process must be analyzed in depth, such as conversion as a function of temperature and time, residence times, pyrolysis temperature, mass ratios, inert atmosphere flows, among others. The selectivity of each product is a function of the operating variables and the raw material to be used. We must consider a quintuple helix approach among research, policy, industries, society, and climate change to achieve decarbonization and avoid irreversible processes in climate change that compromise our generation and future ones.

Acknowledgements Óscar Piña thanks CONAHCyT for doctoral fellowship. The Authors thank IMP for financial support to undertake this work.

References

1. Escalante J, Wei-Hsin C, Tabatabaei M, Anh TH, Eilhann EK, Kun-Yi AL, Ayyadurai S (2022) Pyrolysis of lignocellulosic, algal, plastic, and other biomass wastes for biofuel production and circular bioeconomy: a review of thermogravimetric analysis (TGA) approach. Renew Sustain Energy Rev 169:1–21. https://doi.org/10.1016/jrser.2022.112914
2. Dada TK, Sheehan M, Murugavelh S, Antunes E (2021) A review on catalytic pyrolysis for high-quality bio-oil production from biomass. Biomass Conv Bioref 1–20. https://doi.org/10.1007/s13399-021-01391-3
3. Ryu HW, Kim DH, Jae J, Lam SS, Park ED (2020) Recent advances in catalytic co-pyrolysis of biomass and plastic waste for the production of petroleum-like hydrocarbons. Bioresour Technol 310:1–8. https://doi.org/10.1016/j.biortech.2020.123473
4. Rasaq WA, Golonka M, Scholz M, Bialowiec A (2022) Opportunities and challenges of high-pressure fast pyrolysis of biomass: a review. Energies 14(5426):1–20. https://doi.org/10.3390/en14175426
5. Van de Beld L (2020) Process steam in a dairy factory via fast pyrolysis bio-oil. IEA Bioenergy 1–28. 20.500.12592/9dbhqv
6. Pamreishang K, Sharma DK, Ahmaruzzaman M (2020) Thermal and catalytic decomposition of waste plastics and its coprocessing with petroleum residue through pyrolysis process. Elsevier, pp 1–25. https://doi.org/10.1016/j.clepro.2020.121639
7. Wijeyekoon S, Torr K, Corkran H, Bennett P (2020) Commercial status, or direct thermochemical liquefaction technologies. IE Bioenergy 1–28
8. Stapf D, Ciceri G, Johansson I (2020) Trends, and drivers in alternative thermal conversion of waste. IEA Bioenergy 1–27
9. Chen D, Cen K, Zhuang X, Gan Z, Zhou J, Zhang Y, Zhang H (2022) Insight into biomass pyrolylisis mechanism based on cellulose, hemicellulose, and lignin: evolution of volatiles and kinetics, alucidation of reaction pathways, and characterization of gas, biochar, and bio-oil. Combust Flame 242:1–17. https://doi.org/10.1016/j.combustflame.2022.112142
10. Wang K, Brown RC (2017) Chapter 1. Prospects for fast pyrolysis of biomass. Green chemistry series no 50. In: Fast Pyrolysis of biomass: advances in science and technology. The Royal Society of Chemistry, pp 2–11. https://doi.org/10.1039/9781788010245-00001
11. Lee N, Joo J, Lin KYA, Lee J (2021) Waste-to-fuels: pyrolysis of low-density polyethylene waste in the presence of H-ZSM-11. Polymers 13:1198. https://doi.org/10.3390/polym13081198
12. Kim Y, Thomas A, Robichaud D, Iisa K, John P, Etz B, Fioroni G, Dutta A, McCormick R, Mukarakate C, Kim S (2020) A perspective on biomass-derived biofuels: from catalysis design principles to fuel properties. J Hazard Mater 2020(400):1–11. https://doi.org/10.1016/j.jhazmat.2020.123198
13. Kirubakaran V, Sivaramakrishnan V, Nalini R, Sekar T, Premalatha M, Subramamian P (2009) A review on gasification of biomass. Renew Sustain Energy Rev 13:179–186. https://doi.org/10.1016/j.rser.2007.07.001
14. Mohamed A, Nuno B, Hasan M, Greg P, Muxina K (2020) A review on advanced catalytic co-pyrolysis of biomass and hydrogen-rich feedstock: insights into synergistic effect, catalyst

development and reaction mechanism. Biores Technol 310:1–14. https://doi.org/10.1016/j.biortech.2020.123457
15. Vilma Quitral R, Carla Morales G, Marcela Sepúlveda L, Marco Schwartz M (2012) Nutritive and healthy properties of seaweed and its potential as a functional ingredient. Chil Nutr Mag 34:1–8. https://doi.org/10.4067/S0717-75182012000400014
16. Morais EG, Silveira JT, Shuler LM, Bastos de Freitas BC, Vieira Costa JA, Morais MG, Ferrer I, Barreira L (2023) Biomass valorization via pyrolysis in microalgae-based wastewater treatment: challenges and opportunities for a circular bioeconomy. J Appl Phycol 35(6):1–21. https://doi.org/10.1007/s10811-023-03104-x
17. Egelja J, Stojkovic T (2022) Energy through pyrolysis method from landfill waste. International Consulting Group. Implementation period: 2019–2022. Organization details: International Consulting Group. Dr. Sandra Glisi´c, Jagodina, Republic of Serbia, p 1
18. Czajczynska D, Anguilano L, Ghazal H, Krzyzynska R, Reynolds AJ, Spencer N, Jouhara H (2017) Potential of pyrolysis in the waste management sector. Therm Sci Eng Prog 3:171–197. https://doi.org/10.1016/j.tsep.2017.06.003
19. Li X, Damartzis T, Stadler Z, Moret S, Meier B, Friedl M, Merechal F (2020) Decarbonization in complex energy systems: a study on the feasibility of carbon neutrality for Switzerland I 2050. Front Energy Res 8:1–17. https://doi.org/10.3389/fenrg.2020.549615
20. International Energy Agency. https://www.iea.org/reports/global-energy-review-2020
21. Basu P (2013) Biomass gasification, pyrolysis, and torrefaction. Practical design and theory, 2nd edn. Elsevier. Book Aid International, pp 1–96. ISBN: 978-0-12-396488-5
22. Castellan GW (1947) Fisicoquímica. Fondo Educativo Interamericano. Segunda edición, pp 715–779. ISBN: 968-444316-1
23. Getman FH, Daniels F (1957) Tratado moderno de Fisicoquímica. Compañía editorial continental, S.A. Segunda edición, pp 342–389
24. Al-Rumaihi A, Shahbaz M, Mckay G, Mackey H, Al-Ansari T (2022) A review of pyrolysis technologies and feedstock: a blending approach for plastics and biomass towards optimum biochar yield. Renew Sustain Energy Rev 167:1–16. https://doi.org/10.1016/j.rser.2022.112715
25. Chen Y, Kuo MJ, Lobo R, Lerapetritou M (2024) Ethylene production: process design, tecno-economic and life-cycle assessments. Green Chem 1–9. https://doi.org/10.1039/d3gc03858k
26. Atkins PW (1985) Fisicoquímica. Fondo educativo interamericano, S.A. de C.V. Segunda edición, pp 807–842. ISBN: 968-858-040-6
27. Kennedy M (2012) Biomass gasification. The East African study. Pisces practical action consulting. pp 1–40
28. Torres-García E (2021) Un acercamiento al análisis termocinético a través de la dependencia de la energía de activación con la conversión y la temperatura. Res Proposal 1–40. https://doi.org/10.13140/RG.2.2.24359.27040
29. Hougen OA, Watson KM (1962) Chemical process principles. Part three. Kinetics and catalysis. Wiley, pp 805–822
30. Levenspiel O (2006) Ingeniería de las Recciones Químicas. Editorial Limusa S.A. Tercera edición. de C.V. 2006, pp 13–79. ISBN: 10 968-18-5860-3
31. Pang S (2019) Advances in thermochemical conversion of woody biomass to energy, fuels, and chemicals. Biotechnol Adv 37:589–597. https://doi.org/10.1016/j.biotechadv.2028.11.004
32. Galano A, Aburto J, Sadhukhan J, Torres-García E (2017) A combined theorical-experimental investigation on the mechanism of lignin pyrolysis: role of heating rates and residence times. J Anal Appl Pyrol 2017(128):208–216. https://doi.org/10.1016/j.jaap.2017.10.009
33. DVGW (2022) Hydrogen Europe. Energie Wasser-Praxis. Kompakt. Pyrolysis. Potential and possible applications of a climate-friendly hydrogen production. 2022, pp 1–40
34. Veksha A, Wang Y, Foo JW, Naruse I, Lisak G (2023) Defossilization and decarbonization of hydrogen production using plastic waste: Temperature and feedstock effects during thermolysis stage. J Hazard Mater 452:1–24. https://doi.org/10.1016/j.jhazmat.2023.131270
35. Sánchez-Bastardo N, Schlogl R, Ruland H (2021) Methane pyrolysis for zero-emission hydrogen production: a potential bridge technology from fossil fuels to a renewable and sustainable hydrogen economy. I&EC Res 60:11855–11881. https://doi.org/10.1021/acs.iecr.1c01679

36. Secretaría del Medio Ambiente y Recursos Naturales (2023). https://apps1.semarnat.gob.mx:8443/dgeia/informe15/tema/cap7.html
37. Instituto Nacional de Estadística y Geografía (2020). https://www.inegi.org.mx/programas/ccpv/2020/
38. Agencia de Protección Ambiental de Estados Unidos (EPA). https://espanol.epa.gov/
39. Caldeira K, Hoeght-Guldberg O, Liss PS, Riebesell U (2005) Ocean acidification due increasing atmospheric carbon dioxide. The Royal Society. The Clyvedon Press Ltd, Cardiff, UK, pp 1–69. ISBN: 0-85403-617-2
40. Ville CA (1996) Biology. McGraw-Hill InterAmerican Editor's, S.A. of C.V. 8th edn, pp 155–183. ISBN: 0-7216-9023-8
41. Biris-Dorhoi ES, Michiu D, Pop CR, Rotar AM, Tofana M, Pop OL, Socaci SA, Farcas AC (2020) Macroalgae—a sustainable source of chemical compounds with biological activities. Nutrients 12:1–10. https://doi.org/10.3390/nu12103085
42. Misurcova L (2011) Chemical composition of seaweeds. Research Gate, pp 1–22. https://doi.org/10.1002/9781119977087.ch7
43. Xu W, Ding K, Hu L (2021) A mini review on pyrolysis of natural algae for biofuel and chemicals. Processes 9:1–11. https://doi.org/10.3390/pr9112042
44. Lage S, Gentili FG (2023) Chemical composition and species identification of microalgal biomass grow at pilot-scale whit municipal wastewater and CO_2 from flue gases. Chemosphere 313:1–11. https://doi.org/10.1016/j.chemosphere.2022.137344
45. Li Z, Zhao Y, Xu X, Han R, Wang M, Wang G (2018) Migration and transformation of dissolved carbon during accumulated cyanobacteria decomposition in shallow eutrophic lakes: a simulated microcosm study. PeerJ 7:1–23. https://doi.org/10.7717/peerj.5922
46. Mostafazadeh AK, Solomatnikova O, Drogui P, Tyahi RD (2018) A review of recent research and developments in fast pyrolysis and bio-oil upgrading. Biomass Conv Bioref 8:739–773. https://doi.org/10.1007/s13390-018-0320-z
47. Huber GW, Iborra S, Corma A (2006) Synthesis of transportation fuels from biomass: chemistry, catalysis, and engineering. Chem Rev 106:4044–4098. https://doi.org/10.1021/cr068360d
48. Lewandowski WM, Ryms M, Kosakowski W (2020) Thermal biomass conversion: a review. Processes 8(516):1–45. https://doi.org/10.3390/pr8050516
49. Kazawadi D, Ntalikwa J, Kombe G (2021) A Review of intermediate pryrolysis as a technology of biomass conversion for coproduction of biooil and adsorption biochar. J Renewable Energy 2021:1–10. https://doi.org/10.1155/2021/5533780
50. AlDayyat EA, Saidan MN, Al-Hamamre Z, Al-Addous M, Alkasrawi M (2021) Pyrolysis of solid waste for bio-oil and char production in refugees´ camp: a case study. Energies 14:3861. https://doi.org/10.3390/en14133861
51. Khuenkaeo N, Phromphithank S, Onsree T, Naqvi R, Tippayawmong N (2021) Production and characterization of bio-oils from flash pyrolysis of tobacco processing wastes in an ablative reactor under vacuum. PLoS ONE 167:1–13. https://doi.org/10.1371/journal.pone.0254485
52. Ajazziza M, Albahanasawi A, Eyvaz M, Maskari T, Nassani DE, Abu Amr SS, Abujazar MSS, Bashir MJK (2023) An overview of green bioprocessing of Algae-Deriv biochar and biopolymer: synthesis, preparation, and potential applications. Energies16:791. https://doi.org/10.3390/en16020791
53. Tursi A (2019) A review on biomass: importance, chemistry, classification, and conversion. Biofuel Res J 2019(22):962–979. https://doi.org/10.18331/BRJ2019.6.2.3
54. Lehninger AL (1995) Bioquímica. Ediciones Omega, S.A. Decimoctava reimpresión
55. Wenzl HFJ (1970) The chemical technology of wood. Academic Press Inc., New York and London. ISBN: 012743450X, 9780127434506
56. Amezcua-Allieri MA, Aburto J (2018) Conversion of lignin to heat and power, chemicals or fuels into the transition energy strategy. INTECH, pp 145–160. https://doi.org/10.5772/intechopen.71211
57. Alam M, Bhavanam A, Jana A, Jaimin KSV, Nageswara RP (2020) Co-pyrolysis of bamboo sawdust and plastic: synergy effects and kinetics. Renewable Energy 149:1133–1145. https://doi.org/10.1016/j.renene.2019.10.103

58. Yuan H, Fan H, Shan R, Mingyang H, Gu J, Yong C (2018) Study of synergistic effects during co-pyrolysis of cellulose and high-density polyethylene at various ratios. Energy Conver Manage 157:517–526. https://doi.org/10.1016/j.enconman.2017.12.038
59. General law for the prevention and comprehensive management of waste. Last reform. Official Gazette of the Federation 05–08–2023. https://www.diputados.gob.mx/LeyesBiblio/pdf/LGPGIR.pdf
60. Nanda S, Beruti F (2021) Thermochemical conversion of plastic waste to fuels: a review. Environ Chem Lett 19:123–148. https://doi.org/10.1007/s10311-020-01094-7
61. BASF (2020) Evaluation of pyrolysis with LCA-3 case studies, pp 1–150. According to ISO 14040:2006 and ISO 14044:2006
62. Sun J, Norouzi O, Masek O (2022) A state-of-the-art review on algae pyrolysis for bioenergy and biochar production. Biores Technol 346:1–18. https://doi.org/10.1016/j.biortech.2021.126258
63. Su G, Ong HC, Gan YY, Chen W-H, Chong CT, Ok YS (2020) Co-pyrolysis of microalgae and other biomass wastes to produce high-quality bio-oil: progress and prospective. Biores Technol 344(12):1–13. https://doi.org/10.1016/j.biortech.2021.126096
64. United Nations Climate Change (2023) https://unfccc.int/news/decarbonization-cannot-wait
65. United States Environmental Protection Agency (2023) https://espanol.epa.gov/la-energia-y-el-medioambiente/descripcion-general-de-los-gases-de-efecto-invernadero
66. National Inventory of Greenhouse Gas and Compound Emissions. National Institute of Ecology and Climate Change (2018). https://www.gob.mx/inecc/acciones-y-programas/inventario-nacional-de-emisiones-de-gases-y-compuestos-de-efecto-invernadero
67. Van de Beld B, Funke A, Lindfors C, Sandstrom L (2019) Country reports. Direct thermo-chemical liquefaction. IEA Bioenergy 1–20. https://doi.org/10.5445/IR/1000118109
68. (2017) ENSYN the world's leading biocrude producer: investing in the bioeconomy. pp 1–12
69. (2023) Historical: Petrobras converts soybean oil into biofuels at a refinery in Brazil. H2BusinessNews
70. Pinheiro A, Arauzo J, Fonts I, Domine M, Fernández A, Garcia ME, Montoya J, Chejne F, Pfromm P, Garcia M (2019) Challenges, and opportunities for bio-oil refining: a review. Energy Fuels 1–39. https://doi.org/10.1021/acs.energyfuels.9b00039
71. Yang Q, Zhou H, Bartocci P, Fantozzi F, Masek O, Agblevor FA, Wei Z, Yang H, Chen H, Chen G, Zheng C, Nielsen CP, McElroy MB (2021) Prospective contributions of biomass pyrolysis to China´s 2050 carbon reduction and renewable energy goals. Nat Commun 12:1–12. https://doi.org/10.1038/s41467-021-21868-z
72. Yang Q, Han F, Chen Y, Yang H, Chen H (2016) Greenhouse gas emissions of a biomass-based pyrolysis plant in China. Renew Sustain Energy Rev 53:1580–1590. https://doi.org/10.1016/j.rser.2015.09.049

Chapter 7
Biocrude Production and Subsequent Upgradation via Hydrothermal Liquefaction, Distillation, and Hydrotreatment of Agro-Forestry Biomass: A Canadian Context

John G. B. Churchill, Venu B. Borugadda, and Ajay K. Dalai

Abstract The use of hydrothermal liquefaction and biocrude upgradation technologies for lignocellulosic residue feeds is a lucrative pathway to meet growing sustainable transport fuel demands while avoiding disruption of critical food supply chains. Hydrothermal liquefaction is gaining attention from corporate and academic spheres for its feedstock flexibility; however, produced biocrude faces challenges in reaching fuel-level stability, acidity, viscosity, and heating value, primarily through high oxygen content. As a result, established upgradation for conventional crude oil is being adapted to biocrude upgradation, such as physical distillation and thermochemical hydrotreatment. The infancy of hydrothermal liquefaction and biocrude upgradation techniques remains the biggest opponent for this pathway, with uncertainty in economic as well as technical feasibility. An emerging fuel production pathway ideal for natural resource-abundant countries, this chapter covers ideal lignocellulosic feedstock in a Canadian context, hydrothermal liquefaction and its parameter effects, as well as established biocrude upgradation techniques to produce transportation fuels.

Keywords Biocrude · Catalytic upgradation · Hydrothermal liquefaction · Lignocellulosic biomass · Sustainable fuels

---

J. G. B. Churchill · V. B. Borugadda · A. K. Dalai (✉)
Catalysis and Chemical Reaction Engineering Laboratories, Department of Chemical and Biological Engineering, University of Saskatchewan, Saskatoon, Canada
e-mail: ajay.dalai@usask.ca

J. G. B. Churchill
e-mail: garret.churchill@usask.ca

V. B. Borugadda
e-mail: vbb123@mail.usask.ca

M. Aslam et al. (eds.), *Production and Biorefining of Biocrude Oil: Current Status and Future Developments*, Advances in Sustainability Science and Technology,
https://doi.org/10.1007/978-981-96-5198-6_7

1 Introduction

The world is well into an era of rapid energy demand through exponential industrialization and population. There are no signs of slowing down in energy consumption with global energy estimated to have risen by 4.6% in 2021 and consequent greenhouse gas emissions recorded higher than ever, rebounding from the Covid-19 pandemic [1]. The accelerating energy consumption of solid, gas, and liquid fossil fuels is an increasingly unsustainable practice, associated with concern due to acid rain, air pollution, and global climate change [2]. From a political aspect, growing tensions from large fossil fuel exporters such as Russia, have additionally put pressure on the energy sector with fuel prices surging in 2022. Rapid consumption with little to no renewability has led to declines in fossil fuel reserves, projected to deplete as early as the second half of the twenty-first century [2]. The dire environmental impact, increased price of fossil fuels, and future decline in fossil fuel supply have pushed researchers to investigate sustainable sources of energy. Likewise, governing bodies have recognized a need to increase renewable energy with goals, initiatives, and legislation committed to reducing non-renewables and increasing sustainability. Countries like Canada are striving for carbon neutrality within 30 years and have passed initial legislation such as the *Renewable Fuels Regulations*, *Pan-Canadian Approach to Pricing Carbon Pollution, Clean Fuels Regulations,* and *Clean Fuels Fund* to increase biofuel demand and reduce fossil fuels [3]. For example, Saskatchewan province has requirements for 7.5%vol. renewable gasoline and 2%vol. renewable diesel at fuel pumps as of 2023, expected to increase in the future [4].

Transportation fuels make up nearly 30% of all global energy consumption [5], while accounting for 37% of CO_2 emissions in 2021, maintaining the highest reliance on fossil fuels of any energy sector, as seen in Fig. 1 [6]. As the transportation industry is falling behind in sustainability, renewable energy solutions in the form of biofuels are seen as more important than ever before. Biomass-derived biofuels are a particularly promising solution, utilizing already well-established fuel infrastructure and engines, with the potential to source fuel from biomass that would otherwise be regarded as waste. Biomass sources for fuels have drawn attention as they sequester carbon while growing to create a renewable carbon cycle when combusted as fuel. First-generation biomass includes food crops (grains, sugar cane, potatoes, etc.), oil seed plants (palm, jatropha, etc.), and grasses (i.e., alfalfa), predominantly used for human consumption. First-generation biomass is some of the earliest sustainable energy pathways explored, however, the demand for these edible biomasses competes directly with vital food and feed industries [7].

Cost and sustainability advantages have led researchers to shift focus towards second-generation feedstock, often a waste product of first-generation, for advanced biomass-to-bioenergy technology in the last 10 years. Second-generation biomass includes non-edibles such as agricultural residues (straws, husks, stalks, hulls, and bagasse), aquatic plants (hyacinth, and algae), green grasses (such as switch grass), forestry residues (including sawdust, pulp waste, and shavings), and biogenic municipal or industrial wastes [9]. Exporting millions of tonnes of grain annually, the

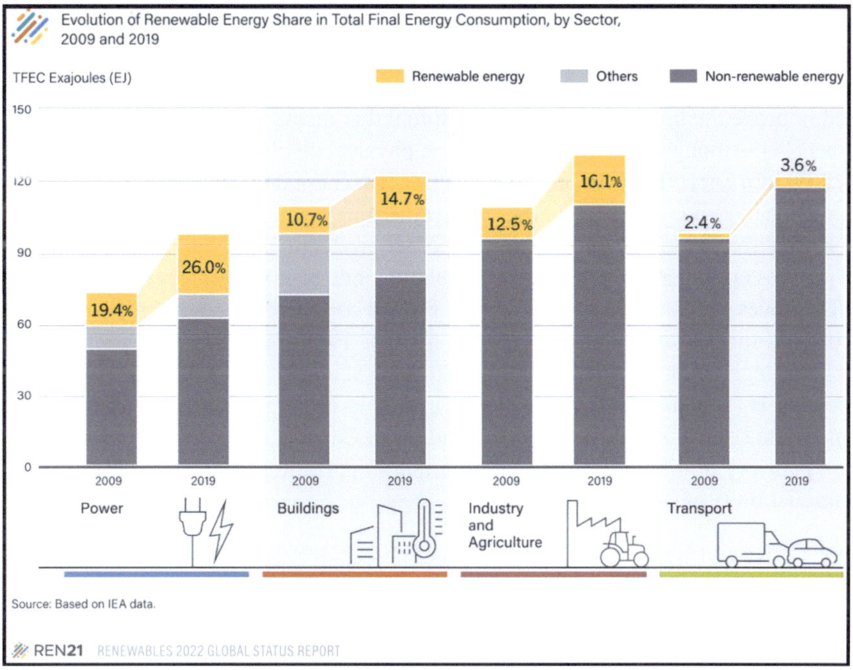

Fig. 1 Reliance of global energy sectors on non-renewable energy—2009–2019 [8]

resource-abundant west accounts for 83% of crop production primarily in the Canadian prairie provinces of Alberta, Saskatchewan, and Manitoba [9]. This agricultural-rich region of Canada has the potential to be a key supplier of biofuels. Despite promising availability, sustainability, and cost over first-generation biomass, second-generation still faces challenges in logistics being a new area of fuel sourcing with little industrial implementation [10].

Hydrothermal liquefaction (HTL) is an emerging process that is suitable for biomass of all water content, allowing flexible use of feeds from dry sawdust to tropical and aquatic biomass with moisture sometimes over 90 wt% [11]. The HTL process uses moderate to high temperatures and pressure in the range of 240–550 °C and 5–25 MPa [10]. Utilizing water as a solvent, HTL exploits water's unique subcritical properties to act as a catalyst and reactant at the same time. Under the subcritical but near-critical HTL conditions, water's bonds are weakened, leading to higher hydroxide and hydrogen ion concentration as well as lower viscosity, surface tension, polarity, and dielectric constant that aid in the complex decomposition, depolymerization, and recombination of biomass into organics such as desired hydrocarbons in the form of a biocrude [12, 13].

With promising experimental results of HTL over other biofuel technology, several drawbacks do exist. HTL biocrude still differs greatly from desired fossil fuel characteristics including composition (primarily oxygen content) associated with higher acidity, higher viscosity, and lower heating value [10, 14]. Besides HTL biocrude production, there is an additional need to individually upgrade or co-process the biocrude with refinery intermediates in order to reduce primarily oxygen content and improve the biocrude to a drop-in biofuel that meets firm transportation regulations [14]. Among upgradation techniques, physical distillation and thermochemical hydrotreatment techniques pioneered by the petroleum industry are commonly used to reduce biocrude oxygen content. As an emerging technology with the lowest technological readiness compared to other biofuel production methods (Fig. 2), the lack of process optimization, pilot-scale production, industrial scaling, and economics for HTL processing is apparent [10, 13, 15]. From a commercial standpoint, Canada has a bright future as an anticipated biofuel hotspot, including HTL. Being a resource-rich and legislation-supporting nation with several universities and corporate biofuel projects, HTL is emerging in the western as well as eastern provinces of Canada [16]. Under the context of lignocellulosic residues in Canada, the relatively new production pathway of biocrude from hydrothermal liquefaction and subsequent biocrude upgradation is presented in this chapter.

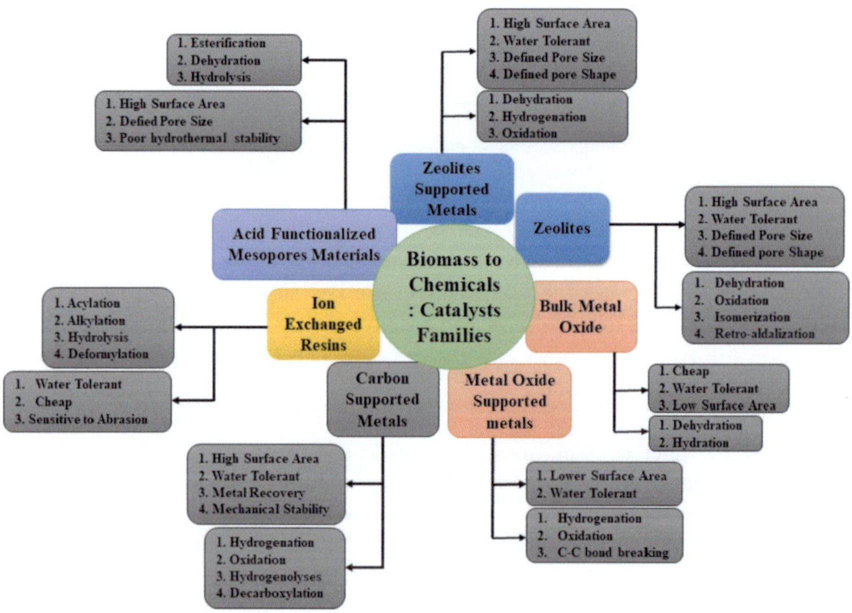

Fig. 2 Technological readiness of biofuel technology [10]

2 Lignocellulosic Biomass

Canada is an international leader when it comes to natural resources. Biomass industries within Canada are well established with forestry and agriculture businesses dating back well over a century and being significant contributors to the nation's economy. Home to vast farmable regions, the Canadian prairies and St. Lawrence lowlands produce millions of tonnes of cereal, pulse, and oilseed crops among many other specialty crops and livestock [17]. It is important to note that Canada's geographic position mostly in the north temperate zone leads to extended and harsh winters with an agricultural season of typically a few months (May–September), limiting biomass to seasonal production. The prairies account for 83% of crop production, a significant region for lignocellulosic residue availability [9]. As the most abundant biomass on land, lignocellulose has been explored as a sustainable alternative for fuel and chemical production [18]. Lignocellulosic biomass is composed primarily of macromolecules known as cellulose, hemicellulose, and lignin, while often containing fewer extractive components such as fats, proteins, and starches, as well as small amounts of ash. These components vary significantly among lignocellulosic biomass and are important factors in HTL biocrude yield and composition [19]. Lignocellulosic biomass contains a complex structure of cellulose and hemicellulose carbohydrate polymers, as well as the rigid lignin aromatic polymer [20]. Cellulose is the most common organic polymer consisting of straight-chain D-glucose molecules with glycosidic bonds in the formula $(C_6H_{10}O_5)_n$, forming ordered bundles capable of degrading to hydrocarbons [20]. Hemicellulose has a large and random amorphous structure consisting of different sugar monomers, containing approximately 500–3000 sugars to cross-link with cellulose and lignin for cell wall strength [18]. Lignin is also amorphous, known to be durable, chemically resistant, and composed of branched long-chain polymers [20]. Utilizing these fibres is considered sustainable as they are renewably produced by sequestering CO_2 and water via sunlight-driven photosynthesis during plant growth. Typical lignocellulosic biomass contains cellulose as the largest component at 33–55 %wt, while hemicellulose and lignin reported at 13–33% wt and 13–32 %wt, respectively [20]. Forestry biomass is typically harder to handle, known to contain higher lignin-dense layers and fewer extractives than agricultural biomass [20].

2.1 Agricultural

Given that western Canada is a massive agriculture producer, second-generation agricultural residues are widely available. These agricultural residues include straws, husks, stalks, and hulls. As second-generation agricultural biomasses, they are non-edible by-products with inherently more availability and lower cost than first-generation. Despite no use for human consumption, it is known that farmers with livestock often use residues for economic feed, soil fertilizer, as well as livestock insulation to get through harsh Canadian winters [21]. Agricultural purposes are the main use of second-generation residues (roughly 95%), however, there is a small construction market that produces biomass-based building materials (roughly 5%) [21]. Despite these applications, annually available residues from key crops were estimated to be 6.6 million tonnes of barley, 37.1 million tonnes of spring wheat, and most abundantly 43.4 million tonnes of canola within the vast prairie regions (Alberta, Saskatchewan, and Manitoba) in 2020 [9].

It should be noted that large-scale use of second-generation agricultural residues for biofuel production has not been implemented in western Canada, with possible impacts predicted. One study hypothesized 12 biorefinery sites across the prairies, each with a predicted annual throughput of 2.42–3.80 million tonnes (MT) for economic feasibility, still only utilizing roughly 40% of the annual available residues in their regions and highlighting abundance [9]. Agricultural biomass has the benefit of being more accessible than forestry residues, as widespread farming equipment like combines, trucks, and bailers exist to bundle and transport the residues from less remote locations. Historically, excess agricultural residues are spread and left on the crop fields as an organic soil-enricher or sometimes even illegally burned if they pose problems, such as fibrous flax straw that can damage combines [22]. Although working agricultural residue back into soil is done out of convenience, there are some positive effects for farmers. Carefoot et al. found that returning no straw to the field over one year reduced soil nitrogen content by 16 kg/ha and organic carbon by 1600 kg/ha, potentially reducing crop yield long-term unless costly fertilizer is used to compensate for nutrient loss [23]. Leaving agricultural residue in the soil not only fertilizes the soil for increased future crop yield but also provides structure and reduces soil erosion, needed for windy regions that the prairies are known for, estimated to need roughly 0.75–1.50 tonne/ha of soil [22]. Given 3.39, 2.62, 1.47, and 1.91 tonne of straw is produced/ha of farmland for durum wheat, barley, canola, and flax, respectively, there is theoretical availability of straw that can be utilized for bioenergy that is otherwise left in the field or disposed of as waste [17]. Given the annual estimated demands for soil (54.8 Mt) and livestock (6.9 Mt) in Canada, an annual 47.9 Mt of agricultural residues was estimated to be available across Canada in a 2012 study, dominated by wheat and secondly barley straw [17].

There may be unrealized impacts on future soil conditions and crop yields if all available straws are utilized in prairies, an area that requires further large-scale study. There is also a large inherent variability in straw availability, predicted to only grow with more unpredictable climate change and regional variation of yields, harvesting technology, soil, and social use of straw in other applications [22]. An estimate of available prairie straw, accounting for soil and livestock needs in 2006, found that the prairies had roughly 15 Mt of annually available straw (5.6 Mt in Alberta, 7.9 Mt in Saskatchewan, and 1.5 Mt in Manitoba) but with a massive variation of 2.3–27.6 Mt depending on crop yields, showing how truly volatile residue availability is [22]. Yet, agriculture production has steadily increased in the prairies over the last few decades with suitable residue availability predicted, promising for utilization in fuel production [9, 22].

2.2 Forestry

Within western Canada, British Columbia is by far the largest producer of forestry products accounting for 38% of Canada's harvested wood in 2020 at approximately 54.5 million cubic metres (Mm^3) [24]. There are also significant logging and pulping industries in the prairie provinces, accounting for nearly a quarter of harvested wood in Canada, with Alberta (29.7 Mm^3), Saskatchewan (4.1 Mm^3), and Manitoba (1.3 Mm^3) along with British Columbia making up 63% of harvested wood in Canada [24]. This harvested wood is primarily used for manufacturing paper, pulp, and wood products. The world's forests are crucial to the environment by providing structure and a form of nutrients for plants as well as animals to thrive in strength to the surrounding soil, and crucial carbon sequestration in the earth's susceptible carbon cycle [25]. With nearly a tenth of the world's forests by area, conservation and management of Canada's forests is crucial [26]. Therefore, direct usage of woody biomass for biofuels creates an ecological footprint and a consequential need to devote resources to maintaining forest sustainability, such as replanting [25]. Forestry biomass directly from the forest also competes with well-established logging and pulping companies, whereas working with these companies to utilize their woody waste is much more economical and logistical. Different sources of forestry waste biomass have been investigated for energy use including sawdust, wood chips, branches, tree bark, wood shavings, low-grade lumber, and rejected trunks, logs, as well as branches from mills [14, 18, 20].

With such a large forest area, Canada has different forest regions each unique with their own ecosystems. Within the prairie provinces, the boreal forest dominates forested regions, north of the prairie provinces. The boreal forest has species diversity when it comes to trees, home to hardwoods like trembling aspen, balsam poplar, and white birch, as well as softwoods like white spruce, black spruce, tamarack, and balsam fir [27]. Towards the Rocky Mountains, western Alberta has 3 types of forests each with their own dominant tree species. The subalpine forest includes mainly softwoods such as Engelmann spruce, subalpine fir, and less but notable amounts of black spruce. The montane forest of western Alberta is known for softwoods such as lodgepole pine, white spruce, Engelmann spruce, Douglas fir, and subalpine fir, as well as hardwood trees such as trembling aspen, balsam poplar, small pockets of white birch. With small parts of the Columbia Forest reaching the Alberta border from British Columbia, this small forest region mainly consists of softwoods including red cedar and western hemlock spruce, as well as lesser amounts of fir, pine, and tamarack [27]. Reflected by prevalence in western Canada, black spruce is the largest tree species in Canada by far, totalling roughly 57.3% of the treed area, while the next common species is trembling aspen and lodgepole pine at 9.8% and 5.9% of the treed area, respectively [28]. As spruce and pine are softwoods, it can be realized that softwoods are drastically more prevalent and harvested than hardwoods, yet milling of just hardwood, as well as both soft and hardwood at the same time, does still occur in Canada. In the year 2021, 55,842 m^3 of softwood was produced by

Canadian lumber companies, while a mere 880 m^3 of hardwood was produced over the same period [29].

Disease, insects, and fire add elements of volatility to the availability of forestry biomass, predicted to only worsen with climate change, similar to other biomass resources like agriculture. This variability can be reflected in the logging and mill industries, with shortages leading to increased prices of wood and waste products while also reducing availability. In 2020 alone, it was estimated 17,768,618 ha of woodland were defoliated or lost to insects, such as the mountain pine beetle or spruce budworm, accounting for roughly 4.5% of wooded land in Canada [24]. Fires are less impactful than insects but still significant, with 6596 wildfires causing the loss of an estimated 4,307,520 ha or roughly 1.1% of wooded land in Canada [24]. Salvaging of wood residues from disease or wildfire has been explored for bioenergy purposes with benefits such as low moisture, similar heating value to unaffected wood, and in some cases degraded lignin, however far greater challenges include transportation, inconsistent availability, high ash content, and increased nitrogen/ sulphur content, an area requiring far more research and development [30, 31]. Statistics on forest waste residue availability from mill and logging industries are not widely available online and are hard to come by, however, models exist for estimating availability through assumptions and back calculating. Sidders et al. [32] reported the annual availability of forest residues to be nearly 49 Mm^3 accounting for harvesting, thinning, and land-clearing residues (roadside), mill residues, and urban residues. A review by Pare et al. [33] summarized the annual availability of Canadian forest biomass residues to roughly range between 30 Mm^3 on the low end and 60 Mm^3 on the high end.

Forestry biomass, like all lignocellulosic biomass, is composed of the 6 components cellulose, hemicellulose, lignin, extractives, minerals, and ash. However extractive, mineral, and ash components are typically minor in the content of forest residues, being notably lower compared to agriculture residues. Within woody biomass, softwoods contain the highest lignin content (25–35 %wt) over hardwoods (16–25 %wt), including far less cellulose, hemicellulose, and extractives. In conflicting studies, this composition difference was reported as a hindrance or a benefit for HTL biocrude yield of high-lignin softwoods [19, 34]. This discrepancy indicates a need for further research comparing varying lignin feeds as no literature could be found directly comparing several woody feedstocks for HTL yields. When considering feedstocks for biofuel applications, availability, HTL yield, and biocrude quality are crucial factors. Given the uncertainty and lack of direct yield comparisons from literature, availability is the initial factor when considering woody feedstocks. Of woody biofuel feedstocks in literature, spruce and pine are commonly used for their availability from forestry industries, while poplar is noted as an ideal source for direct growing/ harvesting as it has the fastest growth rate of trees and is available in less rural non-boreal forest dominated areas like the prairies [35].

3 Hydrothermal Liquefaction

From a liquid fuel perspective, thermochemical pyrolysis and HTL are the main pathways for second-generation biomass. Comparing literature, the two processes with the same feeds have shown that HTL produces biocrude with higher heating value (HHV) and carbon content as well as lower density, oxygen, and nitrogen content, creating a crude with closer desirable characteristics to fossil fuel refinery streams [36]. Not only has HTL been shown to produce conclusively higher quality biocrude than pyrolysis, but the process has also reported comparable, if not higher, yields and energy conversion rates [15]. An emerging techno-economic assessment comparing fast pyrolysis and HTL fuel production found HTL had significantly lower production costs (in USD) such as 0.72 $/L versus 0.85 $/L for jet fuel, 0.97 $/L versus 1.14 $/L for gasoline, and 1.02 $/L versus 1.20 $/L for diesel [15]. From the viewpoint of biocrude quality, mass and energy yield, as well as economics, HTL has been shown to outperform pyrolysis as a promising technology in biocrude production for fuel applications.

The main objective of HTL is to break down the macromolecules of biomass to yield smaller organic molecules, including hydrocarbons that biocrude is comprised of. Biomass's complex macromolecules, cellulose, hemicellulose, and lignin are broken down and converted to energy-rich hydrocarbons through equally complex reaction pathways, with some summarized in Table 1. The moderate temperature and high pressure coupled with the strong subcritical catalytic properties of water create extreme conditions to degrade the biomass into hydrocarbons and other organic compounds, mimicking the production of fossil fuels via natural geological processes [37]. Primarily depolymerization and decomposition reactions first occur with the biomass to produce highly reactive compounds, dissolved in the subcritical water [38]. Depolymerization occurs by the chemical breakdown of the macromolecules into monomers, such as cellulose and hemicellulose breaking down to simple sugars like glucose and xylose as well as lignin breaking down into mono-phenolics [39].

Following depolymerization, decomposition primarily occurs through dehydration, decarboxylation, deamination, and general cleavage reactions where hydroxyl, carboxyl, and amino groups are removed from the biomass, reducing oxygen and nitrogen content, as depicted in Table 1, Eqs. 1–4 [38]. The initial decomposition causes reactive and often water-soluble components to form, participating in various reactions such as hydrolysis, isomerization, defragmentation, dehydration, Maillard, rearrangement, and recombination to form several components, represented in Table 1 with Eqs. 5–10 [40]. Interestingly, the recombination and repolymerization reactions (Table 1, Eq. 10) are reverse reactions of the initial depolymerization and decomposition reactions, primarily driven by a lack of hydrogen molecules and excess free radicals, forming higher molecular weight compounds [37]. These components produced from the various reactions produce a combination of HTL end products, yielding gas, char, aqueous, and biocrude phase products [37]. The exact pathways and mechanisms of HTL are unreported in literature due to the complexity of feeds, end-products, and near-endless intermediate reactions highly influenced

Table 1 General reaction equations of prominent HTL mechanisms [37, 38, 40]

#	Reaction type	Reaction equation
1	Dehydration	$H - R - OH \rightarrow R + H_2O$
2	Decarboxylation	$O = R - OH \rightarrow R - H + CO_2$
3	Deamination	$R - NH_2 + H_2O \rightarrow R - OH + NH_3$
4	General cleavage	$R_1 - R_2 \rightarrow R_1 + R_2$
5	Hydrolysis	$R_1 - R_2 + H_2O \rightarrow R_1 - H + R_2 - OH$
6	Glucose isomerization	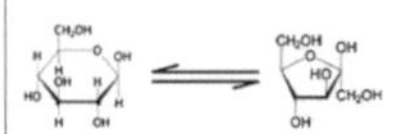
7	Retro-aldol (*defragmentation example)	$OH - R_1 - R_2 = O + OH^- \rightarrow R_1 = O + R_2 = O + H_2O$
8	Maillard reaction	$C_6H_{12}O_6(\text{Glucose}) + R - NH_2 \rightarrow C_6O_5H_{12} = N - R + H_2O$
9	Carbon double bond rearrangement (*rearrangement example)	$R_1 - C = C - C - R_2 \rightarrow R_1 - C - C = C - R_2$
10	Recombination/ Repolymerization/ Condensation*	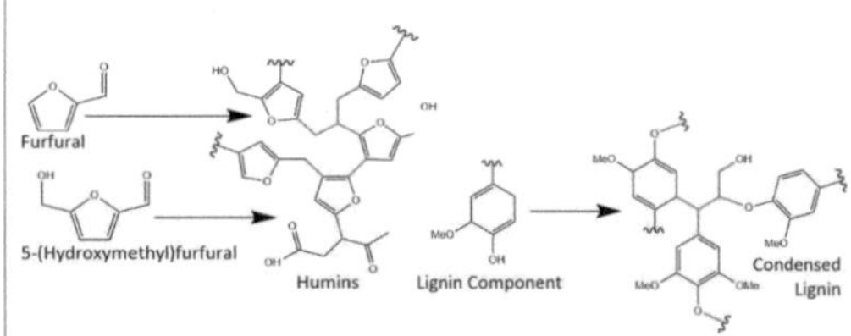
11	Boudouard gas reaction	$C + CO_2 \rightarrow 2CO$

*Example reaction equations are one of many possible equations that can occur during this reaction

by the conditions used. A schematic of the possible complex reaction pathways identified for the common contents of agricultural biomass (lignin, hemicellulose, cellulose) is shown in Fig. 3.

Of the products produced, char has found use as a bio-fertilizer and catalyst support due to its porous nature, low cost, and dense functional groups. The aqueous phase contains organic biochemicals rich in phenols and furfurals for biochemical applications. The high content of these biochemicals has been shown to be inhibiting for reuse in anaerobic digestion applications [37], but success has been found with recycling the aqueous phase within the HTL process to reduce water requirements and improving biocrude yield without sacrificing quality [41]. The gas phase is often small (under 5 %wt) and contains marketable gases such as primarily CO_2 as well as CO, CH_4, and H_2. The biocrude phase is primarily composed of phenols, aromatic hydrocarbons, alcohols, aldehydes, esters, ketones, acids, as well as nitrogen- and sulphur-containing compounds [38]. During the complex HTL reactions, nitrogen is mainly converted to a gaseous phase in the form of N_2O and N_2, oxygen is rapidly oxidized or mineralized to form CO_2 and H_2O, and contents like sulphur, chlorine,

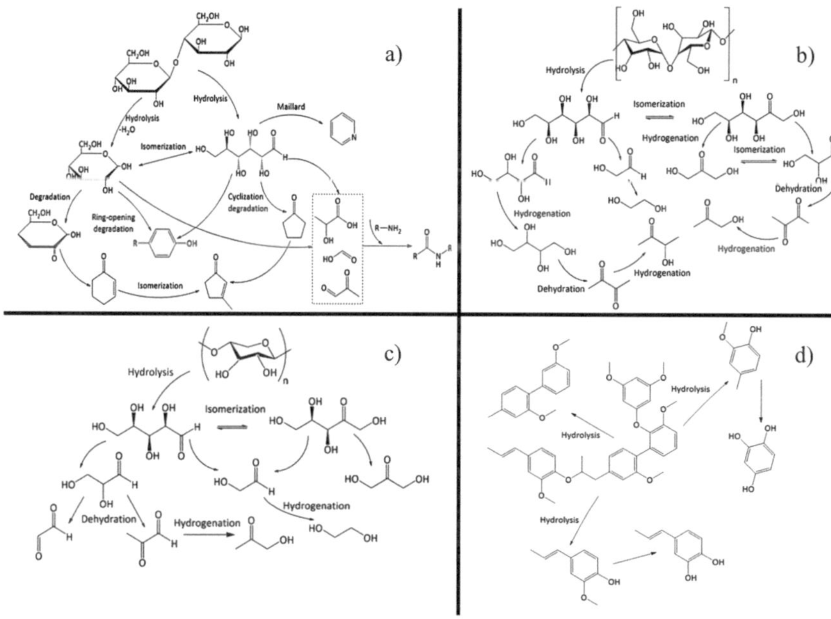

Fig. 3 Possible reaction pathways during HTL of: **a** carbohydrates, **b** cellulose, **c** hemicellulose, and **d** lignin (adapted from Hao et al. [40])

and phosphorus oxidize to inorganic acids or salts, depending on the pH conditions [11]. Compared to biomass and other bio-oils, HTL biocrude has reduced heteroatom and impurity content, being closer in quality to petroleum-based oils [37]. However, this biocrude still differs from conventional petroleum crude oil, primarily through oxygen and nitrogen content, causing undesired properties such as lower heating value, higher acid number, and higher viscosity [11]. For an outline of significant HTL studies see Table 2, summarizing experiments from the two primary types of lignocellulosic feed: agricultural and woody biomass.

3.1 Hydrothermal Liquefaction Among Biofuel Processes

As an emerging energy pathway, different processes have been explored to produce bioenergy from biomass such as pyrolysis, fermentation, direct transesterification, gasification, and HTL among other treatments [10]. Gasification has shown promise in creating drop-in gaseous fuels, as well as pyrolysis in creating gaseous and liquid bioenergy-intermediates, however in recent, hydrothermal liquefaction has shown advantages over these recognized processes [12]. Gasification of biomass is effective but energy-intensive by reaction with a gasifying agent (such as air, steam, or water) at high temperatures of 500–1200 °C, primarily creating H_2, CH_4, CO, and CO_2, among

Table 2 Summary of significant agro-forestry HTL studies in literature

Feedstock	Catalyst	Operating conditions	Key findings	Reference
Agriculture				
Canola straw	10 %wt KOH or NaOH	180–320 °C, W/B 12.5:1, 10 min	KOH produced lower oxygen biocrude at optimal temperature of 280 °C	[42]
Reed/Corn stover	5–20 %wt anhydrous aluminum chloride	210–260, W/B 1:1, 60–240 min	Optimum operating temperature of 260 °C, 5 %wt catalyst and 150 min (biocrude yield not disclosed)	[43]
Barley straw	10 %wt K_2CO_3	280–400 °C, W/B 6.67:1, 15 min	Optimum yield was at 300 °C (34.9 %wt) but greater temperatures had less oxygen and greater HHV. Aqueous phase recycle improved yield	[44]
Corn straw	10–50 %wt CuO + 1 M NaOH	190–270 °C, 2–8 MPa, 30–150 min	CuO + NaOH had synergistic catalyst effect. Optimal yield of 38.2 %wt at 230 °C and 60 min	[45]
Flax straw	5–15 %wt γ–alumina, H-ZSM-5, silica-alumina	200–325 °C, W/B 9:1, 0–120 min, 0–60 bar	γ–alumina > H-ZSM-5 > silica-alumina for biocrude yield. Optimal 66 %wt yield at 325 °C	[46]
Rice, corn, peanut, soybean straw	–	320 °C, W/B 2.67:1, 60 min	Soybean straw (15.8 wt%) > rice straw (15.1) > peanut straw (14.6) > corn straw (7.9) for biocrude yield	[47]
Forestry				
Various softwoods	10 %wt K_2CO_3	280–360 °C, Water-to-Biomass (W/B) 12.5:1, 10 min	K_2CO_3 and 340 °C had maximum yield of 32.2 %wt higher lignin content leads to higher catalyst effect and higher optimum temperature	[48]
Jack pine	5 %wt FeS, 5 %wt $FeSO_4$	200–350 °C, 2–10 MPa, ethanol-to-biomass 10:1, 20–40 min	$FeSO_4$ had most significant effect on yield increasing from 44 to 63 %wt 300 °C optimal but 20–40 min had less effect	[49]
Bamboo	5 %wt HCl	180 °C, W/B 10:1–2:1, 120–1080 min	Optimum yield of 80 %wt (biochar + biocrude) at 840 min	[50]
White pine	–	250–350 °C, 2 MPa, methanol/water-to-biomass 10:1, 15 min	300 °C optimum temperature for biocrude yield (65 %wt)	[51]
Waste wood furniture	–	180–300 °C, 20 MPa, W/B 10:1–2:1, 0–60 min	Optimum biocrude yield was 12.7 %wt at 280 °C, W/B 6:1, and 15 min	[35]

(continued)

Table 2 (continued)

Feedstock	Catalyst	Operating conditions	Key findings	Reference
Soft and hardwoods	1, 5, 10 %wt K_2CO_3	300 °C, W/B 10:1, 30 min	K_2CO_3 enhanced yield to 27.7 %wt from 24, higher oil conversion in softwood	[52]
Eucalyptus	0.5 %wt. NaOH, KOH, Pd/C	260–300 °C, W/B 10:1, 30 min	Optimum temperature at 260 °C with NaOH (yield of 61.2 %wt)	[53]

other desired gaseous products. As for liquid fuels, pyrolysis is also a thermochemical process where feedstock undergoes decomposition due to elevated heat in the absence of oxygen, producing a bio-oil typically with higher oxygen content and lower yield than HTL [15]. Similar to HTL, pyrolysis produces significant amounts of biocrude, gas, and char products depending on temperature and time.

3.2 *Temperature*

Temperature is considered one of the dominant factors in HTL, having a strong influence on the biocrude yield and properties, as many of the degradation and repolymerization reactions have activation energies that are only effectively exceeded with increased temperature [40]. Temperature causes thermal conversion of the lignocellulosic macromolecules to the desired energy-rich hydrocarbons, while allowing water to reach a subcritical or supercritical condition, where it can act as a reactant and catalyst in the process [37]. The main macromolecules differ in atmospheric degradation temperatures with hemicellulose requiring temperatures in the short and low range of 220–315 °C, cellulose slightly higher at 300–400 °C, and lignin being the toughest with a very large and severe range of 150–900 °C [54]. The majority of HTL studies have been conducted at temperatures between 200 and 400 °C and moderate to high pressures of 25–35 MPa, reporting optimal trends of increase with biocrude yield from initial temperature increase before reaching a yield limit and declining in yield at higher temperatures across different sources [49, 55].

With production yields differing due to temperature, depicted in Fig. 4, at lower temperatures the macromolecules are degraded significantly less and the biocrude-producing reactions (Table 1, Eqs. 1–9) are partially or not initiated, yielding a higher hydrochar and lower biocrude yield with undesirably high oxygen content and low HHV due to the lack of these degradation and deoxygenation reactions [44, 56]. With higher HTL temperatures approaching the supercritical level of water at 373 °C and 22.1 MPa, the high thermal energy and near supercritical properties of water cause increased gas formation through further secondary decompositions like decarboxylation and Bourdard gas reactions (Table 1, Eq. 2 and 11), while hydrochar yield slightly increases through high free radical-driven reactions such as recombination

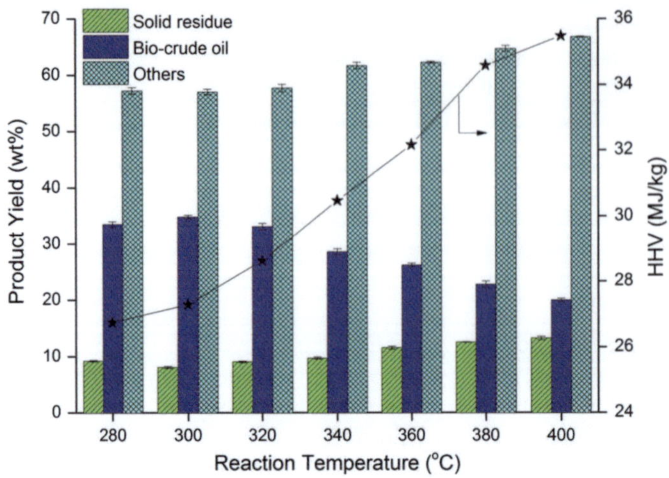

Fig. 4 Yields (left) of hydrochar residue, biocrude (oil), and other products (aqueous and gas products) from HTL at different temperatures as well as biocrude HHV (right) [44]

and condensation (Table 1, Eq. 10) [11, 37]. The increase in gas and hydrochar phase causes a decrease in biocrude yield at high temperatures, indicating optimal operation temperature before these reactions become dominant. The range of 300–330 °C is recommended by reviews, but the exact optimal temperature depends on feed, catalysts, solvents, and other operation parameters [11, 37]. For example, a higher quality biocrude (higher HHV) may be valued much more than yield, indicating a temperature optimum closer to 400 °C should be favoured. When considering the high energy input needed to reach higher temperature reactions, it is more economical to run HTL reactions at the lower temperatures of the optimal range while also reducing heating/cooling times that become prevalent with scale-up [37].

3.3 *Pressure*

Pressure is a critical parameter to HTL, ensuring the effective production of biocrude. A moderate to high pressure of HTL maintains the water solvent in a single-phase at the subcritical or supercritical conditions reached in the reaction. By maintaining a moderate to high reaction pressure above the vapour pressure of water at the reacting temperature, the single-phase solvent avoids large energy losses to the enthalpy of phase change, maintaining a more energy-efficient HTL reaction [57]. Secondly, elevated reaction pressure promotes the decomposition (Table 1, Eqs. 1–4) and extraction of biomass through improved hydrolysis (Table 1, Eq. 5) due higher dissolution rates as well as water penetration from the increased solvent density. However, when maintained near or just above supercritical levels (7–12 MPa) it has been reported that pressure has little influence on the solvent properties and biocrude yield [57, 58].

Too high of pressure in the supercritical region of water can decrease biocrude yield, as the increased water density is hypothesized to create a cage effect that inhibits C–C cleavages and blocks active sites of catalysts [37]. Beyond maintaining pressure above the solvent saturation pressure at the reaction temperature, there is little to no significant effect on biocrude yield, until very high detrimental pressures are reached. Pressure is relatively proportional to temperature and both are detrimental to yield as well as energy demand at high levels, recommending a moderate pressure above the solvent vapour pressure [11].

3.4 Residence Time

Residence time is defined as the time that the reaction maintains the target temperature, not including heating and cooling times that can vary with equipment. The residence time has a comparable effect to temperature, with several papers citing low residence times yielding incomplete degradation and lower yield, while longer residence times also reduce yield primarily causing polymerization of biocrude to solids or further degradation to gases [37, 57, 58]. With short and long residence times reducing biocrude yield, many studies have investigated optimization. Kaur et al. [59] as well as Wang et al. [60] found that 60 min was an optimal residence time for biocrude yield at their previously determined optimal temperatures of 300 °C for castor residues and 290 °C for *Litsea cubeba* seeds, respectively. However, optimal residence time has been shown to differ from as little as 1 to 840 min, depending on production scale, temperature, and feed [50]. For example, a residence time of 15 min was determined optimal for pig manure, 30 min for corn stalk, and 120 min for pulp powder, suggesting different compositions have significant effects [37]. Although the effect of residence time appeared less significant in comparison to temperature, lower residence times had slightly larger aqueous and solid products due to incomplete decomposition reactions (Table 1, Eqs. 1–4), while gas products primarily increase at longer residence times with solids and aqueous phase slightly increasing or remaining constant [60].

The quality of the biocrude was also affected by residence time, with temperature and residence time affecting primarily the elemental composition and subsequent heating value. Deoxygenation, primarily through decarboxylation (Table 1, Eq. 2) and other oxygen-reducing reactions (such as Table 1, Eqs. 1, 7, 8), requires certain thermal energies and reaction lengths, where lower temperatures or residence times can result in the reactions not being complete, yielding higher oxygen content and lower heating value [57]. Generally higher temperatures and residence times promote biocrude quality, further supported by cracking and condensation reactions at high values, outside of optimal yield results over the same range [57]. As quality and yield are independent, this often creates a trade-off between biocrude yield and quality above the optimum yield. In these cases, higher quality biocrude with a lower oxygen content can be achieved but at the much higher expense of harsher HTL conditions, lower yield, and higher energy requirements [57].

3.5 Catalyst

The hydrothermal liquefaction process can be improved using catalysts, with several studies utilizing acidic, basic, metallic as well as homogenous and heterogenous types for improved biocrude yield and quality [37, 58]. The exact mechanisms of hydrothermal liquefaction are not fully understood, including the catalysis of these reactions, with many explanations hypothesized. Catalysts typically reduce secondary char-forming reactions such as condensation and polymerization (Table 1, Eq. 10), increasing the biocrude yield that would otherwise be char [37]. The catalysts can also reduce side reactions, increasing the main biocrude reaction rates, while also reducing optimal reaction temperature and pressure to produce a higher yield and lower oxygen content biocrude [58]. The type and amount of catalyst loaded are important factors for the HTL reaction, where too little catalyst becomes diluted by the water and ineffective, while too much alkali or acidic catalyst has an overwhelming inhibition effect due to change in pH [58]. Studies using homogeneous alkaline catalysts have successfully used catalyst loading as little as 1.0 %wt and typically no larger than 10.0 %wt of the biomass input weight [37]. Large amounts of alkaline or acidic catalyst loading, aside from having an unnecessarily high cost, also increase maintenance frequency due to corrosion and fouling. As catalysts have been developed for decades in well-established industries like petroleum refining or the biochemical industry, many catalysts have been repurposed for HTL or biocrude hydrotreatment such as zeolites, metal oxides, acid-functionalized, etc., summarized with advantages and disadvantages in Fig. 5 [37]. The most popular heterogeneous catalysts include metals like Pd, Fe, Ti, Ni, Ru, Co, Mo, Ni, and Pt supported by SiO_2, Al_2O_3, or zeolites that increase yield and heating value by reducing heteroatom content, also insignificantly affecting acid value [37, 57]. However, homogeneous catalysts are favoured over heterogeneous due to a lack of catalyst poisoning, sintering, and higher mass transfer limitations [57].

The selection of homogenous catalysts has gained popularity for HTL of lignocellulosic biomass due to ease of handling/preparation as a powder and success at increasing biocrude yield as well as quality at an experimental level [12]. However, like the overall HTL reaction mechanisms described in Table 1, the catalytic mechanisms of HTL are not fully understood. In particular, the impact of different bases and salts on HTL biocrude yield has been studied finding K_2CO_3 > KOH > Na_2CO_3 > NaOH, with K_2CO_3 increasing the overall yield by nearly 25 %wt while NaOH increased yield by only 14 %wt from non-catalytic [37]. In general, alkaline carbonates are more effective than their hydroxide counterparts due to the salts reacting in water to form base and biocarbonate ions that have reducing catalytic effects, while also promoting lower viscosity biocrude due to the formation of lower molecular weight hydrocarbons [12]. Higher molecular weight and less popular catalysts such as $Ba(OH)_2$ and Rb_2CO_3 have also been shown to specifically increase phenolic/neutral oil content of biocrude while increasing gas yield and reducing solids yield, but only for low lignin feedstocks like agricultural residues [12]. Weak acids such as formic, acetic, and phosphoric are favoured as catalytic solvents over strong acids

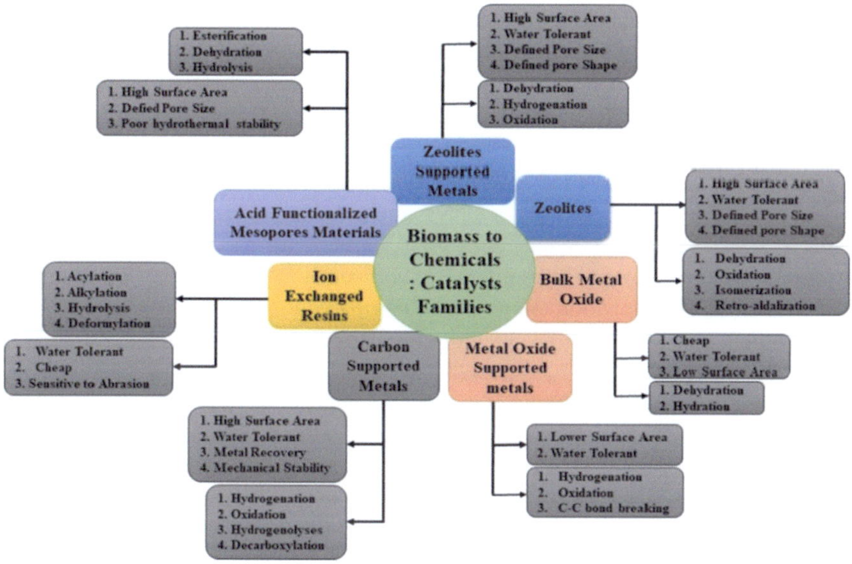

Fig. 5 Thermochemical catalysts with respective reactions and considerations [37]

like perchloric, hydrochloric, and sulphuric acid due to less corrosiveness, increased yield and undesirably higher oxygen content [57]. Iron-based catalysts have shown success such as FeS, $FeSO_4$, and Fe, weakly acidic in nature, assisting in degradation (Table 1, Eqs. 1–5) as well as redox reactions to increase yield by > 15 %wt in some cases [37]. Several HTL reviews have regarded K_2CO_3 as the most effective catalyst for lignocellulosic conversion due to its dual alkaline and carbonate catalytic effect that promotes biocrude formation via intermediate degradation reactions (Table 1, Eqs. 1–9), also reducing polymerization and condensation reactions (Table 1, Eq. 10) that form chars [37, 57].

3.6 *Water-to-Biomass (W/B) Ratio*

As discussed in this chapter's introduction section, water plays a crucial role in HTL by participating as a catalyst and/or reactant in the complex reaction system, helping produce the various HTL products through degradation and hydrolyzing large molecules (Table 1, Eqs. 3 and 5), also repolymerizing small molecules, among other reactions that can favour biocrude over gas and solid yields [37, 57]. Water also plays a crucial role as a solvent for the reaction, suspending lignocellulose initially, aiding in intermediates dissolution during the reaction, and separating unwanted aqueous phase organics from the immiscible biocrude product after the reaction [57]. When it comes to biocrude quality, water's hydrogen ions can effectively stabilize free radicals, reducing oxygen and improving oxidative stability of the biocrude, yielding a more desired biocrude [57]. From an ease of handling standpoint, especially with larger-scale processes that utilize pumps and tough lignocellulosic biomass, a higher W/B ratio makes non-dense biomass easier to transport compared to more dense slurries. A sufficiently high W/B ratio is essential for HTL and biomass handling, differentiating the process from pyrolysis pathways, producing more economical, lower oxygen, and higher yield biocrudes as a result.

Despite the essential role water plays in the HTL process, there are negative impacts to excess water content, indicating the need for an optimum. A high W/B ratio is uneconomical as it requires a significantly higher amount of energy to heat to the high HTL reaction temperatures compared to lower water ratios. A high-water content can cause unprecedentedly high reaction pressures for the system at the harsh conditions of HTL, especially at supercritical conditions, reducing the catalytic and reactant capabilities of water to decrease biocrude yield, described in Sect. 3.3 [37]. Studies have found that an increase in the W/B ratio above 3:1 significantly prevented hydrochar formation from biocrude due to favourable cross-linking between aromatic compounds [57]. Despite this, the same review also reported that these studies reached a limit where biocrude yield remains unchanged or decreases with water ratio increase, shown in Fig. 6. Biocrude yield is hypothesized to level off due to repolymerization reactions (Table 1, Eq. 10) becoming favoured under dilute conditions, competing with biocrude-forming hydrolysis pathways (Table 1, Eq. 5) [57]. Too high of a W/B ratio not only has an increasingly lower impact on yield but also becomes more detrimental to energy usage and its associated costs, creating an optimal ratio for yield, energy, and subsequent economics of the process. Studies have utilized ratios as low as 1:1 [60] or higher than 20:1 [43] for effective biocrude production, however, optimization studies have recommended ratios between 16 and 6:1 (~ 6–16 %wt biomass) for desired yield and quality of plant residue biocrude [61, 62].

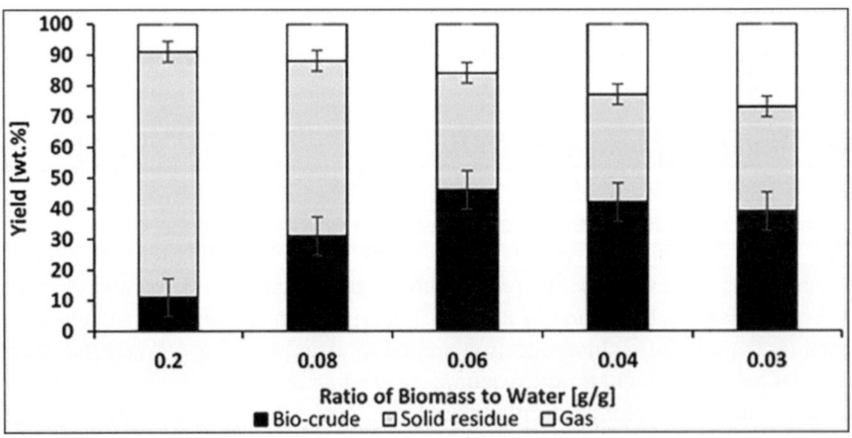

Fig. 6 Effect of W/B ratio on hyacinth plant matter HTL products [62]

4 Biocrude Upgradation Techniques

There are several physical and chemical upgradation techniques used to improve the properties of biofuels and their intermediate biocrudes. Despite better quality than pyrolysis bio-oil through properties like HHV and lower oxygen content, the biocrude often has poor properties such as high viscosity, high heteroatom content, and lower HHV compared to fossil fuels as well as refinery streams. Fortunately, several upgradation techniques can be employed to make biocrude closer to fossil fuel properties, primarily reducing oxygen content to increase HHV and reduce acid number as well as viscosity [37]. Physical methods include previously discussed distillation, blending, and emulsification with other petroleum intermediates, solvents, and/or emulsifiers [63, 64]. Chemical methods are often more complex, including different forms of hydrotreatments, esterification (only for high lipids), as well as catalytic cracking [64, 65]. A summary of the main physical and chemical methods of biocrude upgradation is seen in Fig. 7.

The biggest challenges associated with HTL biocrudes are viscosity, stability, and acidity that stem from oxygen content, while solvent addition can help homogeneously improve these properties [36]. Solvents utilized for extraction and mixing with biocrude range from non-polar solvents like hexadecane, decane, hexane, benzene, pentane, and cyclohexane to polar solvents like alcohols (methanol or ethanol), acetone, ethyl acetate, chloroform, dichloromethane, and tetrahydrofuran [64]. The complex nature of HTL biocrude varies based on the feedstock used, including ranges of components that are polar or non-polar. A study by Garcia-Perez et al. [66] confirmed that the majority of lignocellulosic biocrude was polar in some capacity, with majority phenolic compounds and emphasizing the use of polar solvents, differing from Hossain et al. [67] who reported high alkane content associated with microalgae biocrudes, emphasizing the use of non-polar solvents. Liu et al. [68] confirmed the effective use of polar solvents such as alcohols or acetone on

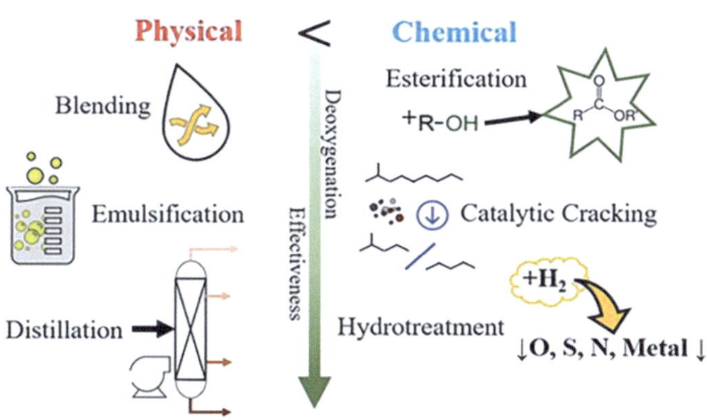

Fig. 7 Summary of the main physical and chemical biocrude upgradation techniques

lignocellulosic biocrude, reducing viscosity and instability when in solution, while diluting moisture content at the expense of reduced HHV. To avoid HHV loss while maintaining/improving the aforementioned properties from solvent blending, emulsifications may form with components that are partially immiscible with biocrude, including hydrocarbons, fuel intermediates, or fuels. Challenges exist with biocrude emulsification such as high-energy inputs through agitation and temperature, challenges in sourcing emulsifying agents, and long-term lack of stability [36]. These physical methods of upgradation, although more convenient than chemical, have reported challenges in practicality if they are effective at all, signifying importance of chemical upgradation, like hydrotreatment instead.

4.1 Distillation of Biocrude

Among techniques to further process raw biocrude, distillation is commonly an initial technique as it is a physical, scalable, and established refinement that is often more economical and reliable than chemical processes [69]. The number of studies distilling HTL biocrude is limited compared to more established petroleum crude distillation and alternative pyrolysis biocrude distillation but HTL distillation has been identified as a potential technique that can upgrade biocrude in combination with other techniques [69]. Distillation helps characterize biocrude's complex composition and produce fractions with more desirable elemental content, viscosity, stability, total acid number (TAN), and HHV [69]. Biocrude distillation has reportedly increased HHV by as much as 46 %wt, reduced oxygen by 95 %wt, and reduced sulphur content by 45 %wt in certain desirable fuel-range fractions [69]. There are variations to the distillation process including simple experimental distillation with 2–3 distillates, steam distillation to separate by lowering the partial pressure of the mixture, and more complex fractionations using higher temperature ranges under atmospheric or a higher equivalent range under vacuum distillation [64, 69].

Since atmospheric and vacuum distillation is utilized in crude oil refinement, there is also the opportunity to co-process biocrude with crude oil for renewable integration to the unsustainable petroleum refinement practice. HTL biocrude contains a majority of heavier-range fractions such as diesel and residue ranges, with independent studies by Hoffmann et al. [63] and Masoumi and Dalai [16] agreeing. These same distillation studies created cuts similar to petroleum crude ranges, finding that the gasoline range ($\lesssim$ 190 °C) of hydrocarbons is the smallest component of biocrude at 3–13 %wt, while the diesel range (~ 190–340 °C) is more than double that at 27–47 %wt, and the vacuum gas oil (~ 340–538 °C) and residue (> 538 °C) fractions combined are often the largest fraction being 40–70 %wt [63]. Comparative to petroleum fractions, these cuts have reportedly similar properties with comparable or slightly lower HHVs, similar viscosity, lower sulphur content but slightly higher oxygen and nitrogen content [63]. The content of these different fractions varies and also corresponds to the carbon number of the hydrocarbons such that the gasoline fraction contains primarily alkanes, aliphatic ketones, and esters in the range of C5-12, while the diesel fraction contains alkanes, aliphatic ketones, cyclics, esters, aromatics, and alcohols in the range of C13-24, and vacuum gas oil residues contain primarily substituted aromatics, esters, and carbonyl groups [63].

4.1.1 Atmospheric Distillation (< 310 °C)

Distillation under atmospheric pressure (1 atm or 1.013 bar) is a common practice in many industries to separate different substances based on their volatility over a temperature range usually no larger than 500 °C [69]. Atmospheric distillation boasts easier handling/operation, less maintenance, less cost, and overall lower energy demand over vacuum distillation, which utilizes a sealed vacuum pump system that can be energy-intensive for the already energy-intensive distillation process [70]. However, at high temperatures, atmospheric distillation becomes more energy intensive due to the harsh conditions needed to maintain > 350 °C, electing for vacuum distillation to reach the equivalent conditions with less energy [70]. When it comes to the complex composition of biocrude containing more than 40 significant compounds and even more minor compounds (< 0.5 %wt) making up to 30 %wt of the total composition, atmospheric distillation is commonly used as an initial characterization method to understand the size of the boiling fractions and separate components into groups for characterization or use in vacuum distillation to further separate heavier, less volatile, components outside the atmospheric range [63, 69].

There are few comparative biocrude distillation studies, but among them, fractions under 310 °C are reportedly similar [63, 69]. Hoffmann et al. [63] have extensively reported enhancement of HHV, density, and elemental content among fractions obtained from both atmospheric and vacuum distillation, aligning with gasoline and diesel range hydrocarbons. Atmospheric distillation is least effective at the high-temperature range often needed for the separation of diesel fractions from vacuum gas oils and residues (> 340 °C) [71]. At high temperatures, atmospheric distillation becomes more energy intensive than vacuum techniques and faces thermal cracking reactions that chemically alter the biocrude composition to form undesired chars/gases [71]. American Society for Testing and Materials (ASTM) 2892 recommends avoiding atmospheric distillation above 310 °C, requiring vacuum for unreactive biocrude distillation of the heavier fractions above this temperature [63]. However, these atmospheric cracking reactions can be favourable to produce lighter hydrocarbons, avoiding the viscosity and oxygen content that is inherent with longer hydrocarbon chains, dominant in HTL biocrude [69]. New techniques of catalytic cracking distillation have emerged for biocrude upgradation, efficiently utilizing the high thermal conditions of atmospheric distillation to thermochemically deoxygenate and crack components while physically fractionating the biocrude components [71].

4.1.2 Vacuum Distillation (> 310 °C)

Vacuum distillation utilizes below atmospheric pressure (< 101.325 kPa) to reduce the resistance of a feed's vapour pressure, allowing vaporization of different volatility components at a lower temperature. The lower the pressure utilized over the process, the higher equivalent atmospheric temperature that is reached [63]. For example, the equivalent atmospheric temperature of 400 °C can be reached at 250 °C under 0.1 kPa vacuum conditions. Therefore, the main advantage of vacuum distillation is the ability to avoid costly energy requirements and thermal degradation temperatures (> 310 °C) of the thermally susceptible biocrude, using less harsh temperatures to separate biocrude at equivalent boiling points [63, 64]. Although some studies have utilized easier-to-operate atmospheric distillation units before feeding heavier components to vacuum distillation, other studies have chosen to strictly use vacuum distillation to reduce process complexity and fully utilize its far larger temperature range than atmospheric distillation [63, 69]. Comparative properties to equivalent fractions, some vacuum distillations have reported some differences in yields and unfavourable mass recovery. At an experimental level, vacuum distillation has reported lower mass recovery (73 %wt) than atmospheric (84 %wt) due to challenges with vacuum-induced fast evaporation and the low theoretical plates of the columns, however, columns with 10–15 theoretical plates and slower pressure change negate this issue [63].

4.2 Hydrotreatment of Biocrude

The high oxygen content of HTL biocrude compared to petroleum crude is a major challenge in integrating biocrude with petroleum intermediates. The high oxygen content (> 10%wt for biocrude vs. < 2 %wt for crude oil) causes major challenges in immiscibility with petroleum streams, higher acidity, higher viscosity, lower HHV, and oxidative instability [64]. Oxygen content of fuel and fuel intermediates can form many bonds with hydrocarbons to create alcohols, ethers, carboxylic acids, ketones, etc., that inherently increase acidity and have lower bond energy to release, yielding a lower HHV. Hydrotreatment is a catalyzed process that utilizes hydrogenation reactions to increase the saturation of hydrocarbons and remove heteroatoms in the presence of a hydrogen gas reactant; a well-established process for denitrification and desulphurization of crude oil, with primary reactions summarized in Table 3. The use of hydrotreatment processes tailored to remove oxygen content, instead of nitrogen and sulphur, is a lucrative idea for HTL biocrude upgradation gaining popularity as a process before biocrude can be marketed or blended [30, 31] The initial upgradation of HTL biocrude using hydrotreatment is limited, with the majority focusing on the upgradation of higher lipid-content microalgae, but lacking investigation of lignocellulosic hydrotreatment of HTL biocrude. However, due to challenges faced with costly harvesting and lack of large-scale microalgae availability, lignocellulose is becoming more explored for HTL biocrude upgradation [30, 31].

Table 3 Hydrogenation reactions during biocrude hydrotreatment [64]

#	Reaction type	Reaction equation
1	Hydrogenation	
2	Cracking	
3	Decarbonylation	
4	Decarboxylation	
5	Hydrocracking	
6	Hydrodeoxygenation	$R—OH + H_2 \rightarrow R—H + H_2O$

The conditions of hydrotreatment can vary from harsh conditions of conventional hydrotreatment that can lead to biocrude cracking, catalyst deactivation, and unwanted char formation, or milder hydrotreatment that utilizes the same catalyst but at milder conditions, keeping the temperature under 300 °C and hydrogen pressure between 0.2 and 0.8 MPa [64]. Mild hydrotreatment has been utilized for a smaller reduction in oxygen content of lower oxygen biocrudes (~ 10 %wt) mainly deoxygenating aldehydes and ketones. For high oxygen or viscous biocrudes containing particularly high carbonyl or phenol/benzyl groups (such as the ones derived from lignin and holocellulose), mild hydrotreatment may be used as an initial step followed by more severe hydrotreatment at 300–405 °C, for 1–5 h, and at a hydrogen pressure from 5 to 15 MPa to remove higher heteroatom content from less reactive and higher hydrogen requiring components primarily derived from lignocellulosic fibres, as outlined in Fig. 8 [30, 31, 64, 72]. Mild hydrotreatment has achieved deoxygenation of ~ 25 %wt whereas severe hydrotreatment has achieved more in the range of 60–99 %wt deoxygenation of biocrudes [64]. Hydrotreatment optimization studies have agreed that temperature has the biggest effect on heteroatom removal and subsequent heating value, with a higher temperature of 405 °C favoured over 300–350 °C [64]. Hydrogen gas pressure and time are also significant factors, with 4 h and 10 MPa favoured over lower values like 1–2 h and 5–7.5 MPa [30, 31]. Scaling up biocrude hydrotreatment to continuous processes requires modification of the catalyst to a packed shape, accounting for throughput and pressure differences. Exploring different pellet shapes such as a sphere, pellet, cylinder, or multi-lobe, limited studies have compared shape effectiveness for biocrude hydrotreatment. Yet, extensive shape investigation for petroleum hydrotreatment has noted the effectiveness of trilobes

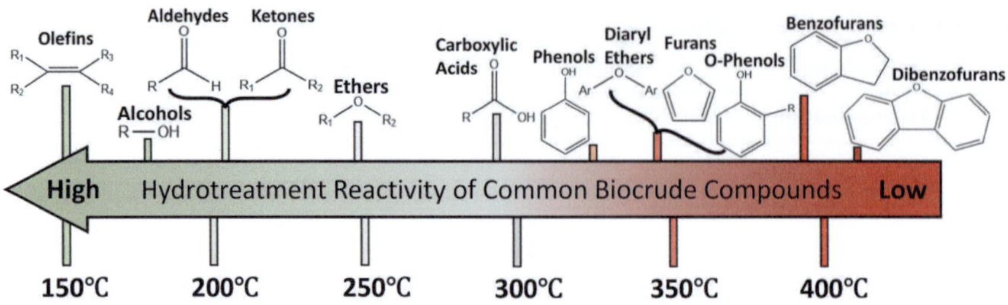

Fig. 8 Reactivity of common lignocellulosic biocrude compounds during hydrotreatment

for mass transfer and desulphurization/denitrogenation over tetralobes, spheres and cylinders [73].

4.2.1 Hydrotreatment Catalyst

Commercial catalysts proven effective for the hydrotreatment of petroleum crude include metal catalysts supported by zeolites such as NiMo/Al_2O_3 and CoMo/Al_2O_3, but the higher water content of biocrudes and tendency of coke formation from alumina-polymerization causes costly deactivation challenges [30, 31]. The use of sulphided catalysts has also been used in petroleum refinement and attempted to be used for HTL biocrude with some success. However, the process unfavourably adds additional sulphur to the low-sulphur biocrude [64]. The investigation of more effective hydrotreatment catalysts specific to HTL biocrude is a novel pursuit with many using petroleum refinement catalysts that can be improved upon [30, 31]. The use of noble metals catalysts such as platinum (Pd) and rhenium (Rh), often deactivated by the high sulphur content in petroleum refinement, have been used for effective mild upgrading (25.5 wt% deoxygenation) of HTL biocrude as it contains much lower sulphur content [64]. An extensive comparison including several types of popular catalyst components (transition/noble active metals, metal promoters, carbon/zeolite/alumina supports) has yet to be reported for biocrude hydrotreatment to the best of the author's knowledge, an area for future research. The use of activated HTL hydrochar as a metal catalyst support is an emerging concept, showing promising results for algae, wood, and non-edible vegetable oil hydrotreatment with lucrative lower cost, high surface area, and thermal stability needed for catalyst development [36]. Of limited biocrude hydrotreatment studies using hydrochar, [72] have found comparable deoxygenation results compared to conventional alumina-supported metal catalysts for non-edible oil feeds. As for catalyst loading, the range of 0–10 %wt has been explored, with the high-end favouring oil production and deoxygenation reactions, however, higher than 10 %wt is often regarded as uneconomical and begins dropping off in performance per mass of catalysts [30, 31, 64]. Limited studies have synthesized complex catalysts through lengthy precipitation and impregnation of metals, yielding bimetallic and trimetallic catalysts that have

Table 4 Examples of HTL biocrude hydrotreatment from lignocellulosic residues in literature

Feedstock	Catalyst	Operating conditions	Key findings	Reference
Aspen wood	NiW/Al_2O_3 $NiMo/Al_2O_3$	350 °C, 7.5 MPa, 2 h	NiMo catalyst outperformed NiW with HDO to < 4.0 %wt	[30, 31]
Pine wood	$NiMo/Al_2O_3$	400 °C, 12.4 MPa H_2	Successful upgradation to drop-in fuels (gasoline, diesel, jet fuel)	[75]
Corn straw	$NiMo/Al_2O_3$	280–370 °C, 1 h, 4 MPa H_2	Highest temperature (370 °C) had largest deoxygenation (74.24 %wt)	[76]
Corn Stover	$CoMo/Al_2O_3$ (sulphided)	400 °C, 0.21 LHSV, 10.3 MPa H_2, 70 h	Direct hydrotreating of biocrude produced acceptable blendstock for diesel, jet, and marine fuel	[77]
Sorghum bagasse	Ru/C (sulphided)	350 °C, 4 h, 3.5 MPa H_2	Effective deoxygenation to fuel levels (further distillation needed)	[78]
Wheat straw/ Cow manure	CoMo or NiMo with Al_2O_3 support	340–400 °C, 8 MPa H_2, 4 h stabilizing, 3 h run-time	340 °C optimal with both catalysts effectively removing oxygen but CoMo showing less carbon build-up	[79]

been shown to improve the deoxygenation of biocrude compared to monometallic alternatives [74]. Notably, there is a lack of agricultural residue-derived biocrude upgradation with algae and woody-sourced biocrudes studied significantly, utilizing both sulphided and non-sulphided catalysts. For an outline of the limited agricultural HTL biocrude hydrotreatment studies and their experimental conditions see Table 4.

5 Production Considerations

5.1 *Biomass-to-Biocrude Through HTL*

When it comes to sourcing the feedstocks for HTL, biomass is viewed as the more challenging material to obtain in bulk compared to abundant water and commercially available catalysts, like potassium carbonate or iron powder. Among HTL biomass feeds, agricultural residues are often seen as the most accessible given the vast rural farmland of North America, let alone around the world. Li et al. [17] analyzed agricultural crop residue in western Canada for cellulosic ethanol production and found that a plant with a processing capacity of 244,800 tonnes/year could be serviced by a surrounding radius of just 37 km, while distances greater than 100 km are noted for some less accessible biomass in other technoeconomic analysis (algae or woody biomass) [10]. Inventory tools are emerging to help estimate and locate ideal locations for processes that utilize biomass such as the biomass inventory mapping and analysis tool used in Canada [80].

Although lack of pretreatment steps like drying is a key advantage of HTL over alternative processes like pyrolysis, grinding of HTL biomass is a necessary step, although energy-intensive and consequently costly. Particle size reduction is necessary to overcome heat and mass transfer limitations within the HTL reaction, shown to improve biocrude yield and handling of the typically low-density biomass residues [37]. When it comes to the particle size of HTL feedstocks, changes in reasonably small particle sizes have reportedly had little to no effect on biocrude yield and quality, being one of the least impactful parameters when under a threshold of a few millimetres [11]. Zhang et al. [55] saw no enhancement of HTL yield when changing particle size between 25.4, 2, and 0.5 mm. Regarded as a less significant parameter, more coarse particles are recommended in literature to reduce grinding costs, noting coarse grinding (~ 3.2 mm) was nearly half the cost of fine grinding (~ 0.8 mm) for various straws, due to the massive inputs of energy required [37]. Across literature, there is no conclusively recommended particle size for HTL pre-treatment, but [11] suggest particle sizes as large as even 4–10 mm for reduced grinding cost at negligible yield effect.

Common experimental product separation is outlined in the flow diagram of Fig. 9. Due to the several products of HTL in different phases, product separation is a particularly significant and several-step process. The gaseous phase is the first to separate, easily able to be vented from the rest of the liquid/solid products. As for the biocrude phase, it is inherently immiscible with the aqueous phase, having a higher affinity for the porous hydrochar phase. Due to this affinity, the aqueous phase is usually separated from the biocrude/hydrochar contents through gravity or filtration methods, where the aqueous phase is a filtrate and the biocrude-containing hydrochar remains a retentate. Separation of the biocrude from the hydrochar phase is typically done by washing with a solvent that biocrude has a higher affinity towards than the hydrochar. Common solvents reported in literature include hexane, toluene, acetone, dichloromethane, ethyl acetate, ethanol, methanol, etc., each with varying properties like polarity and volatility, having no consensus on the optimal solvent [37]. During extraction of the biocrude some heating is typically employed to the mixture ~ 50–60 °C under constant agitation to ensure effective extraction of biocrude from the hydrochar and overcome mass transfer limitations. High temperatures (>80 °C) are usually avoided during extraction to avoid costly solvent losses. After mixing the solvent-extracted biocrude can be easily separated from the solid hydrochar via filtration methods. Given the much higher volatility of the solvent compared to biocrude, this difference can be exploited to remove the solvent through distillation/evaporator techniques. The standardized method of ASTM D2892 is a useful method for reference in removing solvents and residual water contaminants, holding the biocrude at an atmospheric equivalent temperature (AET) of 130 °C until no bubbling occurs [14]. The evaporated solvent can be recycled for washing or biocrude extraction purposes to reduce costs and waste, contributing to this more sustainable practice.

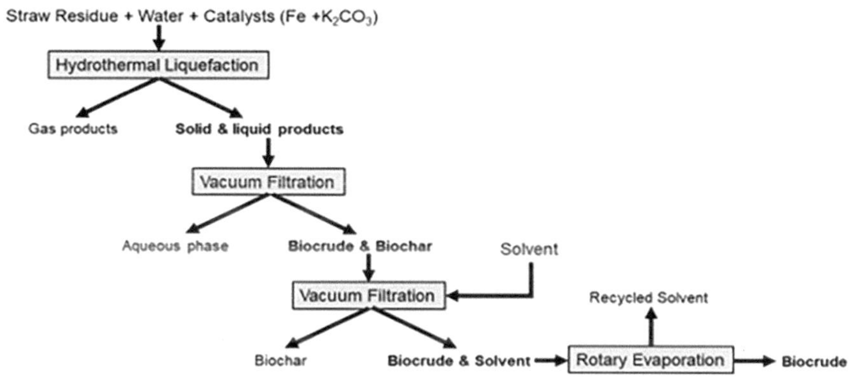

Fig. 9 Flow diagram of experimental product separation for HTL [81]

5.2 *Biocrude Distillation Practices*

The biocrude has relatively similar properties to crude oils and can be distilled by similar practices. Thus, refined biocrude can be distilled as per recommendations of ASTM D2892 (for atmospheric) and ASTM D1160 (for vacuum conditions) [82]. When it comes to distillation of biocrude a unit equivalent to at least 15 theoretical plates is needed to reach the separation desired for the multiple fractions of crude distillates [82]. At the same time, minimal holdup within the column is usually desired to reduce valuable biocrude losses, particularly for smaller batch scales, utilizing spinning band distillation towers with minimal holdup but high theoretical plate counts. Initially, the pressure of the column is kept atmospheric (~ 101.325 kPa) while the pot is heated at a gradual rate, avoiding rapid boil up such as bumping that can damage the column and cause contamination of fractions [82]. An initial first fraction is taken until < 100 °C, containing primarily undesired volatiles (water, solvent, etc.) as a contamination removal step. A second fraction collected at atmospheric pressure between 100 and 190 °C, is known as the gasoline-range fraction. Beyond the gasoline range vacuum conditions are typically utilized to reach the higher boiling temperatures of the diesel fraction by AET. To avoid thermochemical decompositions of the biocrude and distillates, such as cracking of the biocrude that leaves behind troublesome coke, a maximum pot of 310 °C and distillation head of 210 °C is vital to always maintain during biocrude distillation [82]. An AET of 190–340 °C, marks the distillation range of renewable diesel-range hydrocarbons, with this fraction sometimes is further separated to include kerosene and some jet fuel range hydrocarbon fractions, depending on prevalence in the biocrude utilized [83]. Typical vacuum distillation units can ensure no cracking occurs while reaching an AET of ~ 538 °C in part by reducing to pressures < 0.2 kPa. This is key as biocrude contains a large portion of hydrocarbons beyond the diesel range, allowing further separation of the significant vacuum gas oil-range components (340–538 °C) from the remaining heavy residues (> 538 °C). It is important to note that the non-volatile heavy residues have low handling ability and require harsh upgradation as they are extremely non-viscous and contain a high concentration of ash, particularly if derived from high-ash biomass such as manures.

5.3 *Biocrude Hydrodeoxygenation Practices*

Hydrotreatment has many options for production in literature sometimes as the first step in upgradation to directly deoxygenate HTL biocrude or as a final step to upgrade previously processed biocrudes such as distillates. One innovative direction includes utilizing co-hydrotreatment of the biocrude with petroleum refinery intermediates at continuous pilot scales that are closer to commercial hydrotreatment; a sustainable route to reduce fossil fuel reliance by integrating HTL biocrudes within petroleum refinement. For example, biocrude and its fractions have been hydrotreated with blends of conventional crude oil, refinery hydrotreater feeds, and refinery hydrocracking feeds [72]. Although the difference in heteroatom content between biocrude and conventional crude oil has been shown to cause challenges in simultaneously removing all heteroatoms, particularly with co-hydrotreatment catalysts [84]. The primary issue is that hydrodesulphurization (HDS) of the prevalent sulphur compounds in crude oil produces H_2S gas as a by-product, shown to inhibit oxygen components from interacting with the catalyst and strongly inhibit HDO of the biocrude components. Similarly, sulphided catalysts have been found to be deactivated by CO and CO_2, significantly formed during decarbonylation and decarboxylation of carboxylic and ester groups prevalent in biocrude [84]. Further, a detrimental effect was observed between HDO and HDS catalytic mechanisms, shown to compete for the same active sites and significantly reduce activity. In this case, solutions have been proposed including catalysts that are less sensitive to CO/CO_2/H_2S, or pretreatment of the biocrude to further remove carboxyl or ester groups, with no implementation to meet this major challenge yet [84]. Nonetheless, deoxygenation of biocrude through catalytic hydrotreatment has potential with + 90 %wt deoxygenation at experimental levels to produce transportation-level fuels, utilizing catalysts with molybdenum as an active site, promoted by either cobalt or nickel, and supported by alumina or carbon [16, 30, 31]. Setbacks for biocrude hydrotreatment include high cost, short catalyst lifetime/deactivation, high consumption of hydrogen gas for unreactive components, and equipment maintenance for sometimes severe reaction conditions, ultimately holding biocrude hydrotreatment back from eventual industrial implementation [37]. Catalyst development, scale-up, and technoeconomic analysis are areas of urgency in future research as current technoeconomic analyses make many assumptions regarding the unknown nature of scaling up the hydrotreatment process as well as the catalyst used.

6 Conclusions

In summary, the emerging technology of HTL and subsequent biocrude upgradation techniques hold significant promise in converting underutilized lignocellulosic residues into alternative crude oil and transportation fuels. The transportation industry's heavy reliance on non-renewable fossil fuels necessitates a shift towards sustainable biofuel production pathways. Unlike other biofuel technologies, HTL stands out due to its non-strict feedstock requirements, making it adaptable to various biomass feeds crucial for the future biofuel industry. Countries with substantial lignocellulosic residue production, such as Canada with its abundant agricultural and forestry industries, are particularly well-suited for HTL and biocrude upgradation. These industries generate millions of tonnes of lignocellulosic agro-forestry waste, providing a valuable resource for biofuel production in a circular bioeconomy.

HTL employs high water content, high pressure, and moderate temperatures to produce biocrude, along with marketable by-products like hydrochar, biogas, and aqueous biochemicals. While kinetic models for the complex HTL process remain elusive, the breakdown and recombination of lignin, hemicellulose, and cellulose components are identifiable overarching mechanisms. Process optimization is critical for HTL, with temperature identified as the most influential factor. Pressure, retention time, catalyst, and water-to-biomass ratio also significantly impact biocrude yield, quality, and process economics. However, even under ideal conditions, HTL-produced biocrude requires further upgrades to meet strict transportation fuel regulations, addressing issues like high oxygen content, viscosity, acidity, and heating value. Hydrotreatment and distillation are the most promising biocrude upgradation techniques, with ongoing developments achieving nearly complete deoxygenation. Despite the potential, HTL and biocrude upgradation face challenges, including economic and technical scaling uncertainties. The pathway's low readiness factor for commercialization underscores its experimental nature, with further research, kinetic studies, optimization, and scaling efforts needed for potential integration into the mainstream fuel market. As global demand for biofuels rises amidst efforts to reduce fossil fuel dependence, ongoing research in HTL and biocrude upgradation is expected to continue, driven by legislative support and environmental considerations.

Acknowledgements The Author would like to express gratitude to the Department of Chemical and Biological Engineering, University of Saskatchewan, Canada for support and necessary facilities. The granted use of select Figures from other pieces of work is acknowledged and greatly appreciated. Funding for this work from Natural Sciences and Engineering Research Council of Canada (NSERC), Canada Research Chair Program (CRC), and Tidewater Renewables is greatly acknowledged.

Compliance with Ethical Standards The authors declare no competing financial interests or personal relationships that could have influenced the research reported in the current manuscript.

References

1. International Energy Agency (2021) Global energy review: CO_2 emissions 2021. https://www.iea.org/reports/global-energy-review-co2-emissions-in-2021-2. Accessed 22 Apr 2023
2. Darkwa K (2001) Sustainable-energy management practices in an energy economy. 4566:165–172. https://doi.org/10.1117/12.443140
3. Government of Canada (2022a) Net-zero emissions by 2050. https://www.canada.ca/en/services/environment/weather/climatechange/climate-plan/net-zero-emissions-2050.html. Accessed 22 Apr 2023
4. IEA Bioenergy (2021) Implementation of bioenergy in Canada—2021 update
5. Energy Institute (2022) Exploring energy topics—transport. https://www.energyinst.org/exploring-energy/topic/transport. Accessed 22 Apr 2023
6. International Energy Agency (2022) Transport—improving the sustainability of passenger and freight transport. https://www.iea.org/topics/transport. Accessed 22 Apr 2023
7. Vevere L, Fridrihsone A, Kirpluks M, Cabulis U (2020) A review of wood biomass-based fatty acids and rosin acids use in polymeric materials. Polymers (Basel) 12:2706. https://doi.org/10.3390/POLYM12112706
8. REN21 (2019) Why is renewable energy important? https://www.ren21.net/why-is-renewable-energy-important/. Accessed 25 Apr 2023
9. Zheng Y, Qiu F (2020) Bioenergy in the Canadian prairies: assessment of accessible biomass from agricultural crop residues and identification of potential biorefinery sites. Biomass Bioenergy 140:105669. https://doi.org/10.1016/J.BIOMBIOE.2020.105669
10. Kargbo H, Harris JS, Phan AN (2021) "Drop-in" fuel production from biomass: critical review on techno-economic feasibility and sustainability. Renew Sustain Energy Rev 135:110168. https://doi.org/10.1016/J.RSER.2020.110168
11. Akhtar J, Amin NAS (2011) A review on process conditions for optimum bio-oil yield in hydrothermal liquefaction of biomass. Renew Sustain Energy Rev 15:1615–1624. https://doi.org/10.1016/J.RSER.2010.11.054
12. Sahu SN, Sahoo NK, Naik SN, Mahapatra DM (2020) Advancements in hydrothermal liquefaction reactors: overview and prospects. Bioreactors 195–213. https://doi.org/10.1016/B978-0-12-821264-6.00012-7
13. Toor SS, Rosendahl L, Rudolf A (2011) Hydrothermal liquefaction of biomass: a review of subcritical water technologies. Energy 36:2328–2342. https://doi.org/10.1016/J.ENERGY.2011.03.013
14. Pedersen TH, Jensen CU, Sandström L, Rosendahl LA (2017) Full characterization of compounds obtained from fractional distillation and upgrading of a HTL biocrude. Appl Energy 202:408–419. https://doi.org/10.1016/J.APENERGY.2017.05.167
15. Rahman WU, Patel M, Kurian V, Kumar A (2022) A comparative techno-economic assessment of fast pyrolysis, hydrothermal liquefaction, and intermediate pyrolysis of municipal solid waste for liquid transportation fuels production. Energy Convers Manag 267:115877. https://doi.org/10.1016/J.ENCONMAN.2022.115877
16. Masoumi S, Dalai AK (2021) Techno-economic and life cycle analysis of biofuel production via hydrothermal liquefaction of microalgae in a methanol-water system and catalytic hydrotreatment using hydrochar as a catalyst support. Biomass Bioenergy 151:106168. https://doi.org/10.1016/J.BIOMBIOE.2021.106168
17. Li X, Mupondwa E, Panigrahi S et al (2012) A review of agricultural crop residue supply in Canada for cellulosic ethanol production. Renew Sustain Energy Rev 16:2954–2965. https://doi.org/10.1016/J.RSER.2012.02.013
18. Zhou CH, Xia X, Lin CX et al (2011) Catalytic conversion of lignocellulosic biomass to fine chemicals and fuels. Chem Soc Rev 40:5588–5617. https://doi.org/10.1039/C1CS15124J
19. de Caprariis B, De Filippis P, Petrullo A, Scarsella M (2017) Hydrothermal liquefaction of biomass: influence of temperature and biomass composition on the bio-oil production. Fuel 208:618–625. https://doi.org/10.1016/J.FUEL.2017.07.054

20. Yousuf A, Pirozzi D, Sannino F (2020) Lignocellulosic biomass to liquid biofuels. Academic Press
21. González-García S, Luo L, Moreira MT et al (2012) Life cycle assessment of hemp hurds use in second generation ethanol production. Biomass Bioenergy 36:268–279. https://doi.org/10.1016/J.BIOMBIOE.2011.10.041
22. Sokhansanj S, Mani S, Stumborg M et al (2006) Production and distribution of cereal straw on the Canadian prairies
23. Carefoot JM, Janzen HH, Lindwall CW (1994) Crop residue management for irrigated cereals on the semi-arid Canadian prairies. Soil Tillage Res 32:1–20. https://doi.org/10.1016/0167-1987(94)90029-9
24. Government of Canada (2022b) Statistical data. Natural Resources Canada. https://cfs.nrcan.gc.ca/statsprofile/overview/CA. Accessed 23 Jun 2023
25. Kissinger M, Fix J, Rees WE (2007) Wood and non-wood pulp production: comparative ecological footprinting on the Canadian prairies. Ecol Econ 62:552–558. https://doi.org/10.1016/J.ECOLECON.2006.07.019
26. CCFM (2023) Healthy forests. Canadian Council of Forest Ministers (CCFM). https://www.ccfm.org/healthy-forests/. Accessed 23 Jun 2023
27. Stanton CR, Bourchier RJ (2006) Forest regions. The Canadian Encyclopedia
28. Hermosilla T, Bastyr A, Coops NC et al (2022) Mapping the presence and distribution of tree species in Canada's forested ecosystems. Remote Sens Environ 282:113276. https://doi.org/10.1016/J.RSE.2022.113276
29. Government of Canada (2023) Lumber production, shipments, and stocks by species, monthly. https://www150.statcan.gc.ca/t1/tbl1/en/tv.action?pid=1610001701. Accessed 23 Jun 2023
30. Yu J, Biller P, Mamahkel A et al (2017a) Catalytic hydrotreatment of bio-crude produced from the hydrothermal liquefaction of aspen wood: a catalyst screening and parameter optimization study. Sustain Energy Fuels 1:832–841. https://doi.org/10.1039/C7SE00090A
31. Yu J, Biller P, Mamahkel A et al (2017b) Publication metadata title: catalytic hydrotreatment of bio-crude produced from hydrothermal liquefaction of aspen wood: a catalyst screening and parameter optimization study. 1:832–841. https://doi.org/10.1039/C7SE00090A
32. Sidders D, Joss B, Keddy T (2008) Project TID8 25B GIS-based inventory and analysis of forestry and agriculture biomass. Biomass Inventory and Mapping Assessment Tool (BIMAT)
33. Paré D, Bernier P, Thiffault E, Titus BD (2011) The potential of forest biomass as an energy supply for Canada. 87:71–76. https://doi.org/10.5558/TFC87071-1
34. Karagöz S, Bhaskar T, Muto A, Sakata Y (2005) Comparative studies of oil compositions produced from sawdust, rice husk, lignin and cellulose by hydrothermal treatment. Fuel 84:875–884. https://doi.org/10.1016/J.FUEL.2005.01.004
35. Jindal MK, Jha MK (2016) Effect of process parameters on hydrothermal liquefaction of waste furniture sawdust for bio-oil production. RSC Adv 6:41772–41780. https://doi.org/10.1039/C6RA02868C
36. Masoumi S, Borugadda VB, Dalai AK (2020) Biocrude oil production via hydrothermal liquefaction of algae and upgradation techniques to liquid transportation fuels. In: Biorefinery of alternative resources: targeting green fuels and platform chemicals. pp 249–270.https://doi.org/10.1007/978-981-15-1804-1_11/TABLES/3
37. Mishra RK, Kumar V, Kumar P, Mohanty K (2022) Hydrothermal liquefaction of biomass for bio-crude production: a review on feedstocks, chemical compositions, operating parameters, reaction kinetics, techno-economic study, and life cycle assessment. Fuel 316:123377. https://doi.org/10.1016/J.FUEL.2022.123377
38. Gollakota ARK, Kishore N, Gu S (2018) A review on hydrothermal liquefaction of biomass. Renew Sustain Energy Rev 81:1378–1392. https://doi.org/10.1016/J.RSER.2017.05.178
39. Questell-Santiago YM, Galkin MV, Barta K, Luterbacher JS (2020) Stabilization strategies in biomass depolymerization using chemical functionalization. Nat Rev Chem 4(6):311–330. https://doi.org/10.1038/s41570-020-0187-y
40. Hao B, Xu D, Jiang G et al (2021) Chemical reactions in the hydrothermal liquefaction of biomass and in the catalytic hydrogenation upgrading of biocrude. Green Chem 23:1562–1583. https://doi.org/10.1039/D0GC02893B

41. Hu Y, Feng S, Yuan Z et al (2017) Investigation of aqueous phase recycling for improving bio-crude oil yield in hydrothermal liquefaction of algae. Bioresour Technol 239:151–159. https://doi.org/10.1016/J.BIORTECH.2017.05.033
42. Shin J, Hong S-G, Kwon S-I et al (2010) Effect of temperatures to crude oil productions with rapeseed straw on application of hydro-thermal liquefaction technology. 18
43. Lu J, Li X, Yang R et al (2014) Liquefaction of fermentation residue of reed- and corn stover-pretreated with liquid hot water in the presence of ethanol with aluminum chloride as the catalyst. Chem Eng J 247:142–151. https://doi.org/10.1016/J.CEJ.2014.02.094
44. Zhu Z, Toor SS, Rosendahl L et al (2015) Influence of alkali catalyst on product yield and properties via hydrothermal liquefaction of barley straw. Energy 80:284–292. https://doi.org/10.1016/J.ENERGY.2014.11.071
45. Chen Y, Dong L, Miao J et al (2019) Hydrothermal liquefaction of corn straw with mixed catalysts for the production of bio-oil and aromatic compounds. Bioresour Technol 294:122148. https://doi.org/10.1016/J.BIORTECH.2019.122148
46. Jaafari L, Ibrahim H, Jaffary B, Idem R (2019) Catalytic production of furfural by pressurized liquid water liquefaction of flax straw. Renew Energy 130:1176–1184. https://doi.org/10.1016/J.RENENE.2018.08.065
47. Tian Y, Wang F, Djandja JO et al (2020) Hydrothermal liquefaction of crop straws: effect of feedstock composition. Fuel 265:116946. https://doi.org/10.1016/J.FUEL.2019.116946
48. Zhong C, Wei X (2004) A comparative experimental study on the liquefaction of wood. Energy 29:1731–1741. https://doi.org/10.1016/J.ENERGY.2004.03.096
49. Xu C, Etcheverry T (2008) Hydro-liquefaction of woody biomass in sub- and super-critical ethanol with iron-based catalysts. Fuel 87:335–345. https://doi.org/10.1016/J.FUEL.2007.05.013
50. Yip J, Chen M, Szeto YS, Yan S (2009) Comparative study of liquefaction process and liquefied products from bamboo using different organic solvents. Bioresour Technol 100:6674–6678. https://doi.org/10.1016/J.BIORTECH.2009.07.045
51. Cheng S, DCruz I, Wang M et al (2010) Highly efficient liquefaction of woody biomass in hot-compressed alcohol-water co-solvents. Energy Fuels 24:4659–4667.https://doi.org/10.1021/EF901218W/ASSET/IMAGES/LARGE/EF-2009-01218W_0007.JPEG
52. Hwang H, Lee JH, Choi IG, Choi JW (2018) Comprehensive characterization of hydrothermal liquefaction products obtained from woody biomass under various alkali catalyst concentrations. 40:1657–1667. https://doi.org/10.1080/09593330.2018.1427799
53. Wu XF, Zhang JJ, Li MF et al (2019) Catalytic hydrothermal liquefaction of eucalyptus to prepare bio-oils and product properties. Energy Convers Manag 199.https://doi.org/10.1016/J.ENCONMAN.2019.111955
54. Waters CL, Janupala RR, Mallinson RG, Lobban LL (2017) Staged thermal fractionation for segregation of lignin and cellulose pyrolysis products: an experimental study of residence time and temperature effects. J Anal Appl Pyrolysis 126:380–389. https://doi.org/10.1016/J.JAAP.2017.05.008
55. Zhang B, von Keitz M, Valentas K (2009) Thermochemical liquefaction of high-diversity grassland perennials. J Anal Appl Pyrolysis 84:18–24. https://doi.org/10.1016/J.JAAP.2008.09.005
56. Sugano M, Takagi H, Hirano K, Mashimo K (2008) Hydrothermal liquefaction of plantation biomass with two kinds of wastewater from paper industry. J Mater Sci 43:2476–2486. https://doi.org/10.1007/S10853-007-2106-8/TABLES/6
57. Xue Y, Chen H, Zhao W et al (2016) A review on the operating conditions of producing bio-oil from hydrothermal liquefaction of biomass. Int J Energy Res 40:865–877. https://doi.org/10.1002/ER.3473
58. Cao L, Zhang C, Chen H et al (2017) Hydrothermal liquefaction of agricultural and forestry wastes: state-of-the-art review and future prospects. Bioresour Technol 245:1184–1193. https://doi.org/10.1016/J.BIORTECH.2017.08.196
59. Kaur R, Biswas B, Kumar J et al (2020) Catalytic hydrothermal liquefaction of castor residue to bio-oil: effect of alkali catalysts and optimization study. Ind Crops Prod 149:112359. https://doi.org/10.1016/J.INDCROP.2020.112359

60. Wang F, Chang Z, Duan P et al (2013) Hydrothermal liquefaction of Litsea cubeba seed to produce bio-oils. Bioresour Technol 149:509–515. https://doi.org/10.1016/J.BIORTECH.2013.09.108
61. Singh R, Balagurumurthy B, Prakash A, Bhaskar T (2015) Catalytic hydrothermal liquefaction of water hyacinth. Bioresour Technol 178:157–165. https://doi.org/10.1016/J.BIORTECH.2014.08.119
62. Egesa D, Mulindwa P, Mubiru E et al (2021) Hydrothermal liquefaction of water hyacinth: effect of process conditions and magnetite nanoparticles on biocrude yield and composition. J Sustain Bioenergy Syst 11:157–186. https://doi.org/10.4236/JSBS.2021.114012
63. Hoffmann J, Jensen CU, Rosendahl LA (2016) Co-processing potential of HTL bio-crude at petroleum refineries—part 1: fractional distillation and characterization. Fuel 165:526–535. https://doi.org/10.1016/J.FUEL.2015.10.094
64. Ramirez JA, Brown RJ, Rainey TJ (2015) A review of hydrothermal liquefaction bio-crude properties and prospects for upgrading to transportation fuels. Energies 8:6765–6794. https://doi.org/10.3390/EN8076765
65. de Santos Brandao JG, Manmohandas HV (2020) A simplified kinetic model for continuous hydrotreating of HTL biocrude
66. Garcia-Perez M, Chaala A, Pakdel H et al (2007) Characterization of bio-oils in chemical families. Biomass Bioenergy 31:222–242. https://doi.org/10.1016/J.BIOMBIOE.2006.02.006
67. Hossain FM, Kosinkova J, Brown RJ et al (2017) Experimental investigations of physical and chemical properties for microalgae HTL bio-crude using a large batch reactor. Energies 10:467.https://doi.org/10.3390/EN10040467
68. Liu R, Fei W, Shen C (2014) Influence of acetone addition on the physicochemical properties of bio-oils. J Energy Inst 87:127–133. https://doi.org/10.1016/J.JOEI.2013.08.001
69. Taghipour A, Ramirez JA, Brown RJ, Rainey TJ (2019) A review of fractional distillation to improve hydrothermal liquefaction biocrude characteristics; future outlook and prospects. Renew Sustain Energy Rev 115:109355. https://doi.org/10.1016/J.RSER.2019.109355
70. Feng TH, Xue R, Zhu L, Zhu YH (2023) Comparison of antioxidant and antibacterial effects of the Eucommia wood vinegar under the refining methods of atmospheric distillation and vacuum distillation. Ind Crops Prod 192:116013. https://doi.org/10.1016/J.INDCROP.2022.116013
71. Cheng LH, Fu YJ, Liao KS et al (2014) A high-permeance supported carbon molecular sieve membrane fabricated by plasma-enhanced chemical vapor deposition followed by carbonization for CO_2 capture. J Memb Sci 460:1–8. https://doi.org/10.1016/J.MEMSCI.2014.02.033
72. Roy P, Jahromi H, Rahman T et al (2022) Understanding the effects of feedstock blending and catalyst support on hydrotreatment of algae HTL biocrude with non-edible vegetable oil. Energy Convers Manag 268:115998. https://doi.org/10.1016/J.ENCONMAN.2022.115998
73. Kolitcheff S, Jolimaitre E, Hugon A et al (2018) Tortuosity and mass transfer limitations in industrial hydrotreating catalysts: effect of particle shape and size distribution. Catal Sci Technol 8:4537–4549. https://doi.org/10.1039/C8CY00831K
74. Pongsiriyakul K, Kiatkittipong W, Adhikari S et al (2021) Effective Cu/Re promoted Ni-supported γ-Al_2O_3 catalyst for upgrading algae bio-crude oil produced by hydrothermal liquefaction. Fuel Process Technol 216:106670. https://doi.org/10.1016/J.FUPROC.2020.106670
75. Van Dyk S, Ebadian M, Su J et al (2019) Assessment of likely technology maturation (ATM) project executive summary assessment of likely technology maturation pathways for biojet production from forest residues
76. Yang T, Jie Y, Li B et al (2016) Catalytic hydrodeoxygenation of crude bio-oil over an unsupported bimetallic dispersed catalyst in supercritical ethanol. Fuel Process Technol 148:19–27. https://doi.org/10.1016/J.FUPROC.2016.01.004
77. Collett JR, Billing JM, Meyer PA et al (2019) Renewable diesel via hydrothermal liquefaction of oleaginous yeast and residual lignin from bioconversion of corn stover. Appl Energy 233–234:840–853. https://doi.org/10.1016/J.APENERGY.2018.09.115

78. Yue Y, Kastner JR, Mani S (2018) Two-stage hydrothermal liquefaction of sweet sorghum biomass—part ii: production of upgraded biocrude oil. Energy Fuels 32:7620–7629. https://doi.org/10.1021/acs.energyfuels.8b00669
79. dos Passos JS, Straka P, Auersvald M, Biller P (2023) Upgrading of hydrothermal liquefaction biocrudes from mono- and co-liquefaction of cow manure and wheat straw through hydrotreating and distillation. Chem Eng J 452:139636. https://doi.org/10.1016/J.CEJ.2022.139636
80. Agri-Food Canada (2023) Biomass inventory mapping and analysis tool—business data. Open Government Portal. https://open.canada.ca/data/en/dataset/1a759d95-3008-4078-87af-5bb1bdf657b3. Accessed 18 Jun 2023
81. Churchill JGB, Borugadda VB, Dalai AK (2024) Biocrude production via hydrothermal liquefaction of Canadian lignocellulosic residues for sustainable transportation: screening, catalytic effect, and modelling. Sustain Energy Fuels 8:5379–5398. https://doi.org/10.1039/D4SE00878B
82. Montemayor RG (2008) Distillation and vapor pressure measurement in petroleum products
83. Olarte MV, Albrecht KO, Bays JT et al (2019) Autoignition and select properties of low sample volume thermochemical mixtures from renewable sources. Fuel 238:493–506. https://doi.org/10.1016/J.FUEL.2018.10.115
84. Lindfors C, Elliott DC, Prins W et al (2023) Co-processing of biocrudes in oil refineries. Energy Fuels 37:799–804. https://doi.org/10.1021/ACS.ENERGYFUELS.2C04238/ASSET/IMAGES/LARGE/EF2C04238_0001.JPEG
85. Rai AK, Al Makishah NH, Wen Z et al (2022) Recent developments in lignocellulosic biofuels, a renewable source of bioenergy. Fermentation 8:161. https://doi.org/10.3390/FERMENTATION8040161
86. Zhang M, Hu Y, Wang H et al (2021) A review of bio-oil upgrading by catalytic hydrotreatment: advances, challenges, and prospects. Mol Catal 504:111438. https://doi.org/10.1016/J.MCAT.2021.111438

Chapter 8
Sustainable Production of Bio-crude Oil from Algal Biomass and Its Upgradation for Enhanced Fuel Properties

Shreya Anand and Padmini Padmanabhan

Abstract Fossil fuels are the globally demanded component for vital processes such as transportation and industrial production. The negative impact of fossil fuel can be observed once it is burnt, while burning, it releases nitrogen oxides resulting in smog, acid rain, and greenhouse gas emissions. This has led to exploring fossil fuel substitutes as alternative fuel sources, opening up the possibility of several renewable sources. To mitigate current and potential dangers in the energy sector, a range of biofuel are generated from biomass as fossil fuel substitutes. Aquatic biomass has attracted much attention due to its use as a feedstock for third-generation biofuels. Microalgae are one of the source of aquatic biomass that produces biofuels due to its growth cycle and yield. The currently used techniques for the production of bio-crude oil need to be upgraded for the enhanced production. This book chapter discusses the sustainable production approaches for producing bio-crude oil using microalgae as algal biomass. It also discusses the upgradation of process of enhanced production.

Keywords Bio-crude oil · Microalgae · Algal biomass · Sustainable production

1 Introduction

Due to its high output and minimal environmental impact, microalgae bio-crude oil has gained popularity as a renewable resource substitute quite quickly. The method creates a substitute to fossil fuel by using naturally found microalgae. Because it is comparatively less expensive than other biofuels and is seen as sustainable, renewable, and environmentally benign, bio-crude oil can be used instead of fossil fuels. More significantly, compared to bio-oil produced from other biomass, microalgae-derived bio-oil (MB) has a lower oxygen content and a higher carbon and nitrogen content [1]. Additionally, MB was shown to have a substantial amount of aromatic

S. Anand · P. Padmanabhan (✉)
Department of Bioengineering and Biotechnology, Birla Institute of Technology, Mesra, Ranchi, India
e-mail: padmini@bitmesra.ac.in

M. Aslam et al. (eds.), *Production and Biorefining of Biocrude Oil: Current Status and Future Developments*, Advances in Sustainability Science and Technology,
https://doi.org/10.1007/978-981-96-5198-6_8

hydrocarbons [2]. A practical and sustainable way to create sustainable biofuels and renewable energy is through the manufacture of microalgae-based bio-oil [3]. Microalgae are a diverse group of small photosynthetic organisms that have gained attention as potential candidates for bio-oil production because of their high lipid content and quick development rates [2].

Microalgae are unique from other algae because of their capacity to fix carbon rates efficiently and develop more quickly than most other types of algae [4]. They capture carbon and stores it during growth and harvest. Because of their great nutritional content, they are well-known for a variety of commercial uses, such as being included into cosmetics, utilized as animal feed, and enriched in food [5]. Furthermore, microalgae are widely recognized for their capacity to be produced in large quantities and absorbed as greenhouse gases, which is why their potential as a feedstock for the synthesis of bio-oil is so high [6].

There are two different ways to turn microalgae into different types of biofuels: biochemical and thermochemical processes [7]. However, the main benefits of the thermochemical process over the biochemical one are found to be higher energy conversion, shorter reaction times, and lower production costs. These findings suggest that the thermochemical processes are among the utmost operative methods for the production of biofuel using microalgae. Globally, sustainable methods for producing biofuels are being established using algal biomass.

The generation of microalgae bio-oil has promise for reduced carbon emissions and sustainable energy. Before they can be extensively used or advantageous for commercial organizations, various challenges must be resolved, such as biomass composition, high cultivation costs, sustainable sourcing, and effective extraction. Continuous research and technological advancements are required to overcome these obstacles and realize the full potential of microalgae-based bio-oil production [8].

Microalgae-based bio-oil synthesis has enormous promise for long-term, resource- and energy-efficient use. It is anticipated that as technology, research, and commercialization endeavors advance, present problems will be resolved and the industry will progress. One such area is the improvement of microalgae growth techniques such as genetically engineered microalgae, it results in economical large-scale production of microalgae which gradually helps in production of biofuels in large amount. New extraction and conversion methods are being investigated by researchers in an effort to boost bio-crude oil production and improve the quality. The combination of all this processes with additional biorefinery processes facilitates the production of valuable byproducts from microalgae, including proteins, carbohydrates, and lipids. In addition, environmentally sound harvesting practices are essential to the future of producing bio-oil from microalgae. The viability of using waste and wastewater as nutrient sources is being studied by scientists in an effort to lessen reliance on synthetic fertilizers and the harm they cause to the environment [8]. This book chapter discusses the algal biofuels, importance of algal biomass for the production of bio-oil/biofuels, techniques of production of biofuels from algal biomass, and upgradation of biofuels.

2 Algal Biofuels

Advanced sustainable fuels made from algae feedstock through various conversion processes are known as algae-derived biofuels. This is a result of this feedstock's oil-rich composition, which may be connected to its ability to photosynthesize in large quantities [9]. The concentrates obtained from algae naturally comprise a variability of dynamic combinations, including minerals, polysaccharides, unsaturated lipids, pigmentary substances, and agents that fight cancer. It has been observed that greater developments can be achieved with different levels of oil ranging from 20 to 50%. Producing lipids at higher growth is vital for biofuel production because it leads to higher productivity of biomass and high content of lipid reduces extraction costs per unit of product. Thus, in order to allow for the constitutive synthesis of significant amounts of lipids without limiting growth, microalgae must undergo metabolic engineering [10]. Lipids also include the fatty acids that are necessary for the synthesis of some biofuels [11]. Despite the existence of biofuels, algae are thought to be the natural producers of synthetic materials that prevent bacterial, fungal, and viral illnesses in addition to being utilized to produce antioxidants [12]. According to some perspectives, microalgae feedstock is preferred for producing biofuels because it can be grown to a few overlays regardless of sporadic conditions, reduces atmospheric CO_2 and requires no fresh water or cultivable land for growth, and doesn't affect the food chain [13].

(a) **Biobutanol**

Biobutanol, which is butanol produced biologically, is similar to petrol and has a number of intriguing uses. The most effective process that has been employed in the past to produce biobutanol is fermentation of acetone–butanol–ethanol (ABE). This important sustainable biofuel can be employed in addition to already existing fuels due to its exceptional enactment and benefits [14]. Microalgae are thought to be the most promising raw material as carbon source for the biofuels production. Many algal polysaccharides may be broken down into monosaccharides, which can then be used in the process of biofuel production. These days, some of the most well-known producers in the world, such as US Technology Corporation, Gevo, Butamax, and Green Biologics, create biobutanol [15].

Anaerobic conditions are present during the biological techniques' butanol synthesis, which is regarded as a stage of ABE fermentation. Butyric acid and acetic acid are produced in the acidogenesis phase of the biphasic acetone–butanol–ethanol fermentation process. Ethanol, butanol, and acetone are among the solvents that are produced when acids are produced and then reassimilated [16]. Nevertheless, the other solvents are synthesized concurrently with the butanol production process, which lowers the selectivity rate of our intended product [17]. Marine macroalgae, such as Ceylon moss, have been demonstrated to be a viable option for the synthesis of butanol. Clostridial strains were utilized to extract the butanol from the macroalgae feedstock [18].

(b) **Biodiesel**

The term "biodiesel" describes a type of biofuel made of mono-alkyl esters. Through the process of transesterification, these esters can be extracted from organic oils, algae, plants, or animals [19]. The transesterification process is an equilibrium approach that uses an alkali that resembles potassium hydroxide and a catalyst to convert algal oil into biodiesel [20]. An enormous amount of biomass from algae is required to generate the biodiesel. Unicellular algal biomass, primarily found in aquatic environments, are the ones that are commonly used in the manufacturing of biodiesel.

Under ideal circumstances, a green alga can double its biomass in less than twenty-four hours [21]. The successfully grown algae strains must be cleaned through downstream processes including centrifugation, flocculation, filtering, electrophoresis, and sedimentation are the most significant techniques currently used for microalgal collection [22]. Normally, after harvesting, the biomass's dry weight should be amplified. However, if the goal is to develop a system for producing biodiesel or biogas, this step can be eliminated because these processes can handle large amounts of moisture content and can move forward with the process right away [23].

(c) **Biohydrogen**

Biohydrogen can be produced by bacteria. The production of biohydrogen by either thermochemical treatment or biological techniques is known as biohydrogen [24]. Additionally, it encourages application of the environmentally friendly method. Among the processes for creating biohydrogen are anaerobic fermentation of biomass, biophotolysis of water, and photofermentation by photosynthetic bacteria [25]. The biophotolysis of water with microalgae is a potential process for producing biohydrogen [26]. According to certain research, macroalgae is the most important biomass feed for the synthesis of biohydrogen [27].

(d) **Bioethanol**

Because of its benefits for the environment, bioethanol is getting a lot of consideration now-a-days. All three generations of feedstocks—plant, lignocellulosic, and algal biomass—can produce bioethanol [28]. Algal biomass is gathered and dried to eliminate moisture that allows easy management of solid material in producing bioethanol. For this reason, before the oil is extracted, the harvested biomass of algae is subjected to a suitable dehydration procedure to lower the amount of water. Usually, a workable drying process eliminates the moisture content. Several methods for drying algae biomass include sun, freeze, and spring drying [29, 30].

This method produces bioethanol from the algal carbohydrates, where the raw material breaks down into simple carbohydrate which can then be combined with water and yeast in bioreactors [31], followed by distillation for removal of water and other impurities from the crude bioethanol to obtain purified from of bioethanol [32]. The use of microalgae in bioethanol production can help address environmental issues where bioethanol produced from conventional feedstock is found to emit more greenhouse gases than fossil fuels due to feedstock manufacture and process applications [33].

3 Why Are Algae Beneficial in Biofuel Production?

Microalgae are prokaryotes, which can be found in a variety of aquatic conditions. More than 40% of the world's carbon fixation is attributed to algae, most of which are marine microalgae that utilize CO_2 efficiently [34, 35]. Algae have a very high rate of biomass production; some species can double in as little as 6 h, and many can double twice a day [36, 37]. Energy-rich oils can be produced by any type of algae, and it has been observed that certain microalgal species spontaneously accumulate large amounts of oil in their total dry biomass [38]. For instance, it has been determined that up to 50% of the dry mass of certain species of Botryococcus is preserved as long-chain hydrocarbons [39]. There is an abundance of genetic diversity among the microalgal species under investigation as possible biofuel crops because they come from groupings whose ancestral ties are notably wider than those of the most varied land plants [40, 41].

The three primary constituents of algal biomass are lipids/natural oils, proteins, and carbohydrates. Microalgae are the majorly used in biodiesel industry because it can produce triacylglycerol, which is very compatible bio-crude oil for production of biodiesel. Under certain growing conditions, certain species of algae have the ability to create hydrogen gas. Algal biomass can be converted into ashes and can be used for production of biogas. Another method for processing algae biomass is pyrolysis, which produces crude bio-oil [42].

4 Production of Bio-crude Biofuel

The available methods for converting biomass obtained from algae into various forms of energy include biochemical and thermochemical conversion, as well as combustion. The algal biomass can be converted into biofuels can be achieved using hydrothermal liquefaction or thermochemical techniques such as gasification, pyrolysis, and superficial liquid extraction. These methods can be used to turn all of the biomass obtained by algae, except carbohydrates and lipids, into biofuels. It is not necessary to apply particular conditions, like nitrogen scarcity during the growing process in the hopes of reaching maximum content of lipid, if the thermochemical approach is used for further processing of algal biomass after cultivation [43].

The conversion of biomass into biofuels by biochemical processes can be divided into four categories: photobiological hydrogen generation, anaerobic digestion, alcohol fermentation, and transesterification [44]. The biomass can be heated and decomposed using oxygen-assisted thermochemical reactions. Thermochemical methods may produce solid, liquid, and gaseous biofuels, unlike biological processes [45]. It's been confirmed that conversion of algal biomass using a thermochemical method is advantageous than biochemical method because of the less conversion of energy, long durations of reaction, and high cost of production is associated with the biological conversion.

Thermochemical reactions happen significantly faster than biological conversion processes [46]. Compared to chemical and biological processes, the thermochemical approach provides a more direct means of producing biofuels. Transesterification necessitates a methanol recycling system, chemical conversion demands the separation or purification of biomass, and the disposal process is challenging since soap is produced [47]. Biochemical conversion techniques, such as fermentation, usually need several days to produce biofuels [6]. By using the full feedstock and converting a variety of feedstocks from biomass, thermochemical conversion, on the other hand, usually eliminates the requirement for chemicals. Producing biofuels with the thermochemical process also takes relatively little time. Thus, the thermochemical conversion process is a vital step in the production of biofuels from microalgae. Direct combustion, gasification, hydrothermal liquefaction, and pyrolysis are the four basic types of thermochemical conversion processes. The two most popular and successful methods for creating bio-oil are pyrolysis and hydrothermal liquefaction [48].

Conventional pyrolysis typically operates at temperature between 450 and 550 °C and at an atmospheric pressure. On the other hand, microwave pyrolysis and catalytic pyrolysis require temperatures of at least 300 °C and 800 °C, respectively [6]. Pyrolysis has several benefits, one of which is the possibility of producing higher yields of bio-oil. Nevertheless, the requirement for dried biomass with the lowest possible moisture content limits the procedure [49]. Typically, the algae contribute between five and fifty percent of the slurry feed during the hydrothermal liquefaction process. High temperatures are needed to produce biofuels from algae. The procedure can be done with or without a catalyst, and auto thermal water is utilized at pressures between 5 and 20 bar [50]. Methane that is in a gaseous condition is converted into bio-oil and other byproducts through liquefaction. Higher yields can be achieved with this reaction, despite its extreme complexity [49]. Fuels can be produced from microalgae using a biochemical process that uses either microorganisms or enzymes to break them down. Biochemical transformations often happen more slowly and use less energy than thermochemical conversion techniques.

Generally, biomass can be used to produce biogas, anaerobic by converting organic matter using enzymes or microbes into biogas. This process also produces CO_2 and trace amounts of H_2S. Anaerobic digestion is often suitable for organic waste, such as algal biomass, that shows a lot of moisture. This process consists of three phases that are completed in the order listed: hydrolysis, acetogenesis, and methanogenesis. An important component that might be achieved through pretreatment is the algal vulnerability to an enzyme attack, since in the anaerobic fermentation, hydrolysis acts as a rate-limiting step. Despite being specifically intended for microalgae, power-driven pretreatment is still a frequent method; nonetheless, thermal pretreatment is still an excellent option [50]. Fermentation is usually carried out using microalgae with a higher starch concentration. This technique entails hydrolyzing the main ingredients, such as sugars and starches, to produce bioethanol. Yeast fermentation, which results in the generation of bioethanol, continues the process. Subsequently, the algal cell wall is subjected to pretreatment procedures that involve either sonication or enzymatic methods to liberate carbohydrates [51]. Unfortunately, the carbohydrate and protein components of the feedstock are not utilized and is restricted to the conversion of lipids only.

5 Upgradation of Bio-crude Oil

Bio-oil is a viscous, dark brown liquid with a unique smell [52, 53]. A complex mixture with hundreds of chemical components, bio-oil is influenced by various factors such as raw material type, thermochemical process, and operational parameters [54]. Bio-oil mostly consists of the following constituents: phenols, acids, alcohols, aldehydes, esters, ketones, and guaiacol [55]. These substances give bio-oil its undesired characteristics. The literature reports that bio-oil can be upgraded by a variety of physical and chemical techniques, such as hydrotreating, catalytic cracking, emulsification, solvent addition, extraction, esterification, steam reforming, and supercritical fluids. A few of the above-mentioned techniques have been discussed below.

(a) **Extraction**

Bio-oil contains a wide range of compounds, some of which are used in industry (phenol for the resin industry, organic acids, alkanes, etc.) [56]. As a result, the chemicals that are removed from bio-oil effectively raise its value. Extraction from bio-oil is the separation technology, much like in petrochemical facilities [57]. Fractionation, distillation, and absorption are all part of the technology. Acetone is used as the adsorption solvent, and there are two types of fractionation: phase separation and aqueous extraction [58]. According to Eboibi et al. 's research [59], there is a rise in HHV and a drop in the O element's concentration. The development of low-cost separation and refining processes is necessary for commercialization [60].

(b) **Solvent Addition**

It has been normal practice for several decades to use polar solvents to lower the viscosity of bio-oils. The normal polar solvents include methanol, ethanol, acetone, and ethyl acetate [61]. The heating value rises when polar solvents are added because they boil at higher boiling point than biofuels. The viscosity of the biofuel decreases due to the chemical interaction of the solvent and the biofuel.

(c) **Emulsification**

For upgrading the production of bio-crude oil, emulsification can also be used along with additional chemicals. Since it is challenging to combine fuels made from petroleum with bio-oil made from biomass, surfactants are frequently utilized. Although bio-oil emulsions have good ignition properties, engine applications show increased corrosion. A quick method of upgrading bio-oil without a chemical reaction is emulsification. The heating value, corrosiveness, and cetane number do not significantly improve using this approach, despite the possibility that it will improve the ignition properties [62]. Emulsification is a somewhat easy method of upgrading bio-oil with diesel or biodiesel, however, it is expensive due to the surfactant addition and significant energy usage.

(d) **Esterification**

Esterification is a common method for producing biodiesel. When fatty acids and alcohol combine at atmospheric pressure to form alkyl ester or biodiesel, acid catalyst is a key component of the process [63]. Typically, the reaction occurs at a temperature below 60 °C, which is lower than methanol's boiling point. It is also possible to carry out the esterification process in supercritical circumstances. It is more efficient to upgrade biofuel in supercritical conditions as opposed to subcritical ones. As a byproduct of the process, there is a lot of water production which can be extracted and removed after distillation, to improve the production of biofuel.

(e) **Supercritical Fluids**

A novel technique for refining bio-oil has garnered significant interest: the use of supercritical fluids. The solvent can sustain a fluid at high pressure and temperature, referred to as the supercritical fluid, above the solvent's critical point, making it challenging to discriminate between the liquid and gas phases. The substance may diffuse through solids like a gas or dissolve in the supercritical fluid. As an alternative to organic solvents, it shows promise. The ability of undissolved materials could be enhanced in the supercritical fluid [64]. The technology offers faster mass and heat transfer rates, gas-like diffusivity and viscosity, and density similar to that of a liquid.

6 Conclusion

The most promising use of the biomass derived from algae species is the production of third-generation biofuels, an improved kind of biofuel. There is no competition between the production of algae and that of food, fiber, or fuel. Algae have been widely used in industrial settings, most notably in the manufacturing of biofuels. The development of renewable fuels, such as nuclear, solar, wind, and biomass energy, is necessary since fossil fuels may run out in a few decades. There is a chance that biomass will be utilized extensively. One crucial technology for converting biomass into useful biofuel is liquefaction. The two primary processes for converting biomass into biofuels are thermodynamic liquefaction and hydrolysis fermentation. Direct liquefaction of biomass is a promising technology. It is possible to improve bio-oil using physicochemical means. Advances in the techniques of isolating algal biomass from water and obtaining oil from biomass will boost the potential of algal oil.

References

1. Oasmaa A, Lehto J, Solantausta Y, Kallio S (2021) Historical review on VTT fast pyrolysis bio-oil production and upgrading. Energy Fuels 35(7):5683–5695. https://doi.org/10.1021/acs.energyfuels.1c00177
2. Wang Y, Akbarzadeh A, Chong L, Du J, Tahir N, Awasthi MK (2022) Catalytic pyrolysis of lignocellulosic biomass for bio-oil production: a review. Chemosphere 297:134181. https://doi.org/10.1016/j.chemosphere.2022.134181
3. Raheem A, Prinsen P, Vuppaladadiyam AK, Zhao M, Luque R (2018) A review on sustainable microalgae based biofuel and bioenergy production: recent developments. J Clean Prod 181:42–59. https://doi.org/10.1016/j.jclepro.2018.01.125
4. Chen WH, Wu ZY, Chang JS (2014) Isothermal and non-isothermal torrefaction characteristics and kinetics of microalga Scenedesmus obliquus CNW-N. Biores Technol 155:245–251. https://doi.org/10.1016/j.biortech.2013.12.116
5. Spolaore P, Joannis-Cassan C, Duran E, Isambert A (2006) Commercial applications of microalgae. J Biosci Bioeng 101(2):87–96. https://doi.org/10.1263/jbb.101.87
6. Chen WH, Lin BJ, Huang MY, Chang JS (2015) Thermochemical conversion of microalgal biomass into biofuels: a review. Biores Technol 184:314–327. https://doi.org/10.1016/j.biortech.2014.11.050
7. Ağbulut Ü, Sirohi R, Lichtfouse E, Chen WH, Len C, Show PL et al (2023) Microalgae bio-oil production through pyrolysis and hydrothermal liquefaction: mechanism and characteristics. Bioresour Technol 128860.https://doi.org/10.1016/j.biortech.2023.128860
8. Ahmed SF, Rafa SJ, Mehjabin A, Tasannum N, Ahmed S, Mofijur M et al (2023) Bio-oil from microalgae: materials, production, technique, and future. Energy Rep 10:3297–3314
9. Adeniyi OM, Azimov U, Burluka A (2018) Algae biofuel: current status and future applications. Renew Sustain Energy Rev 90:316–335. https://doi.org/10.1016/j.rser.2018.03.067
10. Tan KWM, Lee YK (2016) The dilemma for lipid productivity in green microalgae: importance of substrate provision in improving oil yield without sacrificing growth. Biotechnol Biofuels 9:1–14. https://doi.org/10.1186/s13068-016-0671-2
11. Javidialesaadi A, Raeissi S (2013) Biodiesel production from high free fatty acid-content oils: experimental investigation of the pretreatment step. APCBEE Proc 5:474–478. https://doi.org/10.1016/j.apcbee.2013.05.080
12. Michalak I, Chojnacka K (2015) Algae as production systems of bioactive compounds. Eng Life Sci 15(2):160–176. https://doi.org/10.1002/elsc.201400191
13. Khan MI, Shin JH, Kim JD (2018) The promising future of microalgae: current status, challenges, and optimization of a sustainable and renewable industry for biofuels, feed, and other products. Microb Cell Fact 17:1–21. https://doi.org/10.1186/s12934-018-0879-x
14. Rathour RK, Ahuja V, Bhatia RK, Bhatt AK (2018) Biobutanol: new era of biofuels. Int J Energy Res 42(15):4532–4545. https://doi.org/10.1002/er.4180
15. Wang Y, Ho SH, Yen HW, Nagarajan D, Ren NQ, Li S et al (2017) Current advances on fermentative biobutanol production using third generation feedstock. Biotechnol Adv 35(8):1049–1059.https://doi.org/10.1016/j.biotechadv.2017.06.001
16. Terracciano JS, Kashket ER (1986) Intracellular conditions required for initiation of solvent production by Clostridium acetobutylicum. Appl Environ Microbiol 52(1):86–91. https://doi.org/10.1128/aem.52.1.86-91.1986
17. Birgen C, Dürre P, Preisig HA, Wentzel A (2019) Butanol production from lignocellulosic biomass: revisiting fermentation performance indicators with exploratory data analysis. Biotechnol Biofuels 12:1–15. https://doi.org/10.1186/s13068-019-1508-6
18. Sun C, Zhang S, Xin F, Shanmugam S, Wu YR (2018) Genomic comparison of Clostridium species with the potential of utilizing red algal biomass for biobutanol production. Biotechnol Biofuels 11:1–15. https://doi.org/10.1186/s13068-018-1044-9
19. Demirbas A (2007) Importance of biodiesel as transportation fuel. Energy Policy 35(9):4661–4670. https://doi.org/10.1016/j.enpol.2007.04.003

20. Chisti Y (2007) Biodiesel from microalgae. Biotechnol Adv 25(3):294–306. https://doi.org/10.1016/j.biotechadv.2007.02.001
21. Demirbas A, Demirbas MF (2011) Importance of algae oil as a source of biodiesel. Energy Convers Manage 52(1):163–170. https://doi.org/10.1016/j.enconman.2010.06.055
22. Ananthi V, Balaji P, Sindhu R, Kim SH, Pugazhendhi A, Arun A (2021) A critical review on different harvesting techniques for algal based biodiesel production. Sci Total Environ 780:146467. https://doi.org/10.1016/j.scitotenv.2021.146467
23. Ananthi V, Viklund SB, Hackl R, Karlsson M, Berntsson T (2014) Algae-based biofuel production as part of an industrial cluster. Biomass Bioenerg 71:113–124. https://doi.org/10.1016/j.biombioe.2014.10.019
24. Mohan SV, Pandey A (2013) Biohydrogen production: an introduction. In: Biohydrogen. Elsevier, pp 1–24. https://doi.org/10.1016/B978-0-444-59555-3.00001-5
25. Kim MS, Kim DH, Lee JK (2011) Biohydrogen. In: Industrial biotechnology and commodity products. Elsevier Inc, pp 115–125. https://doi.org/10.1016/B978-0-08-088504-9.00166-5
26. Nagarajan D, Lee DJ, Kondo A, Chang JS (2017) Recent insights into biohydrogen production by microalgae—from biophotolysis to dark fermentation. Biores Technol 227:373–387. https://doi.org/10.1016/j.biortech.2016.12.104
27. Kumar MD, Kavitha S, Tyagi VK, Rajkumar M, Bhatia SK, Kumar G, Banu JR (2021) Macroalgae-derived biohydrogen production: biorefinery and circular bioeconomy. Biomass Convers Biorefinery 1–23.https://doi.org/10.1007/s13399-020-01187-x
28. Baeyens J, Kang Q, Appels L, Dewil R, Lv Y, Tan T (2015) Challenges and opportunities in improving the production of bio-ethanol. Prog Energy Combust Sci 47:60–88. https://doi.org/10.1016/j.pecs.2014.10.003
29. Desmorieux H, Decaen N (2005) Convective drying of spirulina in thin layer. J Food Eng 66(4):497–503. https://doi.org/10.1016/j.jfoodeng.2004.04.021
30. Porpatham E, Ramesh A, Nagalingam B (2012) Effect of compression ratio on the performance and combustion of a biogas fuelled spark ignition engine. Fuel 95:247–256. https://doi.org/10.1016/j.fuel.2011.10.059
31. Singh A, Nigam PS, Murphy JD (2011) Renewable fuels from algae: an answer to debatable land based fuels. Biores Technol 102(1):10–16. https://doi.org/10.1016/j.biortech.2010.06.032
32. Bibi R, Ahmad Z, Imran M, Hussain S, Ditta A, Mahmood S, Khalid A (2017) Algal bioethanol production technology: a trend towards sustainable development. Renew Sustain Energy Rev 71:976–985. https://doi.org/10.1016/j.rser.2016.12.126
33. Mahmood T, Hussain N, Shahbaz A, Mulla SI, Iqbal HM, Bilal M (2023) Sustainable production of biofuels from the algae-derived biomass. Bioprocess Biosyst Eng 46(8):1077–1097. https://doi.org/10.1007/s00449-022-02796-8
34. Falkowski PG, Barber RT, Smetacek V (1998) Biogeochemical controls and feedbacks on ocean primary production. Science 281(5374):200–206. https://doi.org/10.1126/science.281.5374.200
35. Parker MS, Mock T, Armbrust EV (2008) Genomic insights into marine microalgae. Annu Rev Genet 42:619–645. https://doi.org/10.1146/annurev.genet.42.110807.091417
36. Sheehan J, Dunahay T, Benemann J, Roessler P (1998) A look back at the US Department of Energy's aquatic species program: biodiesel from algae. Nat Renew Energy Lab 328:1–294
37. Huesemann MH, Hausmann TS, Bartha R, Aksoy M, Weissman JC, Benemann JR (2009) Biomass productivities in wild type and pigment mutant of Cyclotella sp. (Diatom). Appl Biochem Biotechnol 157:507–526. https://doi.org/10.1007/s12010-008-8298-9
38. Rodolfi L, Chini Zittelli G, Bassi N, Padovani G, Biondi N, Bonini G, Tredici MR (2009) Microalgae for oil: strain selection, induction of lipid synthesis and outdoor mass cultivation in a low-cost photobioreactor. Biotechnol Bioeng 102(1):100–112. https://doi.org/10.1002/bit.22033
39. Kojima E, Zhang K (1999) Growth and hydrocarbon production of microalga Botryococcus braunii in bubble column photobioreactors. J Biosci Bioeng 87(6):811–815. https://doi.org/10.1016/S1389-1723(99)80158-3

40. Deschamps P, Moreira D (2009) Signal conflicts in the phylogeny of the primary photosynthetic eukaryotes. Mol Biol Evol 26(12):2745–2753. https://doi.org/10.1093/molbev/msp189
41. Reeb VC, Peglar MT, Yoon HS, Bai JR, Wu M, Shiu P et al (2009) Interrelationships of chromalveolates within a broadly sampled tree of photosynthetic protists. Mol Phylogenet Evol 53(1):202–211.https://doi.org/10.1016/j.ympev.2009.04.012
42. Zhiyou W (2019) Algae for biofuel production. Farm Energy. https://farm-energy.extension.org/algae-for-biofuel-production/
43. Hallenbeck PC, Grogger M, Mraz M, Veverka D (2016) Solar biofuels production with microalgae. Appl Energy 179:136–145. https://doi.org/10.1016/j.apenergy.2016.06.024
44. Azizi K, Moraveji MK, Najafabadi HA (2018) A review on bio-fuel production from microalgal biomass by using pyrolysis method. Renew Sustain Energy Rev 82:3046–3059. https://doi.org/10.1016/j.rser.2017.10.033
45. Biller P, Sharma BK, Kunwar B, Ross AB (2015) Hydroprocessing of bio-crude from continuous hydrothermal liquefaction of microalgae. Fuel 159:197–205. https://doi.org/10.1016/j.fuel.2015.06.077
46. Kumar PK, Krishna SV, Verma K, Pooja K, Bhagawan D, Srilatha K, Himabindu V (2018) Bio oil production from microalgae via hydrothermal liquefaction technology under subcritical water conditions. J Microbiol Methods 153:108–117. https://doi.org/10.1016/j.mimet.2018.09.014
47. Morais MG, Cassuriaga APA, Cruz CG, Moraes L, Costa JAV (2022) Metabolism of microalgae and metabolic engineering for biomaterial applications. In: Biomass, biofuels, and biochemicals. Elsevier, pp 1–20. https://doi.org/10.1016/B978-0-323-96142-4.00010-5
48. Nagi GK, Minhas AK, Gaur S, Jain P, Mandal S (2021) Integration of algal biofuels with bioremediation coupled industrial commodities towards cost-effectiveness. Front Energy Res 9:735141. https://doi.org/10.3389/fenrg.2021.735141
49. Juneja A, Ceballos RM, Murthy GS (2013) Effects of environmental factors and nutrient availability on the biochemical composition of algae for biofuels production: a review. Energies 6(9):4607–4638. https://doi.org/10.3390/en6094607
50. Raheem A, Azlina WW, Yap YT, Danquah MK, Harun R (2015) Thermochemical conversion of microalgal biomass for biofuel production. Renew Sustain Energy Rev 49:990–999. https://doi.org/10.1016/j.rser.2015.04.186
51. Ward AJ, Lewis DM, Green FB (2014) Anaerobic digestion of algae biomass: a review. Algal Res 5:204–214. https://doi.org/10.1016/j.algal.2014.02.001
52. Zhu X, Lobban LL, Mallinson RG, Resasco DE (2011) Bifunctional transalkylation and hydrodeoxygenation of anisole over a Pt/HBeta catalyst. J Catal 281(1):21–29. https://doi.org/10.1016/j.jcat.2011.03.030
53. Abdullah N, Gerhauser H (2008) Bio-oil derived from empty fruit bunches. Fuel 87(12):2606–2613. https://doi.org/10.1016/j.fuel.2008.02.011
54. Alvarez J, Lopez G, Amutio M, Bilbao J, Olazar M (2014) Bio-oil production from rice husk fast pyrolysis in a conical spouted bed reactor. Fuel 128:162–169. https://doi.org/10.1016/j.fuel.2014.02.074
55. Lian X, Xue Y, Zhao Z, Xu G, Han S, Yu H (2017) Progress on upgrading methods of bio-oil: a review. Int J Energy Res 41(13):1798–1816. https://doi.org/10.1002/er.3726
56. Cao JP, Zhao XY, Morishita K, Li LY, Xiao XB, Obara R et al (2010) Triacetonamine formation in a bio-oil from fast pyrolysis of sewage sludge using acetone as the absorption solvent. Bioresour Technol 101(11):4242–4245.https://doi.org/10.1016/j.biortech.2010.01.031
57. Wang S, Gu Y, Liu Q, Yao Y, Guo Z, Luo Z, Cen K (2009) Separation of bio-oil by molecular distillation. Fuel Process Technol 90(5):738–745. https://doi.org/10.1016/j.fuproc.2009.02.005
58. Elkasabi Y, Mullen CA, Boateng AA (2014) Distillation and isolation of commodity chemicals from bio-oil made by tail-gas reactive pyrolysis. ACS Sustainable Chem Eng 2(8):2042–2052. https://doi.org/10.1021/sc5002879
59. Eboibi BEO, Lewis DM, Ashman PJ, Chinnasamy S (2014) Hydrothermal liquefaction of microalgae for biocrude production: improving the biocrude properties with vacuum distillation. Biores Technol 174:212–221. https://doi.org/10.1016/j.biortech.2014.10.029

60. Schulzke T, Conrad S, Westermeyer J (2016) Fractionation of flash pyrolysis condensates by staged condensation. Biomass Bioenerg 95:287–295. https://doi.org/10.1016/j.biombioe.2016.05.022
61. Pidtasang B, Udomsap P, Sukkasi S, Chollacoop N, Pattiya A (2013) Influence of alcohol addition on properties of bio-oil produced from fast pyrolysis of eucalyptus bark in a free-fall reactor. J Ind Eng Chem 19(6):1851–1857. https://doi.org/10.1016/j.jiec.2013.02.031
62. Jiang X, Zhong Z, Ellis N, Wang Q (2011) Aging and thermal stability of the mixed product of the ether-soluble fraction of bio-oil and bio-diesel. Chem Eng Technol 34(5):727–736. https://doi.org/10.1002/ceat.201000441
63. Xu Y, Wang T, Ma L, Zhang Q, Liang W (2010) Upgrading of the liquid fuel from fast pyrolysis of biomass over MoNi/γ-Al_2O_3 catalysts. Appl Energy 87(9):2886–2891. https://doi.org/10.1016/j.apenergy.2009.10.028
64. Tang Z, Zhang Y, Guo Q (2010) Catalytic hydrocracking of pyrolytic lignin to liquid fuel in supercritical ethanol. Ind Eng Chem Res 49(5):2040–2046. https://doi.org/10.1021/ie9015842
65. Dismukes GC, Carrieri D, Bennette N, Ananyev GM, Posewitz MC (2008) Aquatic phototrophs: efficient alternatives to land-based crops for biofuels. Curr Opin Biotechnol 19(3):235–240. https://doi.org/10.1016/j.copbio.2008.05.007

Chapter 9
Biorefinery Catalysts: Current Scenario and Future Developments

Francisco Simão Neto, Kaiany Moreira dos Santos, Mauricio Quintas Salamba, Jessica Lopes da Silva, José Roberto de Matos Filho, Letícia da Costa Silva, and José Cleiton Sousa dos Santos

Abstract Excessive dependence on non-renewable fossil fuels has led to numerous economic, social, and environmental problems. These include resource scarcity and energy crises, significant emissions of environmental pollutants, global climate change, and unsustainable development. A biorefinery is a processing facility that uses biomass as a base to produce a variety of products. Examples of biomass that can be converted into profitable products include wood, straw, starch, sugars, waste or algae. These can be transformed into chemicals, biomaterials, energy, electricity, oils, food, feed, and much more. In this context, the present work aims to apply advanced bibliometrics to study catalyst applied to biorefinery. This study will address primary research using catalysts applied to biorefinery, the main countries/regions involved in production, areas of research related to the topic and future trends in these areas. The study observed cooperation between China, the USA, and India, with the Department of Aeronautics (USA) being the main institution to publish on the subject. Finally, it can be concluded that new catalysts offer significant opportunities to increase efficiency, reduce costs, and promote sustainability in biorefinery processes.

Keywords Biofuel · Biorefinery · Catalyst · Bibliometric analysis

F. S. Neto · J. C. S. dos Santos (✉)
Department of Chemical Engineering, Federal University of Ceara, Fortaleza, Brazil
e-mail: jcs@unilab.edu.br

F. S. Neto
e-mail: fcosimao@aluno.unilab.edu.br

K. M. dos Santos · M. Q. Salamba · J. L. da Silva · J. R. de Matos Filho · L. da Costa Silva · J. C. S. dos Santos
Engineering and Sustainable Development Institute, University of International Integration of Afro-Brazilian Lusophony, Redenção, Brazil

M. Aslam et al. (eds.), *Production and Biorefining of Biocrude Oil: Current Status and Future Developments*, Advances in Sustainability Science and Technology,
https://doi.org/10.1007/978-981-96-5198-6_9

1 Introduction

Numerous economic, social, and environmental problems have been caused by excessive dependence on non-renewable fossil fuels. These include resource scarcity and energy crisis, significant emission of environmental pollutants, global climate change, and unsustainable development [1, 2]. Growing concerns are driving the search for viable and sustainable alternative resources. Biomass available from lignocellulose, agricultural residues, animal fats, and crops is seen as a promising substitute for fossil fuels as it is the most abundant and renewable natural reserve [3, 4].

Catalysts are substances that speed up a chemical reaction without consuming it [5]. The catalysis panorama can be divided into two categories: heterogeneous catalysis, which uses solid catalysts, and homogeneous catalysis, which uses molecularly defined complexes dissolved in a liquid medium [6]. As catalysis allows obtaining "new products" of high yield, selectivity, and conversion under more favorable operating conditions, it plays an important role in the conversion of biomass [7].

Currently, trends favor heterogeneous catalysis due to the possibility of recovery and reuse of the catalyst. However, acid hydrolysis and other lignocellulosic biomass conversion processes have also been studied through homogeneous catalysis [8]. Due to the low use of bioenergy to provide heat and energy in the industrial, and transport sectors around the world, one of the most studied questions is how to transform biomass into biofuels through catalytic processes [9] Furthermore, studies were carried out on high added-value compounds created by heterogeneous catalysis for use in the cosmetic, pharmaceutical, and chemical sectors [10]. Therefore, catalysis is essential to (i) bring industry and academia closer together about the use of biomass, (ii) improve biorefinery projects, (iii) develop new biomass conversion pathways, and (iv) increase sustainability of the processes [11].

A biorefinery is a processing facility that uses biomass as a base to produce a variety of products [12]. Wood, straw, starch, sugars, waste, or algae are examples of biomass that can be transformed into various profitable products such as chemicals, biomaterials, energy, electricity, oils, food, feed, and much more [13, 14]. The word "biorefinery" emerged as a response to four industrial trends in 1990. The first was an improvement in the profitability and competitiveness of pulp mills; the second was the production of ethanol from cellulose and wood; the third was the gasification of biomass and greater attention to the production of starch for energy use; and the fourth was the increase in interest [15, 16]. Biorefinery products are renewable carbon sources that allow unlimited fuel, material, and energy needs to be met. They also reduce the threat of climate change [17].

Biorefineries are designed based on a comprehensive study of raw materials and promising technologies available [18]. Biorefineries are generally called "first or second generation" according to the raw material used. 1ª generation (1G) biorefineries use food crop resources such as vegetable oil and corn sugar, while 2ª generation (2G) biorefineries process non-food materials such as wood, agricultural waste, and energy crops generally rich in lignocellulose [19–21]. Biorefineries include thermochemical, biotechnological, chemical, and physical processes that allow the production of various compounds. Thus, the application of catalysts can be present in each of these processes [22].

Biorefineries are an essential part of the transition to a sustainable, low-carbon economy. They contribute to reducing dependence on fossil fuels, reducing greenhouse gas emissions, and promoting the economic use of renewable resources. To improve the financial and environmental viability of biorefinery systems, continuous innovation is needed [23, 24]. The objective of this research is to carry out a complete bibliometric analysis of catalysts used in biorefineries from 2014 to 2024. The study aims to identify trends and progress in the use of heterogeneous and homogeneous catalysts in biomass conversion processes, as well as to evaluate the efficiency, sustainability, and impact of these catalysts on the development of biorefinery technologies. Furthermore, the study aims to show how industry and academia collaborate, emphasizing the main scientific and technological changes that have driven the sector.

2 Methodology

2.1 Data Source

To prepare this paper, a bibliographic survey was carried out on the Web of Science-Core Collection platform, where a database was set up with the results of the research. The material analyzed was selected using the tools and delimitations developed for the advanced bibliometric analysis method. Access to the platform was authorized by the Coordenação de Aperfeiçoamento de Pessoal de Nível Superior (CAPES), through its Periódicos Capes domain [25].

The material collected will be the basis for the analysis carried out in this study, where its composition was defined by the following delimiters: The term "Biofuel" was inserted into the search line with the "All fields" filter. A new line was then inserted with the term "2014–2024" and the filter "Year of Publication." In addition, this material was followed up with other delimitations to refine the material and ensure that the data was faithful to the objectives and terms analyzed.

The delimitations were made as follows: Types of documents, in which the limitation was restricted to articles, conference papers, and review articles; Language, in which only works in English would be considered. Finally, the material was reduced to 46,000 articles which will be used in the advanced bibliometric analysis method. Next, the initial material generated two smaller databases that would be used to analyze two other topics relevant to the research topic.

The first database was delimited based on the analysis of the catalyst used in biorefineries. Thus, a new line was added to add a new term: "Biorefinery," and the other filters and terms were maintained to give continuity to the 46,000 articles already defined. After this new delimitation, the base was reduced to 4128 articles related to the terms searched (i.e., biorefinery catalyst) from 2014 to 2024. The last base was defined taking into account the production of biocrude. Therefore, the new line used the term "Catalyst" with the filter "All Fields." A return of 1812 articles related to the topic searched was obtained, out of the 46,000 articles in the database.

2.2 *Data Analysis*

Three software programs were used for analysis in this work: CiteSpace (version 5.8.R6, Philadelphia, PA, USA) and VOSviewer (version 1.6.17, Leiden, South Holland, Netherlands). In order of application, in VOSviewer and CiteSpace, data visualization was carried out to allow better understanding and analysis of the data.

3 Results

3.1 *Trend of Global Publications and Citations*

A data search conducted in the Scopus database yielded various research-related information, including the total number of publications (1800) and citations (80,000) from 2014 to June 2024. These data are presented in Fig. 1, which depicts a bar and line graph representing the evolution of the number of publications and citations over this period. Substantial growth is observed in both the number of publications and citations related to the biorefinery catalyst theme between 2014 and 2021. Comparing 2014 with 2023, a significant difference is noted. In 2014, the number of publications was approximately twenty-five, while in 2023, this number increased to about 375. This represents a 1400% increase in the number of publications over this period. Similarly, the number of citations also showed significant growth, rising from about 500 in 2014 to approximately 17,000 in 2023, equating to a 3300% increase.

This substantial growth can be attributed to technological advances, increased research funding, greater international collaboration, and a growing interest in the field of biorefinery catalysts, where catalysis plays a significant role in the development of new and improved biorefineries [24]. The significant increase in citations suggests that research published during this period has had a significant impact on the scientific community. However, in 2024, there is a significant decline, with publications decreasing to about two hundred and citations to approximately 12,000. It is worth noting that these data only reflect the first half of 2024. Therefore, it is expected that the number of publications and citations will increase, surpassing the values of 2023 by the end of the year.

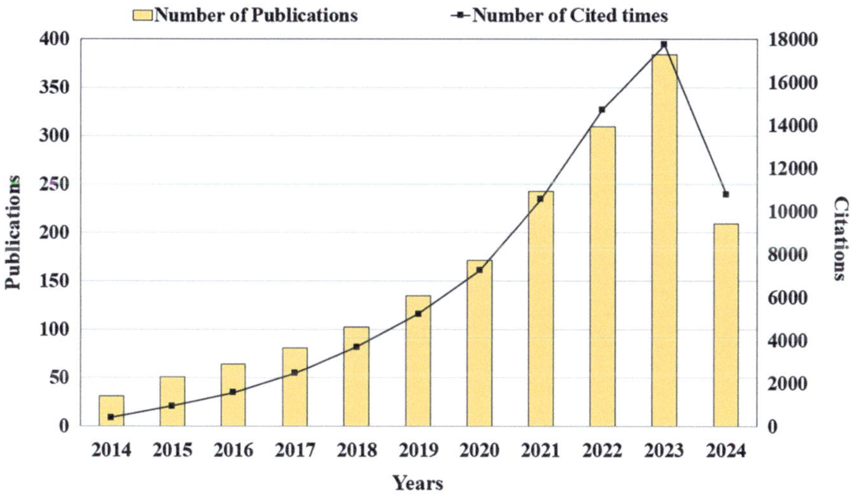

Fig. 1 Bar and line graph show the evolution of the number of publications (yellow bars) and the number of citations (black line with dots) related to the topic of biorefinery catalysts from 2014 to June 2024

3.2 Contributions of Countries/Regions

Figure 2 is a graphical representation of a network of connections between countries, highlighting each one's contribution to the academic publications landscape on biorefinery catalysts. The inclusion of a country in the graph was conditioned by a minimum criterion of twenty-five published works. Among the eighty-nine countries evaluated, thirty-one met this criterion. The lines connecting the countries in the graph vary in thickness, with thicker lines indicating stronger and more frequent connections. Additionally, it is possible to observe a dense web of interconnections within each grouping and between them, illustrating a global network of interdependencies and research cooperation. This visualization highlights China as the most central, interconnected node, represented by red, which is the most prominent on the map, indicating that this country has an extensive network of relationships with several other nations. This highlight reflects China's position as one of the main global actors in terms of research on the proposed topic. China's centrality indicates its significant leadership and influence in the international academic landscape. Following closely behind China, the United States (purple) and India (yellow) also appear as central nodes, albeit with fewer connections. This demonstrates that both countries play important roles, contributing to the global research network, albeit on a smaller scale compared to China.

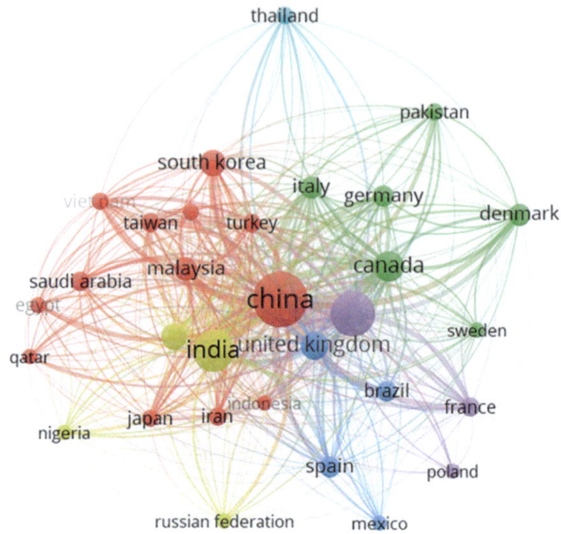

Fig. 2 Map of the collaboration network among countries using twenty-five works as a basis and with a total of thirty-nine participating countries; lines and colors indicate degrees of cooperation between the period of 2014 to June 2024 based on data extracted from the Scopus database, generated via VOSviewer

3.3 Contributions of Institutions

Figure 3 presents a map of the institutions that have contributed the most to publications on biorefinery catalysts from 2014 to June 2024. Among 4422 organizations, only thirty-five were selected based on a minimum criterion of six publications. The institutions were separated into groups of contribution and collaboration, distinguished by distinct colors. The most interconnected ones are highlighted in green, red, and blue. With the highest number of publications, the Department of Aeronautics (green) stands out with twenty works and 2295 citations, followed by the Institute of Nuclear (orange) with sixteen works and 525 citations, and the Department of Agricultural (yellow) with twelve publications and 696 citations. It is important to note that, although the Institute of Nuclear has a higher number of publications compared to the Department of Agriculture, its works are fewer cited. It is worth mentioning that despite some institutions having a lower number of publications, they may still have an important level of citations. Some of these are the Innovative Green (blue) with eleven publications and 934 citations, and the Department of Mechanical with eight works and 919 citations, suggesting that despite fewer works, their impact levels are significant.

3.4 Contributions of Authors

A broad interest in the subject is noticeable, as evidenced in Fig. 4 by the identification of 6665 authors, of which only thirty were analyzed. The minimum number

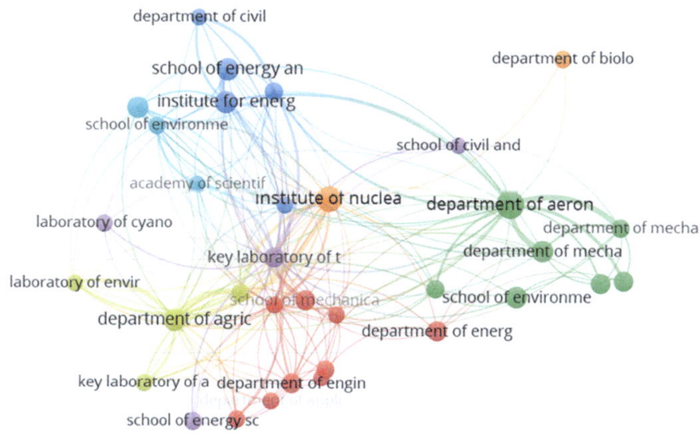

Fig. 3 Collaboration network map among organizations using six publications as the basis, involving thirty-five participating institutions. Lines and colors indicate degrees of cooperation between the period of 2014 to June 2024, based on data extracted from the Scopus database and generated via VOSviewer

of publications for these authors was twelve works. Although some authors belong to groups of distinct colors, there is significant interaction among all, highlighting the high participation and co-authorship of the green, red, and yellow groups. The authors who stand out the most are Zhang Yuanhui (yellow), with 22 documents and 1170 citations; Xu Donghai (green), also with 22 publications and 1133 citations; Liu Zhidan (yellow), with 21 works and 1410 citations; and Chen Wei-Hsin (red), with 20 publications and 2265 citations, which may indicate the significant impact of their research and a possible interest in their work within the academic community. Among the authors mentioned above, it is possible to perceive that out of the four authors with the highest contribution, two are part of the yellow group, which evidences a significant research effort in the area.

3.5 *Journal Analysis*

Figure 5 presents the thirty journals that contributed the most to the proposed theme between 2014 and June 2024. Out of three hundred sources found, only thirty met the inclusion criteria of fifteen documents. The predominant colors on the map are red, green, and blue. In the red cluster, notable journals include "Bioresource Technology," with 106 documents and 7393 citations, and "Journal of Cleaner Production," with fifty-five papers and 1899 citations. The green cluster contains journals such as "Renewable and Sustainable Energy Reviews," with seventy-two publications and 12,152 citations, and "Green Chemistry," with thirty-nine articles and 2181 citations. In the blue cluster, journals like "Fuel" stand out with 113 publications and 3659 citations, and "Energy and Fuels" with seventy-five works and 558 citations. It is

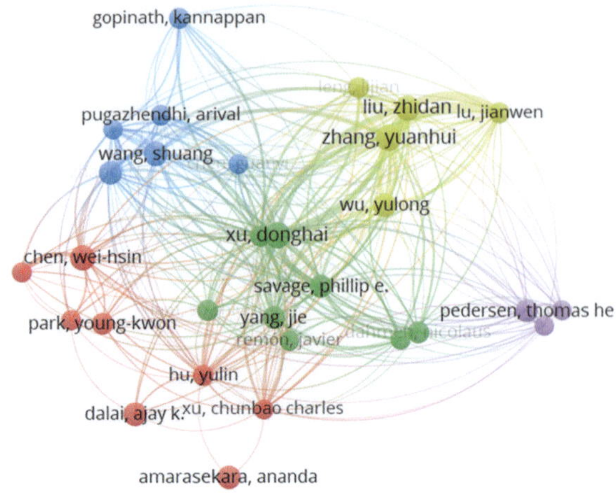

Fig. 4 Collaboration and co-authorship network map using twelve publications as a basis and involving thirty participating authors, with lines and colors indicating degrees of cooperation, spanning from 2014 to June 2024 using data extracted from the Scopus database, generated via VOSviewer

worth noting the high degree of cooperation among all clusters, highlighting the interconnectedness and intense collaboration among different journals and research areas.

Table 1 details the top ten journals that contributed the most from 2014 to June 2024, displaying the interest in the current theme. These journals were extracted from the Scopus database and represent 35.1% of the total publications. The United Kingdom leads the ranking with three journals, representing 12.9% of the contribution, and possesses the journal with the highest Impact Factor (IF). "Renewable Sustainable Energy Reviews" leads with an IF of 16.7, followed by "Energy Conversion and Management" with 12.1, and "Green Chemistry" with 10.6. Despite the UK journals having the highest IF, the most contributions come from the Netherlands, with the journal "Fuel" contributing 6.4%, totaling 113 works.

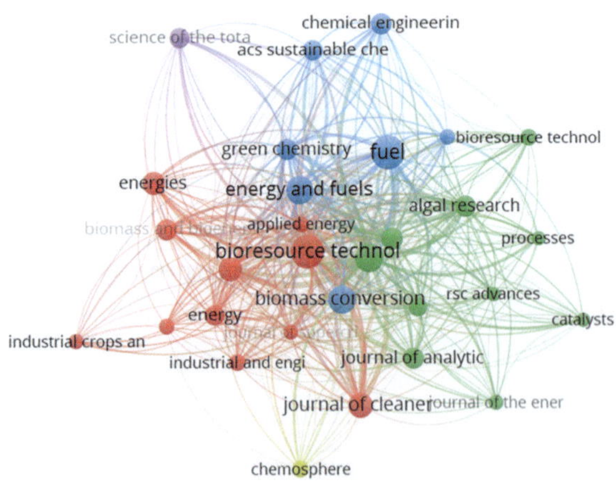

Fig. 5 Map of the network of newspapers and magazines using fifteen publications as a base and totalling thirty sources; lines and colors indicate degrees of cooperation

Table 1 Top ten scientific journals that published the most about refinery catalysts from 2014 to June 2024

Rank	Journal title	Count	Percentage (%)	IF	Quartile in category	H-index
1	Fuel	113	6.4	6.7	Q1	243
2	Bioresource Technology	106	5.9	11.2	Q1	312
3	Energy and Fuels	75	4.2	4.654	Q2	223
4	Renewable Sustainable Energy Reviews	72	4.0	16.799	Q1	420
5	Biomass Conversion and Biorefinery	71	4.0	4.452	Q2	44
6	Journal of Cleaner Production	55	3.0	11.072	Q1	245
7	Energy Conversion and Management	50	2.8	12.132	Q1	228
8	Energies	45	2.5	3.252	Q2	133
9	Renewable Energy	41	2.3	8.634	Q1	197
10	Green Chemistry	39	2.1	10.675	Q1	267

Data extracted from the Scopus database, generated via VOSviewer. IF: Impact Factor

4 Keywords Analysis

Table 2 highlights the classifications of the top ten most used keywords, identified from the Scopus database. Initially, 9954 words were found, but only twenty-four met the minimum repetition criterion of 150 times. A keyword analysis is crucial in bibliometrics as it allows visualizing knowledge networks, identifying relationships between concepts, and discerning emerging trends in the scientific literature [26]. In this context, the keywords with the highest occurrence are “Biomass” (775 citations), “Liquefaction” (713 citations), and “Hydrothermal” (634 citations). These words also have the highest Total Link Strength (TLS), with “Liquefaction” leading with a TLS of 3578, followed by “Biomass” with 3300, and “Hydrothermal” with 3202, demonstrating strong connections with other keywords and indicating their relevance and centrality in the proposed theme.

In Fig. 6, the frequency over the years in which keywords appear and are used is observed. The words “Biomass,” “Liquefaction,” and “Hydrothermal Liquefaction” stand out not only for their frequent use but also for co-occurrence, evidenced by the thickness of the lines connecting them. These words are often used together, reflecting the interrelation between these topics. The line colors are also indicative, showing that these words prominently appeared in the same specific period, dated

Table 2 Ranking table of the twenty-four most prominent keywords mentioned in the analyzed articles

Rank	Keyword	Occurrences	TLS	Rank	Keyword	Occurrences	TLS
1	Biomass	775	3300	**11**	Biorefinery	235	968
2	Liquefaction	713	3578	**12**	Catalysis	235	1295
3	Hydrothermal liquefaction	634	3202	**13**	Enzyme	235	1279
4	Biofuels	496	2652	**14**	Feedstocks	234	966
5	Pyrolysis	374	1483	**15**	Lignocellulosic	197	1348
6	Algae	286	1554	**16**	Temperature	185	1205
7	Micro-algae	269	1598	**17**	Lignin	180	719
8	Catalysts	265	1246	**18**	Article	179	1168
9	Bio-oil	249	1375	**19**	Carbon	174	799
10	Microorganisms	237	1520	**20**	Catalyst	168	1139

2021. Additionally, newer appearing words such as "Bio-oils" in 2022 and "Catalyst" in 2023 indicate current trends and focus on catalysts for biorefinery. On the other hand, words that appeared earlier, like "Micro-algae" in 2020 and "Algae" in 2018, show the evolution of interest and research in the area over time.

Figure 7 distinctly presents keywords, organizing them into three clusters and showing the frequency with which they appear in each group. The first cluster, marked in red, highlights words like "Biomass," "Pyrolysis," and "Catalysts." In the green cluster, we find terms like "Biofuels," "Biofuel," and "Algae." The last cluster, in blue, consists of only three words but with a high usage index, such as "Liquefaction," "Hydrothermal Liquefactions," and "Hydrothermal Liquefaction." Additionally, it is worth mentioning some words that span across all clusters, as mentioned above, indicating that despite being in diverse groups, they still have significant connections.

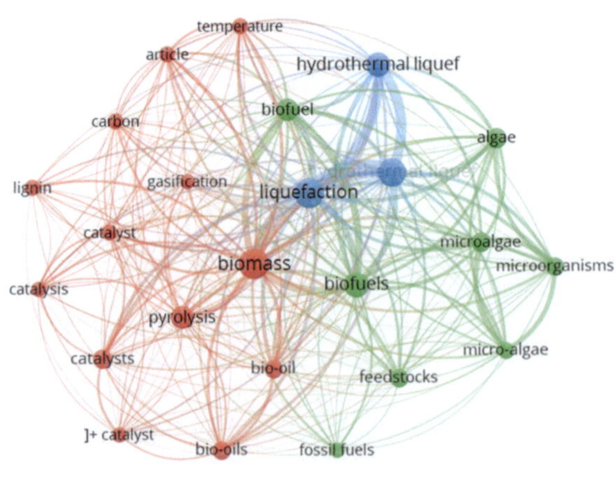

Fig. 6 Overlay visualization of the analysis of the twenty-four co-occurring keywords

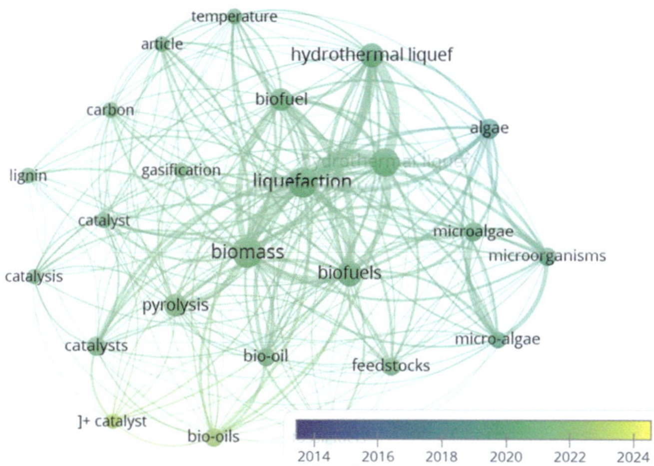

Fig. 7 Visualization network of the clustering of the twenty-four keywords related to co-occurrence analyses from 2014 to June 2024

This interconnection reinforces the importance of these terms within the analyzed context and suggests that they play a fundamental role in the addressed theme.

Figure 8 shows the density map of the twenty most frequent keywords and provides a visual representation of the interactions between these keywords. The color shading on the map reflects the strong thematic connections and interactions among these keywords, showing how they are interconnected. This pattern of interactions demonstrates how frequently these keywords repeat in scientific works in the field.

5 Advanced Analysis

5.1 Types of Catalysts

Considering the scenarios discussed so far, it is crucial to optimize and guarantee the efficiency of converting biomass, such as forest and agricultural residues, into biofuels and high-value chemicals in biocrude biorefineries [27–34]. This process not only promotes sustainability but also supports the circular economy by reducing energy consumption, a globally relevant issue that affects everything from the transportation sector to various industrial segments [27, 33, 35–37]. This is why catalysts play a key role not only in accelerating chemical reactions but also in facilitating recycling in multiple use cycles [38–43]. Following this approach aligns with the principles of a sustainable economy, where efficiency and the reduction of waste generation are essential to ensure responsible economic and environmental practices.

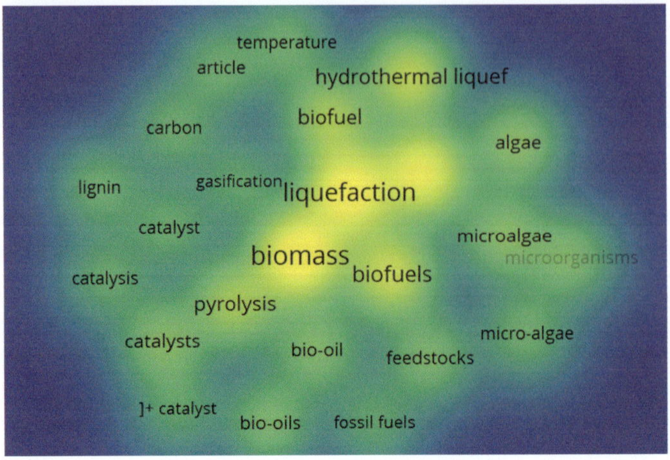

Fig. 8 The density map of the twenty most frequent keywords indicates strong thematic interactions generated via VOSviewer, with data extracted from the Scopus database

Below, we will analyze the two main categories of catalysts: homogeneous and heterogeneous. Our aim is to understand their specific characteristics, their different applications and the impact they have on biomass conversion processes.

High catalytic activity and selectivity in processes such as transesterification and esterification are essential for to production biodiesel from vegetable oils and animal fats, so as they have these two characteristics, homogeneous catalysts are widely used in these procedures [38, 40, 44, 45]. One particularity is that these catalysts are usually found in the same phase as the reactants, with the liquid phase being the most common [45–48]. They offer the advantage of operating under moderate temperature and pressure conditions [49–51], which not only reduces operating costs but also allows precise control of the composition of the product, essential for applications that require a high degree of purity [52–56]. According to analyses by Dias et al. [57], the transesterification of vegetable oils with homogeneous alkaline catalysts, such as NaOH and KOH, can achieve conversion rates of up to 97% with pure oils. However, the presence of high levels of free fatty acids (FFA) in waste oils can reduce this efficiency to around 92%, due to the formation of soap which complicates the separation of the products [57]. Thus, removing these catalysts after the reactions is a challenge due to their miscibility with the reaction medium, requiring complex separation and purification methods. This increases costs, prolongs processing time, and limits the useful life of the catalysts, as well as restricting their applicability on a large scale due to the need for specific reagents and the generation of waste that requires appropriate environmental treatment.

On the other hand, heterogeneous catalysts operate in a phase separate from the reactants, usually in the solid state, while the reactants are in the liquid or gaseous phases [58, 59]. These catalysts are highly valued in biocrude biorefineries because they facilitate simple separation and recovery after the reaction [58–60]. Some examples of these catalysts are materials such as zeolites [61, 62], supported transition metals [63, 64], and solid acids [65, 66]. Heterogeneous stimulants such as $Ba(OH)_2$ and CaO can achieve biodiesel yields of 92% and 84% [67], respectively, even when the feedstock is waste oil, due to their ability to manage high levels of free fatty acids without generating excessive soap. Furthermore, in the study by Abhishek Guldhe et al. [67], the chromium–aluminum mixed oxide catalyst (CRAL) showed a higher biodiesel conversion of 98.28% compared to the homogeneous catalyst (96.47%).

These catalysts offer large surface areas and adjustable properties, ideal for a variety of catalytic processes, such as cracking [68], isomerization [69], hydrogenation [70], and deoxygenation [71]. The ability of these catalysts to be reused in multiple reaction cycles not only improves economic efficiency by reducing the need for new catalysts but also minimizes the amount of waste generated and simplifies purification processes [72, 73]. For example, CaO derived from eggshells has demonstrated excellent catalytic activity and sustainability [74]. However, the effectiveness of these catalysts can be affected by impurities in the biomass, requiring advanced synthesis techniques and treatments to maximize their performance.

Determining the appropriate catalysts for biocrude biorefineries is a multifaceted task that involves analyzing the biomass source, understanding the specific reactions, and considering the economic and environmental aspects of the project [59]. While homogeneous catalysts provide precise control and selectivity in reactions, heterogeneous catalysts are more commonly used due to their environmental benefits and positive impact on the circular economy.

Advancements in catalytic materials and process optimization are key to enhancing both the technical performance and economic feasibility of biocrude biorefineries. Given the global push toward energy transition, it's essential to keep investing in the research and development of catalysts. Such investment will lead to innovative, environmentally friendly ways to use biomass more efficiently, significantly contributing to a more sustainable energy future, in addition to generating new possibilities to produce biofuels and other chemical products from biological resources.

5.2 Catalyst Development

Currently, there are several catalysts used in biorefineries, the main ones being: zeolites [75], metal oxides [76], bifunctional catalysts [77], and transition metals [78]. In the latter category, typical catalysts are low molecular weight acidic or basic compounds and transition metal complexes used in solutions. Additionally, catalysis [79] can improve yields and conversions through the addition of solid catalysts, providing a reliable and efficient alternative to promote the use of biomass instead of non-renewable resources [24].

It is noticeable that all the analyses conducted assume the development of catalysts. The focus is on the development of more effective and long-lasting catalysts, aiming at greater resistance to deactivation [80]. In a futuristic development, catalytic processes are being implemented to improve waste streams, minimizing the environmental impact of biorefineries while simultaneously increasing economic performance with platform molecules, products derived from catalytic processes to enhance the sustainability of biorefineries and reduce dependence on crude oil [24].

It is already possible to affirm that zeolites are one of the most important catalysts in the chemical and petrochemical industry, facilitating the development of quantitative synthesis–structure–function relationships for zeolites that incorporate the effects of arrangement [81]. The potential of this increasingly important class of materials can still be widely explored. Zeolites and other microporous molecular sieves with analogous structures can be modified by methods that include ion exchange [81], isomorphic substitution during crystallization or after the synthesis of the material [82]. Besides the appropriate pore size, the use of a given zeolite in catalysis should also be evaluated by the affinity between the polarity of the solvent, reagent, and product with the chemical composition of the microporous material. Zeolites are efficient catalysts because the forced proximity between reagent molecules under the influence of the strong electrostatic potentials existing within the channels and cavities lowers the activation energy required for the phenomenon of catalysis [82].

The leaching of the redox metal is the main obstacle in the use of redox sieves as catalysts for liquid-phase hydrocarbon oxidation [82]. The results showed that bifunctional catalysts are also of paramount importance for use in biorefineries, given that they can perform simultaneous reactions [77]. The prospects for the use of catalysts in biorefineries seem promising and encompass various areas of research and development. The development of more efficient and durable catalysts will continue to be a priority, focusing on increasing resistance to deactivation, essential to improving process efficiency and reducing operational costs [80] However, the number of research studies must continue rising, as shown in Fig. 1, the increase in studies in this specific area can potentially enhance and expand the variety of publications addressing the proposed topic further.

5.3 Future Innovations

Conventional methods and technologies used in biorefineries present various challenges, such as the search for more efficient catalysts, high energy consumption for recovery after the reaction, high costs and difficulties in the availability of raw materials, and the lifespan of these catalysts [83, 84]. However, for the implementation of new biorefinery projects, many studies have been conducted considering the role of more selective green catalysts capable of directing with greater efficiency, yield, and purity toward specific products, as well as the use of advanced characterization techniques and computational modeling to better understand reaction mechanisms and optimize catalyst design [85, 86].

Huirong Li et al., claim that the unique properties of core–shell structures are promising for the use of selective catalytic NO_x reduction technology in conjunction with ammonia (NH3-SCR) [87]. Furthermore, Guo et al. highlight that in the hydrogenation of acetylene, catalysts drive the reaction toward the formation of target products, thus preventing the occurrence of undesired reactions [88]. Catalytic hydrogenation occurs under different parameters and using different catalysts. To explore the theory of quantum mechanics with the Monte Carlo method, Che et al. investigated the catalytic mechanism of the acetylene hydrogenation reaction, identifying that the increased silver content of the catalyst leads to increased ethylene desorption on the catalyst surface, which culminates in the inhibition of ethylene formation. Finally, this inhibition facilitates the hydrogenation of acetylene [89].

A strong emerging trend for catalytic reactions is the use of artificial intelligence and other types of computational modeling to identify important information about the catalytic site. Techniques, such as XANES (X-ray Absorption Near Edge Structure), used to define the electronic properties and local bonding environment of catalysts, and others like EXAFS, used to reflect bond lengths and types of atoms around it, as well as electronic structure calculations, have been widely used for the development of the structure–activity relationship to gather crucial information about the active site of the catalyst and disclose the physical and chemical [90–94]. Moreover, studies such as Ahn et al. [95] and others provide insights for the development of new technologies to be used in catalytic reactions, as well as reflect the current landscape and provide futuristic predictions of advanced techniques of reaction mechanisms [95–99].

6 Final Considerations

Regarding the origin of the journals that deal with catalysts applied in biorefineries, a notable diversity of publications on the topic was observed. The variety of journals suggests a high applicability for these catalysts, since each journal has distinct areas of interest. Another highlight is cooperation between countries. The analysis reveals that collaboration between organizations within each country has generated promising academic results. Furthermore, there is significant cooperation between nations, where borders have become bridges for the exchange of information and knowledge. This exchange of information, research, and technology is crucial for developing countries, especially in the field of catalysts applied to biorefineries.

The use of catalysts is an important alternative to develop sustainable methods and solve problems related to inadequate waste disposal. Therefore, it can be stated that the use of catalysts in biorefineries represents an ecologically correct approach, aiming to make processes more sustainable.

Acknowledgements The authors would like to express their gratitude for the support that was provided through the Instituto de Engenharias e Desenvolvimento Sustentável (IEDS) at the Universidade da Integração Internacional da Lusofonia Afro-Brasileira (UNILAB). We gratefully acknowledge the following Brazilian Agencies for Scientific and Technological Development: Fundação Cearense de Apoio ao Desenvolvimento Científico e Tecnológico (FUNCAP) (PS1-0186-00216.01.00/21; 31052.000417/2023; UNI-0210-00537.01.00/23), Conselho Nacional de Desenvolvimento Científico e Tecnológico (CNPq) (311062/2019-9; 440891/2020-5; 307454/2022-3), and Coordenação de Aperfeiçoamento de Ensino Superior (CAPES) (finance code 001).

Declaration of Competing Interest
The authors declare that they have no known competing financial interests or personal relationships that could have appeared to influence the work reported in this paper.

Data Availability Data will be made available on request.

References

1. Hou Q, Qi X, Zhen M et al (2021) Biorefinery roadmap based on catalytic production and upgrading 5-hydroxymethylfurfural. Green Chem 23:119–231. https://doi.org/10.1039/D0GC02770G
2. Chen S, Wojcieszak R, Dumeignil F et al (2018) How catalysts and experimental conditions determine the selective hydroconversion of furfural and 5-hydroxymethylfurfural. Chem Rev 118:11023–11117. https://doi.org/10.1021/acs.chemrev.8b00134
3. Ramos R, Abdelkader-Fernández VK, Matos R et al (2022) Metal-supported biochar catalysts for sustainable biorefinery, electrocatalysis, and energy storage applications: a review. Catalysts 12:207. https://doi.org/10.3390/catal12020207
4. Narayan R (1992) Biomass (renewable) resources for production of materials, chemicals, and fuels. pp 1–10
5. Somorjai GA, Kliewer CJ (2009) Reaction selectivity in heterogeneous catalysis. React Kinet Catal Lett 96:191–208. https://doi.org/10.1007/s11144-009-5531-7
6. Iemhoff A, Vennewald M, Palkovits R (2023) Single-atom catalysts on covalent triazine frameworks: at the crossroad between homogeneous and heterogeneous catalysis. Angew Chem Int Ed 62.https://doi.org/10.1002/anie.202212015
7. Ahorsu R, Constanti M, Medina F (2021) Recent impacts of heterogeneous catalysis in biorefineries. Ind Eng Chem Res 60:18612–18626. https://doi.org/10.1021/acs.iecr.1c02789
8. Karuppasamy K, Theerthagiri J, Selvaraj A et al (2023) Current trends and prospects in catalytic upgrading of lignocellulosic biomass feedstock into ultrapure biofuels. Environ Res 226:115660. https://doi.org/10.1016/j.envres.2023.115660
9. Khemthong P, Yimsukanan C, Narkkun T et al (2021) Advances in catalytic production of value-added biochemicals and biofuels via furfural platform derived lignocellulosic biomass. Biomass Bioenergy 148:106033. https://doi.org/10.1016/j.biombioe.2021.106033
10. Deng F, Amarasekara AS (2021) Catalytic upgrading of biomass derived furans. Ind Crops Prod 159:113055. https://doi.org/10.1016/j.indcrop.2020.113055
11. Yan P, Wang H, Liao Y, Wang C (2023) Zeolite catalysts for the valorization of biomass into platform compounds and biochemicals/biofuels: a review. Renew Sustain Energy Rev 178:113219. https://doi.org/10.1016/j.rser.2023.113219
12. Poveda-Giraldo JA, Solarte-Toro JC, Cardona Alzate CA (2021) The potential use of lignin as a platform product in biorefineries: a review. Renew Sustain Energy Rev 138:110688. https://doi.org/10.1016/j.rser.2020.110688
13. Alonso DM, Bond JQ, Dumesic JA (2010) Catalytic conversion of biomass to biofuels. Green Chem 12:1493. https://doi.org/10.1039/c004654j
14. Shahid MK, Batool A, Kashif A et al (2021) Biofuels and biorefineries: development, application and future perspectives emphasizing the environmental and economic aspects. J Environ Manage 297:113268. https://doi.org/10.1016/j.jenvman.2021.113268

15. Esteban J, Ladero M (2018) Food waste as a source of value-added chemicals and materials: a biorefinery perspective. Int J Food Sci Technol 53:1095–1108. https://doi.org/10.1111/ijfs.13726
16. De Buck V, Polanska M, Van Impe J (2020) Modeling biowaste biorefineries: a review. Front Sustain Food Syst 4.https://doi.org/10.3389/fsufs.2020.00011
17. Esposito D, Antonietti M (2015) Redefining biorefinery: the search for unconventional building blocks for materials. Chem Soc Rev 44.5821–5835. https://doi.org/10.1039/C4CS00368C
18. Palmeros Parada M, Osseweijer P, Posada Duque JA (2017) Sustainable biorefineries, an analysis of practices for incorporating sustainability in biorefinery design. Ind Crops Prod 106:105–123. https://doi.org/10.1016/j.indcrop.2016.08.052
19. Parajuli R, Dalgaard T, Jørgensen U et al (2015) Biorefining in the prevailing energy and materials crisis: a review of sustainable pathways for biorefinery value chains and sustainability assessment methodologies. Renew Sustain Energy Rev 43:244–263. https://doi.org/10.1016/j.rser.2014.11.041
20. Guo M, Song W, Buhain J (2015) Bioenergy and biofuels: history, status, and perspective. Renew Sustain Energy Rev 42:712–725. https://doi.org/10.1016/j.rser.2014.10.013
21. Kumar B, Verma P (2021) Biomass-based biorefineries: an important architype towards a circular economy. Fuel 288:119622. https://doi.org/10.1016/j.fuel.2020.119622
22. Shah AA, Sharma K, Haider MS et al (2022) The role of catalysts in biomass hydrothermal liquefaction and biocrude upgrading. Processes 10:207. https://doi.org/10.3390/pr10020207
23. Mombeshora ET, Stark A (2023) Graphene oxide applications in biorefinery catalysis to chemical commodities: critical review, prospects and challenges. Biomass Convers Biorefin 13:4619–4638. https://doi.org/10.1007/s13399-021-01499-6
24. Solarte-Toro JC, Ortiz-Sanchez M, Inocencio-García P-J, Cardona Alzate CA (2023) Sustainable biorefineries based on catalytic biomass conversion: a review. Catalysts 13:902. https://doi.org/10.3390/catal13050902
25. Simão Neto F, Fernandes de Melo Neta MM, Sales MB et al (2023) Research progress and trends on utilization of lignocellulosic residues as supports for enzyme immobilization via advanced bibliometric analysis. Polymers 15:2057. https://doi.org/10.3390/POLYM15092057
26. Cavalcante IO, Simão Neto F, da Silva Sousa P et al (2024) Evolving sustainable energy technologies and assessments through global research networks: advancing the role of blue hydrogen for a cleaner future. RSC Sustainability 2:348–368. https://doi.org/10.1039/D3SU00309D
27. Dessbesell L, Xu C, Pulkki R et al (2017) Forest biomass supply chain optimization for a biorefinery aiming to produce high-value bio-based materials and chemicals from lignin and forestry residues: a review of literature. Can J For Res 47:277–288. https://doi.org/10.1139/cjfr-2016-0336
28. Klein BC, Scheidemantle B, Hanes RJ et al (2023) Economics and global warming potential of a commercial-scale delignifying biorefinery based on co-solvent enhanced lignocellulosic fractionation to produce alcohols, sustainable aviation fuels, and co-products from biomass. Energy Environ Sci 17:1202–1215. https://doi.org/10.1039/d3ee02532b
29. Gielen D, Boshell F, Saygin D et al (2019) The role of renewable energy in the global energy transformation. Energ Strat Rev 24:38–50. https://doi.org/10.1016/j.esr.2019.01.006
30. Elliott DC, Biller P, Ross AB et al (2015) Hydrothermal liquefaction of biomass: developments from batch to continuous process. Bioresour Technol 178:147–156. https://doi.org/10.1016/j.biortech.2014.09.132
31. Kim S, Kwon EE, Kim YT et al (2019) Recent advances in hydrodeoxygenation of biomass-derived oxygenates over heterogeneous catalysts. Green Chem 21:3715–3743. https://doi.org/10.1039/c9gc01210a
32. Cho EJ, Trinh LTP, Song Y et al (2020) Bioconversion of biomass waste into high value chemicals. Bioresour Technol 298:122386. https://doi.org/10.1016/j.biortech.2019.122386
33. Ginni G, Kavitha S, Yukesh Kannah R et al (2021) Valorization of agricultural residues: different biorefinery routes. J Environ Chem Eng 9:105435. https://doi.org/10.1016/j.jece.2021.105435
34. Zaky AS (2021) Introducing a marine biorefinery system for the integrated. Processes 9:1841

35. Hailemariam A, Erdiaw-Kwasie MO (2023) Towards a circular economy: implications for emission reduction and environmental sustainability. Bus Strategy Environ 32:1951–1965. https://doi.org/10.1002/bse.3229
36. Tiwari S, Si Mohammed K, Mentel G et al (2024) Role of circular economy, energy transition, environmental policy stringency, and supply chain pressure on CO_2 emissions in emerging economies. Geosci Front 15:101682. https://doi.org/10.1016/j.gsf.2023.101682
37. Yang M, Chen L, Wang J et al (2023) Circular economy strategies for combating climate change and other environmental issues. Environ Chem Lett 21:55–80. https://doi.org/10.1007/s10311-022-01499-6
38. Guczi L, Erdöhelyi A (2011) Catalysis for alternative energy generation. Catal Altern Energy Gener 9781461403:1–536. https://doi.org/10.1007/978-1-4614-0344-9
39. Feliczak-Guzik A (2018) Hierarchical zeolites: synthesis and catalytic properties. Microporous Mesoporous Mater 259:33–45. https://doi.org/10.1016/j.micromeso.2017.09.030
40. Bender TA, Dabrowski JA, Gagné MR (2018) Homogeneous catalysis for the production of low-volume, high-value chemicals from biomass. Nat Rev Chem 2:35–46. https://doi.org/10.1038/s41570-018-0005-y
41. Lin L, Han X, Han B, Yang S (2021) Emerging heterogeneous catalysts for biomass conversion: studies of the reaction mechanism. Chem Soc Rev 50:11270–11292. https://doi.org/10.1039/d1cs00039j
42. De S, Dutta S, Saha B (2016) Critical design of heterogeneous catalysts for biomass valorization: current thrust and emerging prospects. Catal Sci Technol 6:7364–7385. https://doi.org/10.1039/c6cy01370h
43. Kim S, Tsang YF, Kwon EE et al (2019) Recently developed methods to enhance stability of heterogeneous catalysts for conversion of biomass-derived feedstocks. Korean J Chem Eng 36:1–11. https://doi.org/10.1007/s11814-018-0174-x
44. Rizwanul Fattah IM, Ong HC, Mahlia TMI et al (2020) State of the art of catalysts for biodiesel production. Front Energy Res 8:1–17. https://doi.org/10.3389/fenrg.2020.00101
45. Khatib SJ, Oyama ST (2015) Direct oxidation of propylene to propylene oxide with molecular oxygen: a review. Catal Rev Sci Eng 57:306–344. https://doi.org/10.1080/01614940.2015.1041849
46. Miao W, Tak HC (2006) Ionic-liquid-supported synthesis: a novel liquid-phase strategy for organic synthesis. Acc Chem Res 39:897–908. https://doi.org/10.1021/ar030252f
47. Hugl H, Nobis M (2008) Multiphase catalysis in industry. Top Organomet Chem 23:1–17. https://doi.org/10.1007/3418_044
48. Zhang X, Tang Z (2020) Molecular nitrogen as an extraordinary accelerator for hydrodeoxygenation. Chem 6:16–18. https://doi.org/10.1016/j.chempr.2019.12.018
49. Natongchai W, Posada-Pérez S, Phungpanya C et al (2022) Enhancing the catalytic performance of group I, II metal halides in the cycloaddition of CO_2 to epoxides under atmospheric conditions by cooperation with homogeneous and heterogeneous highly nucleophilic aminopyridines: experimental and theoretical study. J Org Chem 87:2873–2886. https://doi.org/10.1021/acs.joc.1c02770
50. Jagtap SA, Bhanage BM (2018) Rhodium/Phosphine catalysed selective hydroformylation of biorenewable olefins. Appl Organomet Chem 32:1–9. https://doi.org/10.1002/aoc.4478
51. Wang Y (2018) Room-temperature conversion of methane becomes true. Joule 2:1399–1401. https://doi.org/10.1016/j.joule.2018.07.013
52. Malabadi RB, Sadiya MR, Kolkar KP, Chalannavar RK (2023) Biodiesel production via transesterification reaction. Open Access Res J Sci Technol 9:010–020. https://doi.org/10.53022/oarjst.2023.9.2.0064
53. Merakeb L, Robert M (2021) Advances in molecular electrochemical activation of dinitrogen. Curr Opin Electrochem 29:100834. https://doi.org/10.1016/j.coelec.2021.100834
54. Huang J, Zhang F, Li H (2012) Organometal-bridged PMOs as efficiency and reusable bifunctional catalysts in one-pot cascade reactions. Appl Catal A Gen 431–432:95–103. https://doi.org/10.1016/j.apcata.2012.04.021

55. Smith DJ (2018) The past, present, and future of sustainable chemistry. Chemsuschem 11:5–10. https://doi.org/10.1002/cssc.201702329
56. Seifert AI, Wegener HW, Brühl K et al (2023) Polymer-grade bio-monomers from oleochemicals by combining homogeneous catalysis and selective product crystallization in an integrated process. Processes 11.https://doi.org/10.3390/pr11102861
57. Dias JM, Alvim-Ferraz MCM, Almeida MF (2008) Comparison of the performance of different homogeneous alkali catalysts during transesterification of waste and virgin oils and evaluation of biodiesel quality. Fuel 87:3572–3578. https://doi.org/10.1016/j.fuel.2008.06.014
58. Mueller JJ, Baum S, Hilterhaus L et al (2011) Simultaneous determination of mono-, di-, and triglycerides in multiphase systems by online Fourier transform infrared spectroscopy. 83:9321–9327. https://doi.org/10.1021/ac2018662
59. Engineering M (2015) Review of the production of biodiesel from waste cooking. J Mech Eng Sci (JMES) 8:1302–1311. https://doi.org/10.15282/jmes.8.2015.5.0127 REVIEW
60. Mandari V, Devarai SK (2022) Biodiesel production using homogeneous, heterogeneous, and enzyme catalysts via transesterification and esterification reactions: a critical review. Bioenergy Res 15:935–961. https://doi.org/10.1007/s12155-021-10333-w
61. Shabani JM, Ameh AE, Oyekola O et al (2022) Fusion-assisted hydrothermal synthesis and post-synthesis modification of mesoporous hydroxy sodalite zeolite prepared from waste coal fly ash for biodiesel production. Catalysts 12.https://doi.org/10.3390/catal12121652
62. Liu Y, Liu Z, Hui Y et al (2023) Rhodium nanoparticles supported on silanol-rich zeolites beyond the homogeneous Wilkinson's catalyst for hydroformylation of olefins. Nat Commun 14.https://doi.org/10.1038/s41467-023-38181-6
63. Hermes ED, Jenness GR, Schmidt JR (2015) Decoupling the electronic, geometric and interfacial contributions to support effects in heterogeneous catalysis. Mol Simul 41:123–133. https://doi.org/10.1080/08927022.2014.926549
64. Huang L, Han B, Xi Y et al (2015) Influence of charge on the reactivity of supported heterogeneous transition metal catalysts. ACS Catal 5:4592–4597. https://doi.org/10.1021/acscatal.5b00689
65. Kolvari E, Koukabi N, Hosseini MM (2015) Perlite: a cheap natural support for immobilization of sulfonic acid as a heterogeneous solid acid catalyst for the heterocyclic multicomponent reaction. J Mol Catal A Chem 397:68–75. https://doi.org/10.1016/j.molcata.2014.10.026
66. Guldhe A, Moura CVR, Singh P et al (2017) Conversion of microalgal lipids to biodiesel using chromium-aluminum mixed oxide as a heterogeneous solid acid catalyst. Renew Energy 105:175–182. https://doi.org/10.1016/j.renene.2016.12.053
67. Miyuranga KAV, Arachchige USPR, Marso TMM, Samarakoon G (2023) Biodiesel production through the transesterification of waste cooking oil over typical heterogeneous base or acid catalysts. Catalysts 13.https://doi.org/10.3390/catal13030546
68. Perras FA, Wang Z, Naik P et al (2017) Natural abundance 17 O DNP NMR provides precise O−H distances and insights into the Brønsted acidity of heterogeneous catalysts. Angew Chem 129:9293–9297. https://doi.org/10.1002/ange.201704032
69. Bayu A, Abudula A, Guan G (2019) Reaction pathways and selectivity in chemo-catalytic conversion of biomass-derived carbohydrates to high-value chemicals: a review. Fuel Process Technol 196:106162. https://doi.org/10.1016/j.fuproc.2019.106162
70. Strekalova AA, Shesterkina AA, Kustov AL, Kustov LM (2023) Recent studies on the application of microwave-assisted method for the preparation of heterogeneous catalysts and catalytic hydrogenation processes. Int J Mol Sci 24.https://doi.org/10.3390/ijms24098272
71. Fu J, Mei D (2021) A theoretical study of propionic acid decarboxylation over hydroxyapatite supported platinum catalysts. Catal Today 365:181–192. https://doi.org/10.1016/j.cattod.2020.05.014
72. Rabie AM, Shaban M, Abukhadra MR et al (2019) Diatomite supported by CaO/MgO nanocomposite as heterogeneous catalyst for biodiesel production from waste cooking oil. J Mol Liq 279:224–231. https://doi.org/10.1016/j.molliq.2019.01.096
73. Zhang HJ, Da QS, Niu XY et al (2014) Metallic nanoparticles immobilized in magnetic metal-organic frameworks: preparation and application as highly active, magnetically isolable and reusable catalysts. Catal Sci Technol 4:3013–3024. https://doi.org/10.1039/c4cy00072b

74. Erchamo YS, Mamo TT, Workneh GA, Mekonnen YS (2021) Improved biodiesel production from waste cooking oil with mixed methanol–ethanol using enhanced eggshell-derived CaO nano-catalyst. Sci Rep 11:1–12. https://doi.org/10.1038/s41598-021-86062-z
75. Rahman MdM, Liu R, Cai J (2018) Catalytic fast pyrolysis of biomass over zeolites for high quality bio-oil—a review. Fuel Process Technol 180:32–46. https://doi.org/10.1016/j.fuproc.2018.08.002
76. Oliveira LCA, Fabris JD, Pereira MC (2013) Óxidos de ferro e suas aplicações em processos catalíticos: uma revisão. Quim Nova 36:123–130. https://doi.org/10.1590/S0100-40422013000100022
77. Ramos LP, Brugnago RJ, da Silva FR et al (2014) Simultaneous esterification and transesterification of acid oils using layered zinc carboxylates as bifunctional catalysts. Quim Nova. https://doi.org/10.5935/0100-4042.20140274
78. Bernardo-Gusmão K, Pergher SBC, dos Santos EN (2017) Um panorama da Catálise no Brasil nos últimos 40 anos. Quim Nova. https://doi.org/10.21577/0100-4042.20170083
79. Ferreira L, Ribeiro R, Madriaga V et al (2021) Uma Breve Revisão Sobre A Catálise Por Átomos Isolados: Conceitos E Aplicações. Quim Nova. https://doi.org/10.21577/0100-4042.20170822
80. Lombardo EA, Cornaglia C, Múnera J (2016) Development of an active, selective and durable water-gas shift catalyst for use in membrane reactors. Catal Today 259:165–176. https://doi.org/10.1016/j.cattod.2015.06.015
81. Cordeiro CS, da Silva FR, Wypych F, Ramos LP (2011) Catalisadores heterogêneos para a produção de monoésteres graxos (biodiesel). Quim Nova 34:477–486. https://doi.org/10.1590/S0100-40422011000300021
82. Luna FJ, Schuchardt U (2001) Modificação de zeólitas para uso em catálise. Quim Nova 24:885–892. https://doi.org/10.1590/S0100-40422001000600027
83. He C, Gong Y, Li S et al (2024) Single-atom alloys materials for CO_2 and CH_4 catalytic conversion. Adv Mater 36
84. Norouzi O, Taghavi S, Arku P et al (2021) What is the best catalyst for biomass pyrolysis? J Anal Appl Pyrolysis 158
85. Sun K, Zhang Z, Shen C et al (2022) The feasibility study of the indium oxide supported silver catalyst for selective hydrogenation of CO_2 to methanol. Green Energy Environ 7.https://doi.org/10.1016/j.gee.2021.05.004
86. Martínez Tejada LM, Munõz A, Centeno MA, Odriozola JA (2016) In-situ Raman spectroscopy study of Ru/TiO_2 catalyst in the selective methanation of CO. J Raman Spectro 47.https://doi.org/10.1002/jrs.4774
87. Li H, Schill L, Fehrmann R, Riisager A (2023) Recent developments of core-shell structured catalysts for the selective catalytic reduction of NO_x with ammonia. Inorg Chem Front 10
88. Guo J, Lei Y, Liu H et al (2023) Recent research advances on catalysts for selective hydrogenation of ethyne. Catal Sci Technol 13
89. Che C, Gou G, Wen H et al (2021) Study on the reaction mechanism of acetylene selective hydrogenation catalysts Pd-Ag/Al_2O_3. Inorg Nano-Metal Chem 51.https://doi.org/10.1080/24701556.2020.1762217
90. Grazianetti C, Martella C, Molle A (2020) The Xenes generations: a taxonomy of epitaxial single-element 2D materials. Physica Status Solidi—Rapid Res Lett 14
91. Hirose M, Shimomura K, Higashino T et al (2020) Nanoscale determination of interatomic distance by ptychography-EXAFS method using advanced Kirkpatrick-Baez mirror focusing optics. J Synchrotron Radiat 27.https://doi.org/10.1107/S1600577519017004
92. Jenner AL, Aogo RA, Davis CL et al (2020) Leveraging computational modeling to understand infectious diseases. Curr Pathobiol Rep 8
93. Tropsha A, Isayev O, Varnek A et al (2024) Integrating QSAR modelling and deep learning in drug discovery: the emergence of deep QSAR. Nat Rev Drug Discov 23.https://doi.org/10.1038/s41573-023-00832-0
94. Zhang H, Sun S, Shang X et al (2022) Ultrafast photonics applications of emerging 2D-Xenes beyond graphene. Nanophotonics 11

95. Ahn S, Hong M, Sundararajan M et al (2019) Design and optimization of catalysts based on mechanistic insights derived from quantum chemical reaction modeling. Chem Rev 119
96. Chen H, Simoska O, Lim K et al (2020) Fundamentals, applications, and future directions of bioelectrocatalysis. Chem Rev 120
97. Lopes TF, Cerone N, Contuzzi L et al (2020) Análise Do Impacto Do Processo Organosolv Para Pré-Tratamento De Biomassa Lenhocelulósica Na Redução Das Emissões De Gases Com Efeito De Estufa (GEE). In: CIES2020. As Energias Renováveis na Transição Energética: Livro de Comunicações do XVII Congresso Ibérico e XIII Congresso Ibero-americano de Energia Solar
98. Morales-García Á, Viñes F, Gomes JRB, Illas F (2021) Concepts, models, and methods in computational heterogeneous catalysis illustrated through CO_2 conversion. Wiley Interdiscip Rev Comput Mol Sci 11
99. Motagamwala AH, Dumesic JA (2021) Microkinetic modeling: a tool for rational catalyst design. Chem Rev 121

Chapter 10
Biocrude Oil Through Hydrothermal and Pyrolysis Processes: A Comparative Assessment of Properties and Downstream Upgradation

Priyanka Padhi, Nilutpal Bhuyan, Neelam Bora, Anuron Deka, Mondita Athparia, Bidisha Borah, and Rupam Kataki

Abstract The cleaner and technologically feasible alternative to fossil fuels is biocrude oil, a liquid fuel derived from renewable biomass. Its production accounts for the sustainable solution against the increasing energy demand and greenhouse gas emission. Technologies developed to produce renewable energy from biomass include: physical, namely drying, pressing, crushing, and palletization; biochemical, including fermentation and anaerobic digestion; and thermochemical, which include pyrolysis, gasification, liquefaction, and combustion pathways. Of these, the thermochemical pathways have gained much attention since they allow for high-energy output and efficiency. Thermochemical processes, especially pyrolysis and hydrothermal liquefaction, show significant potential in the conversion of biomass into biocrude. Pyrolysis is the thermal decomposition of organic material in the absence of oxygen that produces bio-oil, biochar, and syngas. In contrast, the hydrothermal liquefaction transforms the biomass into a liquid product in an aqueous environment under high temperature and pressure. Furthermore, pyrolysis produces a higher yield of bio-oil, exhibiting lower heating value and stability due to its high oxygen content, whereas hydrothermal liquefaction produces biocrude with lesser oxygen content and heating value, thus more similar to conventional liquid fuels. The process, however is highly energy-intensive and, also requires downstream upgradation to meet fuel quality standards. This chapter reflects the current state of research and development with emphasis on important issues, opportunities, recent technological developments, and techno-economic feasibility of biomass pyrolysis and hydrothermal liquefaction to produce biocrude oil.

P. Padhi · N. Bhuyan · N. Bora · A. Deka · M. Athparia · B. Borah · R. Kataki (✉)
Biofuel Laboratory, Department of Energy, Tezpur University, Tezpur, Assam, India
e-mail: rupam@tezu.ernet.in

N. Bhuyan
Department of Chemistry, D.C.B. Girl, College, Jorhat, Assam, India

M. Aslam et al. (eds.), *Production and Biorefining of Biocrude Oil: Current Status and Future Developments*, Advances in Sustainability Science and Technology,
https://doi.org/10.1007/978-981-96-5198-6_10

Graphical Abstract

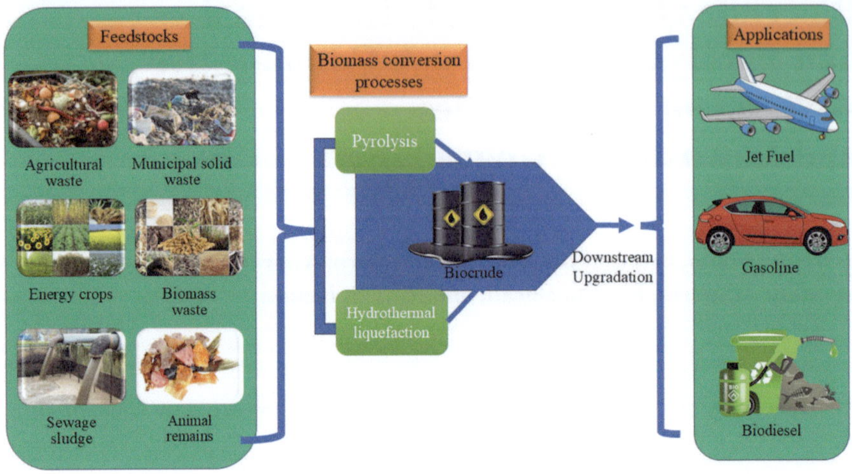

Keywords Biocrude · Hydrothermal liquefaction · Pyrolysis · Feedstock · Biomass · Conversion

1 Introduction

Biocrude oil from biomass has emerged as an important source of renewable fuel that can replace fossil fuels [1]. Biocrude oil is a liquid fuel produced from biomass through several forms of thermochemical processing, such as pyrolysis or hydrothermal liquefaction (HTL). It can be a potential alternative to traditional fossil fuels with several environmental and economic advantages. Biocrude oil has the potential for the reduction of the energy reliance on non-renewable sources and eliminates greenhouse gases [2]. As a versatile feedstock, biocrude oil could be further processed to include various transportation fuels, chemicals, and high-value products [3]. Besides these advantages, the production could alleviate many problems associated with the waste management issue, especially in using organic waste material along with agricultural residues. Moreover, the production of biocrude oil gives new market sources for rural economies and generates additional employment opportunities in the area of bioenergy [4]. Source diversification through advancement in biocrude oil technology will increase energy security while reliance on imported petroleum is reduced [5], further supporting the global drive toward a sustainable, circular economy. However, challenges persist concerning the effectiveness of the production processes of biocrude oil, reducing the costs, and optimizing its quality to achieve the standards of the industry.

Two dominant methods for converting biomass into useful energy feedstock are compared in this chapter, HTL and pyrolysis (Table 1). Both of these methods of technologies produce biocrudes of varying properties, therefore dictating the needs for subsequent upgrading. Although the quality of biocrude obtained from HTL is higher, due to lower oxygen content and higher energy density [6], pyrolysis provides process flexibility and can accept various input materials [7]. Biocrude oil quality is further enhanced through downstream upgradation by catalytic and refining techniques [8]. These include deoxygenation, hydrogenation, and cracking reactions that minimize oxygen levels while maximizing the hydrogen-to-carbon ratio and the maximization of molecular weight distributions [9]. Catalytic hydrotreating is used primarily for the removal of heteroatoms and improvement in stability [10]. The choice between HTL and pyrolysis would depend on type of feedstock, desired product, and economic considerations [11]. Both have scaling issues and are not yet cost-competitive with conventional fossil fuels. Additional research is being conducted toward increasing process efficiency, development of catalysts, and integration into existing refinery infrastructure to make biocrude oil more commercially viable [8]. HTL is a more energy-efficient process since it accepts wet biomass, thus enabling lower overall emissions of greenhouse gases [12]. The apparent advantage of pyrolysis over HTL is with regard to carbon sequestration through the production of biochar [13]. This chapter describes properties of biocrude oils prepared both by HTL and pyrolysis to assess the respective merits and demerits along with assessing the necessity for further downstream upgrading. It details characteristics of the feedstock and processing parameters that may affect the quality and yield of the biocrude oil. It covers the chemical and physical properties of biocrudes from HTL and pyrolysis, including viscosity, heating values, and molecular weight distribution. The paper also assesses the effectiveness of fuel upgrading techniques such as hydrotreating, hydrocracking, and isomerization. It presents the possibility of making transportation fuels and chemicals as a co-product from downstream processing of biocrude. Further discussion is devoted to economic and environmental considerations for both processes, aiming at identification of possible areas for improvement and future research directions. Finally, the life cycle impacts, as well as potential market applications, of biocrude oils from different feedstocks and conversion pathways are discussed.

2 Hydrothermal Conversion of Biomass to Biocrude

Hydrothermal conversion is one of the advanced thermochemical processes to convert biomass or organic matter under hot, pressurized conditions with water acting both as a carrier and as a reactant. The process occurs in a range of specific hydrothermal conditions, typically between 250 and 374 °C (523–647 K) and pressures in a range between 4 and 22 MPa. These conditions facilitate the breakdown of large polymeric biomass structures into smaller molecular pieces over a suitable period of reaction time. The process, in this case, involves a series of reactions, including hydrolysis,

Table 1 Comparison of properties and downstream upgradation of HTL biocrude and bio-oil

Property	HTL biocrude	Pyrolysis bio-oil	References
Heating value	Higher	Lower	[14]
Oxygen content	Lower	Higher	[15]
Viscosity	Higher	Lower	[16]
Water content	Higher	Lower	[14]
Downstream upgradation	Requires less severe hydrotreating	Requires more severe hydrotreating	[17, 18]
Final products	Similar to petroleum-based fuels	Requires additional processing for fuel compatibility	[19, 20]

dehydration, and decarboxylation, which produce soluble intermediates, while subsequent condensation reactions can lead to insoluble products such as liquid biofuels, known as biocrude, and solid fuels, known as hydrochar [19, 21].

The classification of the hydrothermal processing depends on the temperature. The hydrothermal treatment performed below 250 °C (520 K) is known as carbonization, mainly yielding hydrochar with characteristics similar to those of low-rank coal [22]. At temperatures above 374 °C (647 K), gasification reactions dominate and synthetic fuel gases such as hydrogen, methane, carbon dioxide, and carbon monoxide are produced. The temperature range of 250–374 °C is suitable for hydrothermal method to yield biocrude in the form of a dense, water-insoluble liquid having a heating value in the range of 30–40 MJ/kg [21, 23]. Biocrude produced through this method is different from many other bio-oils, because it contains much less oxygen: between 10 and 20% by weight. In contrast, oxygen contents of up to 50% have been reported for fast pyrolysis bio-oils [24]. This is a significant advantage, but the process is still rather energetic to achieve the desired reaction temperatures in heating of high-moisture-content biomass slurries [25].

Hydrothermal conversion method is somewhat similar to pyrolysis but performed in a hot and pressed water medium. Classical hydrothermal method usually operates without any catalyst, but recent studies focused on the catalytic approaches, which can effectively improve the process efficiency and quality of products. Alkali catalysts, sodium or potassium hydroxide, have been studied extensively since they can alter the ionic character of the medium with the reaction environment in support of base-catalyzed reactions. This might favor the production of desired aromatic oils rather than solid product formation more typical under acid-catalyzed conditions through polymerization reactions [26]. Hydrothermally produced biocrude is complex oxygenated organic mix of wide molecular weight range. It consists of a variety of chemical groups such as acids, alcohols, ketones, phenols, and more complex aromatic structures of methoxyphenols, which are generated from the lignin in softwood biomass, naphthols, and benzofurans. The degree of composition in biocrude arises from the variety of chemical reactions involved in the method, including dehydration, decarboxylation, repolymerization, among others [19, 27].

The conversion mechanism of hydrothermal method is feedstock-dependent, and therefore, the process mechanism has been classified for two kinds of feedstock: (i) Lignocellulosic biomass (dry biomass with the major compositions of cellulose, hemicellulose, and lignin), and (ii) Algal biomass (wet biomass composed of lipids, proteins, carbohydrates, as well as unique biopolymers like algaenans. Such variations require customized approaches in these conversion processes for the optimal yield of biocrude [28].

2.1 Hydrothermal Conversion of Lignocellulosic Biomass

In hydrothermal conversion for lignocellulosic biomass, two types of reactors are usually used: batch and continuous flow reactors. The choice of reactor design influences significantly the process conditions in terms of efficiency and scalability. In industry, continuous reactors are used, usually integrated with a certain feed system adaptable to high-pressure feed conditions. Pumps in most cases used are slurry pumps, while others use lock and hopper mechanisms for allowing the biomass particles. Furthermore, pretreatment of biomass is also performed in continuous hydrothermal method. This is usually performed to break the particles into smaller sizes and remove contaminants in the product for efficient pumping and stabilization of the slurry. Such chemical methods for enhancement of breakdown during hydrothermal reaction, include alkali treatment to the biomass [29].

The hydrothermal conversion of lignocellulosic biomass that has been pre-treated at high temperature and pressure results in phase separation: gaseous stream (mainly CO_2), solid residues (biochar), liquid biocrude, and an aqueous phase containing dissolved organics. Water recycling and overall bio-oil yield are of paramount importance in post-processing the aqueous phase. The methane-rich or hydrogen-rich syngas produced by such a phase can use such techniques as anaerobic digestion or catalytic hydrothermal gasification, thus adding efficiency in the energy process. The remaining solid residue could be directly applied as a soil amendment or else as a precursor for activated carbon, whereas the biocrude usually requires further upgrading via hydrotreatment to obtain a low oxygen and other heteroatoms concentration product with improved quality for direct employment as a liquid fuel [30] (Fig. 1).

2.2 Hydrothermal Conversion of Wet/Algal Biomass

The hydrothermal process for wet biomass, like microalgae, is different from that of lignocellulosic materials because the feedstock itself has different properties. For wet biomass, the pretreatment step that often has to be adopted when using lignocellulose is usually bypassed for microalgae due to their small size. Furthermore, the dried algae slurry usually has around 20% of its total solid concentration after dewatering

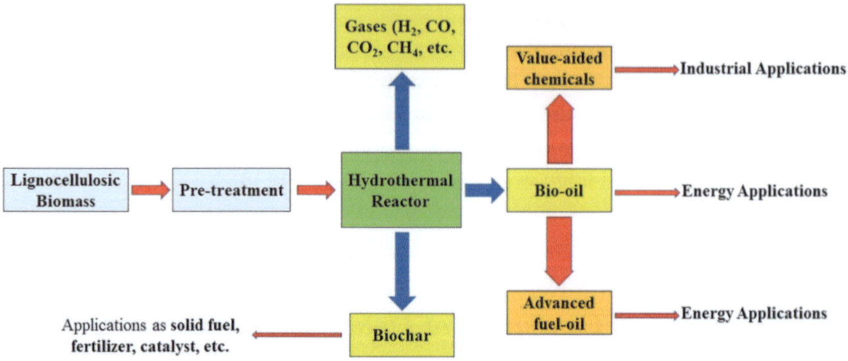

Fig. 1 Schematic diagram of hydrothermal conversion of lignocellulosic biomass into its by-products

and therefore can be directly fed into the hydrothermal reactor without any additional size reduction or other pretreatment activities. In this hydrothermal process, biocrude is produced as the first liquid product along with an aqueous phase concentrated in dissolved organic compounds and nutrients. The water phase can be reused and fed back to grow more algae, and the CO_2 that is produced during cultivation can be supplied to spur photosynthesis and encourage growth in other cultures of algae. Such closed-loop approaches reduce resource usage in general and facilitate the overall process to be more sustainable [30].

Biocrude generally must undergo hydrotreatment to recover properties suitable for hydrocarbon fuel use. Upgrading is primarily aimed at removing oxygen, nitrogen, and other heteroatoms to enhance the energy density and stability of the fuel. At the same time, nutrients recovered in the aqueous phase can be recycled for algae cultivation in a closed-loop nutrient recovery system, thereby enhancing the appeal of hydrothermal conversion as a route for wet biomass in terms of economic and environmental concerns [29, 31] (Fig. 2).

2.3 *Effect of Process Parameters on Biocrude Yield and Properties*

The yields and quality of biocrudes produced using hydrothermal conversion are heavily dependent on several process parameters, including temperature, pressure, reaction time, and catalyst. The effect of all these parameters determines the efficiency of conversion, composition of biocrude, and general sustainability in the process.

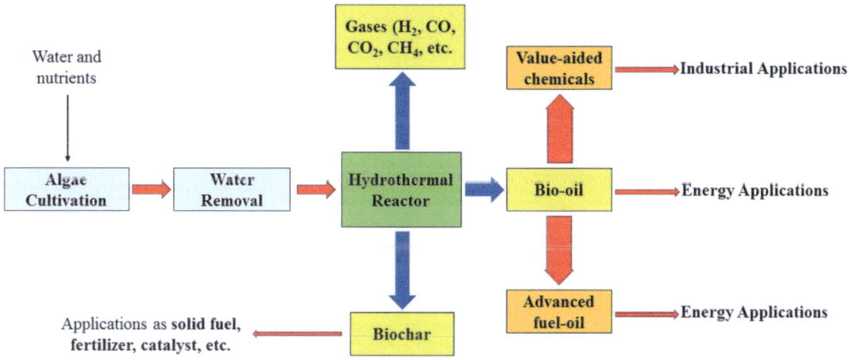

Fig. 2 Schematic diagram of hydrothermal conversion of algal or wet biomass into its by-products

2.3.1 Temperature

Temperature is one of the most significant factors in hydrothermal conversion. Usually, hydrothermal process is performed between the temperature of 250 and 350 °C. When the temperature is applied between 300 and 350 °C, the yield of biocrude would be higher, since the high temperature increases the reaction kinetics by decomposing the components of biomass like cellulose, hemicellulose, and lignin into smaller organic molecules.

When the temperature exceeds 350 °C, cracking, overgasification, and charring occur and, therefore, lead to a yield reduction in biocrude. In such conditions, organic intermediates probably would break into gaseous products such as CO_2, H_2, and methane or form solid residues (char) during repolymerization reactions. According to literatures, the optimum temperature for most types of biomass is between 300 and 350 °C, at which a balance between biocrude production and minimal gas/char formation could be achieved [32].

2.3.2 Pressure

The pressure in hydrothermal method determines the state of water (subcritical or supercritical) and its properties as a solvent and reactant. The average pressure values between 10 and 20 MPa are often enough to keep water in the subcritical state, where it can dissolve organic compounds substantially and promote hydrolysis reactions and others. The solubility of organic matter increases at such pressures, thereby enhancing reaction rates and yields of biocrude. At supercritical conditions of pressure over 22 MPa, organic solubility could be raised further, which would increase the potential conversion efficiency. Under these conditions, gasification reactions would preferentially occur in place of liquefaction reactions. At those higher yields of gas compared to biocrude, the liquid fuel yield may decrease, and hence, supercritical hydrothermal conversion may not always be the ideal method for obtaining maximum liquid fuel yield [32, 33].

2.3.3 Residence Time

Residence time, i.e., the time that biomass remains in the reactor, is another factor that affects the output of hydrothermal conversion of biomass to biocrude. The longer residence time usually results in higher conversion of the biomass to biocrude, since the longer reaction time gives the complex biomass biopolymers more exposure to heat and pressure. This leads to sets of extensive reaction mechanisms, including depolymerization, dehydration, and decarboxylation.

However, extensive longer residence time is not preferable, which can result in over-thermal degradation of the biocrude that ultimately results more gas and char. The optimum residence time is mainly dependent on biomass type and product quality. Wet biomass feedstocks such as algae require shorter residence times, while long residence times favor lignocellulosic feedstocks [34].

2.3.4 Application of Catalyst

Catalysts play a big role in the potential improvement of efficiency as well as selectivity of hydrothermal conversion method. The use of most potent alkali catalysts (e.g., KOH or Na_2CO_3) changes the ionic environment of the reaction medium favoring base-catalyzed reactions that further would contribute to the depolymerization of biomass components and the formation of aromatic compounds in the biocrude [35].

Transition metal catalysts, such as ruthenium (Ru/C), nickel (Ni), or cobalt (Co), are more effective for hydrodeoxygenation, a process that reduces the oxygen content of the biocrude and improves its heating value and stability. These can further significantly improve the quality of the biocrude product to result in a lower oxygen-containing product than could be obtained through non-catalytic hydrothermal conversion method. The use of catalysts also allows for lower temperature and pressure conditions to achieve similar conversion efficiencies, potentially reducing overall energy consumption [36].

2.4 Characterization of Biocrude Oil Obtained from Hydrothermal Conversion

Biocrude oil obtained through hydrothermal conversion of biomass is a very complex mixture of organic compounds. Therefore, investigation of its composition, physical, and chemical properties forms part of characterization of biocrude. The quality of such a product plays an important role in deciding if it may serve as a good renewable fuel or be a source for further refining processes. The characterization techniques used for biocrude oil primarily include GC–MS for determining composition, while an elemental analysis provides the carbon, hydrogen, and oxygen ratios [37].

2.4.1 Composition of Biocrude Oil

The composition of biocrude oil is heterogeneous and heavily reliant upon the type of biomass used together with specific hydrothermal conditions such as temperature, pressure, residence time, and catalysis. The major components of biocrude are the various oxygenated hydrocarbons, including:

(i) Fatty acids and alcohols: These compound species are often extracted from the lipid content in the biomass, especially in wet feedstocks like algae.
(ii) Ketones, aldehydes, and esters: These compound classes in the biocrude are formed due to thermal breakdown of carbohydrates and proteins of biomass.
(iii) Phenolic compounds: These compounds are found in biocrude are formed by the dissolution of lignin in lignocellulosic biomass [38, 39].
(iv) Nitrogen-containing compounds: The compounds like amides and heterocyclic nitrogenous compounds are rich in protein content of the feedstock.
(v) Aromatic hydrocarbons: The biocrude contains the aromatic compounds including polycyclic aromatic hydrocarbons (PAHs) and alkylbenzenes, which are formed when higher temperatures and longer residence times are employed in the process [40].

2.4.2 Properties Biocrude Oil

Biocrude properties are crucial indicators of its potential as a fuel or chemical feedstock. The major properties of biocrude include:

(i) Heating Value: Biocrude oil typically has a higher heating value (HHV) in the range of 30–40 MJ/kg, which is closely comparable to fossil crude oil. On the other hand, the energy content is high due to aliphatic and aromatic hydrocarbons and possibly low due to oxygenates.
(ii) Viscosity: The viscosity of biocrude is often higher than that of traditional crude oil, which is influenced by the molecular weight and extent of polymerization. Therefore, the direct utilization of biocrude oil as a fuel is always a challenge.
(iii) Density: The density of biocrude typically is in between 0.9 and 1.2 g/cm^3 as the feedstock and conversion conditions make a difference with heavier molecules and residual water content with the higher densities.
(iv) Water and Oxygen Content: Oxygen content in biocrude is about 10–20% by weight, which is lesser than that of the bio-oil (50%) from pyrolysis. Biocrude exhibits a significant amount of water content that may affect stability and calorific value of the biocrude [41, 42].

2.4.3 Quality of Biocrude Oil

The quality of biocrude is evaluated in terms of chemical composition as it can be upgraded to be used as a direct drop-in replacement for petroleum fuels. Some important factors that affect the quality of it are:

(i) Oxygen Content: Higher concentration of oxygen in biocrude usually lowers down its stability and calorific value. Hydrodeoxygenation process is typically needed to upgrade biocrude quality from extra oxygen removal [43].
(ii) Nitrogen and Sulfur Content: Biocrude will contain nitrogen (up to 5%) and sulfur species, especially if the feedstocks are protein-rich, such as algae. Such elements could form pollutants during combustion and have to be removed as part of the upgrading process [44].
(iii) Stability: Biocrude is unstable and prone to polymerization due to its complex and heterogeneous nature. It has to be stabilized for a long shelf-life applying techniques such as hydrotreatment or blending with other fuels [45].
(iv) Metal Content and Ash: The presence of metals like sodium or potassium and residual ash can be detrimental for subsequent processing. Such impurities lead to fouling and corrosion in the refining equipment so that further purification is required [46].

3 Pyrolytic Conversion of Biomass to Biocrude

Lignocellulosic biomass is the dry plant material that is primarily comprised of hemicellulose, cellulose, and lignin. The composition of cellulose, hemicellulose, and lignin in biomass varies by plant type. The availability of lignocellulosic biomass makes it a promising feedstock for pyrolysis [47]. Cellulose is a primary structural component of plant cell walls and the most abundant organic polymer. It is a linear chain polysaccharide of D-glucose units linked by β (1–4) glycosidic bonds. Due to its long and linear molecules, cellulose is insoluble in water. Cellulose chains aggregate to form cellulose fibers, which are interconnected by van der Waals and hydrogen bonds, creating microfibrils. These long microfibrils are arranged in a crisscross pattern within the cell wall, providing strength and structure [48]. Lignin and hemicellulose surround the cellulose microfibrils. The three hydroxyl groups on the pyranose rings form inter and intramolecular hydrogen bonds. These hydrogen bonds create a crystalline structure in cellulose, contributing to its mechanical strength and chemical stability. Crystalline structure imparts strength and rigidity to the roots, leaves, and stems. Hemicelluloses are heteroglycans composed of various monomers, including glucose, arabinose, glucuronic acid, galactose, xylose, and mannose. The monomers form polysaccharides like xylan, glucuronoxylan, arabinoxylan, glucomannan, and xyloglucan. Hemicellulose is the second most abundant component of lignocellulosic biomass, constituting one-fourth to one-third of plant matter. Unlike cellulose, hemicellulose has an amorphous structure [49]. Lignin is a highly branched polymer composed of phenylpropane units linked by alkyl-aryl ether bonds. On the basis of number of R–O–CH_3, phenylpropane monomers can be categorized into syringyl, guaiacyl, and p-hydroxyphenyl that are bonded with each other by C–C, ether, or ester bonds. Lignin, an aromatic and hydrophobic biopolymer, diffuses into the matrix of hemicellulose and cellulose, thereby conveying structural inelasticity and firmness to the biomass [50]. Biomass pyrolysis is a complex phenomenon in comparison with other solid-state reactions due to the complicated nature of biomass. Decomposition of biomass components takes place over various temperature ranges, reflecting distinct reaction mechanisms and exhibiting different rates of decomposition. Numerous reactions, including depolymerization, dehydration, aromatization, isomerization, decarboxylation, as well as charring, occur simultaneously during the pyrolysis process [51]. Cellulose thermally decomposes within the temperature range of 310–405 °C by cleaving the β (1–4) glycosidic linkages. During the initial phase, cellulose undergoes depolymerization reaction, producing oligosaccharides, which later decompose to form anhydro-monosaccharides. Levoglucosan (LGA) is the preliminary product of cellulose decomposition, which subsequently undergoes isomerization and dehydration reactions to produce products such as 1,6-anhydro-β-D-glucofuranose, levoglucosenone, and 1,4:3,6-dianhydro-α-D-glucopyranose [52]. Anhydrosugars formed may undergo further reactions to form glyceraldehyde, hydroxyacetone via retro-aldol condensation or fragmentation reactions and furans via dehydration reaction. Decarboxylation and decarbonylation reactions form CO_2 and CO, and polymerization reactions yield char [53]. Hemicellulose decomposition takes place within 180–350 °C. The principal degradation products of xylan include furfural, acetic acid, hydroxyacetone, CO, CO_2, water. Low temperature conditions cause separation of o-acetyl groups on the xylan polymer, yielding acetic acid. Degradation of xylan proceeds through three mechanisms: (a) hydrolytic cleavage to monosaccharides and oligosaccharides, (b) dehydration reactions producing pyran ring and furan intermediates, and (c) ring cleavage to yield low-molecular-weight oxygenates [48]. Lignin pyrolysis is a complex process occurring between 200 and 400 °C, and secondary reactions, such as char formation dominating at temperatures beyond 400 °C. Aromatic groups exhibit thermal stability during the pre-pyrolysis stage. But they become reactive within the range of 400–450 °C, leading to cleavage and formation of phenolic compounds. The primary volatile products produced from G-lignin pyrolysis comprises of coniferyl aldehyde, coniferyl alcohol, vanillin, isoeugenol, and acetovanillone. As the pyrolysis temperature rises from 400 °C to 450 °C, secondary reactions take place, leading to the transformation of syringols/guaiacols into simpler aromatic compounds such as pyrogallols, o-cresols, phenols. Above 700 °C, the pyrolysis process favors the production of polycyclic aromatic hydrocarbons (PAHs). Simultaneously, o-cresols and phenols which are relatively stable at high temperatures are also detected in the product mixture [48].

3.1 *Effect of Process Parameters on Biocrude Yield*

The yield of biocrude is a function of operating parameters which include heating rate, temperature, holding time, and pressure. These parameters govern the yield of biocrude as well as product distribution, including the quality of the pyrolysis oil and gas. A thorough discussion of the influence of these parameters is vital, as the optimization of operating parameters focuses on maximizing the yield of a specific product.

3.1.1 Effect of Final Temperature and Heating Rate

The two principal factors affecting biocarbon yield include heating rate and temperature. During elevated pyrolysis temperatures, the resultant biocrude possesses enhanced calorific value and increased carbon content. Heating rate governs the pyrolysis regime, namely slow, flash, or fast pyrolysis, and significantly impacts the product distribution. Higher heating rate promotes the formation of volatile products, whereas slower heating rate favors char formation. Biocrude yield increases significantly when the reaction temperature is raised from 360 to 490 °C. As the reaction temperature increases, the yield of biocrude and char decreases, while the yield of gas increases. The increased gas yield at higher temperatures can be attributed to secondary cracking reactions that incorporate thermal decomposition of pyrolysis vapours and char. The discrepancies in composition and pyrolysis product yields can be ascribed to variations in reaction parameters, feedstock properties, reactor configuration, and experimental uncertainties [53].

3.1.2 Effect of Particle Size

Particle size influences the pyrolysis product distribution. Larger particles (due to their lower surface area-to-volume ratio), exhibit slower heat transfer rates, resulting in reduced biocrude yield and thereby increased char formation. Minimized particle size leads to enhanced gas yield due to higher heat transfer coefficient. Particle size distribution significantly impacts the product yield distribution, emphasizing the significance of uniform temperature profiles within the reactor. Smaller particle size facilitates rapid heating and complete devolatilization during the pyrolysis process. The enhanced heat transfer associated with small-sized particles causes production of more volatiles, which subsequently condenses to form biocrude, leading to enhanced biocrude yield [54].

3.1.3 Effect of Carrier Gas Flow Rate

As the gas flow rate is increased, the biocrude yield initially increases, but subsequently reduces. During low gas flow rates, incomplete fluidization impeded the completion of pyrolysis process and longer vapor residence time facilitated secondary reactions like repolymerization and cracking. High gas flow rate reduced the yield of biocrude as volatiles were swept out of the reactor prior to condensation, leading to increased gas formation. The carrier gas flow rate influences the holding time of vapors produced during pyrolysis. By rapidly sweeping out these vapors from the hot zone of the reactor, secondary cracking reactions are minimized. Additionally, increasing the carrier gas flow rate further intensifies the bubbling motion, thereby enhancing mixing between the biomass and bed material [55].

3.1.4 Effect of Residence Time

Longer residence time leads to increased cracking of larger molecules into smaller ones, thereby lowering biocrude yield and enhancing gas yield. On the contrary, shorter residence time minimizes secondary reactions, leading to increased biocrude yield. However, it may result in incomplete pyrolysis and reduction in the overall conversion of biomass [55].

3.2 Characterization of Biocrude Obtained from Pyrolytic Conversion

Biocrude is a mixture of organic compounds (alcohols, acids, esters, aldehydes, sugars, ketones, lignin-derived compounds, and phenols) and water that is collected during post pyrolysis process. The physico-chemical characteristics of biocrude encompasses viscosity, density, acid number, water content, and calorific value. The water content of biocrude is quantified using the Karl Fischer titration method. Moisture content adversely affects the flame temperature and heating value and causes ignition delay [56]. Based on the data obtained from elemental analysis, biocrude possesses higher hydrogen and carbon content, and lower oxygen content, in comparison with the original feedstock. Heating value of a fuel, encompasses lower heating value (LHV) and higher heating value (HHV), is a measure of the energy released during complete combustion. Biocrude possesses a higher calorific value due to an increase in carbon content and reduction in oxygen content [57]. Table 2 depicts the typical physico-chemical profile of biocrude.

Table 2 Chemical compound profile present in biocrude [61]

Major components	Wet weight (%)
Acetic acid, formic acid, propanoic acid	2–27
Glycolaldehyde, acetol	3–26
Acetaldehyde, formaldehyde	3–18
Phenol, catechol, methyl phenol, dimethyl phenol	3–13
Furfurol, hydroxymethyl furfural, furfural	2–11
Acetone	4–6
Methanol, ethanol	2–6

3.2.1 pH

The presence of volatile acids (such as acetic acid and formic acid) and diluted water in biocrude contribute to its low pH, within the range of 2–3. The acidic nature of biocrude renders it corrosive during storage and handling equipment. Henceforth, biocrude must undergo certain upgrading processes to mitigate its corrosive properties along with improvement in fuel quality.

3.2.2 Flash Point

Flash point of a fuel is the minimum temperature at which it evaporates to form an ignitable mixture in air. Flash point is determined using flash-point analyzer in which the test temperature is increased at a heating rate of 5.5 °C per minute within the temperature range of 30–80 °C [58].

3.2.3 Ash

Ash content (AC) of biocrude is determined gravimetrically according to ASTM D482. The typical range of AC is from 0.004 to 0.03 wt% and is further influenced by the mineral content of the feedstock and operating conditions. Ash deposits can accelerate corrosion, erosion, and fouling of engine components [59].

3.2.4 Density

Density of biocrude is determined using a pycnometer, in accordance with ASTM D 4052. Density typically ranges from 1.1 to 1.3 kg/m^3. The specific density can vary depending on the pyrolysis conditions and feedstock employed [60].

3.2.5 Physico-chemical Analysis

Gas chromatography-mass spectrometry (GC–MS) has been the most extensively used technique for analyzing the individual components present in biocrude. Biocrude primarily consists of furfural as furan derivative, acetic acid as the major acidic component, and ketones in the chemical form of cyclopentadiones, 1-hydroxy-2-propanone, and cyclopentanones [58]. Levoglucosan, a product of cellulose pyrolysis, signifies the partial degradation of carbohydrate biopolymers. Acetic acid is primarily produced through two mechanisms: the cleavage of acetyl groups linked to the C-2 position of xylose units and the ring-scission of uronic acid residues, involving the elimination of carbonyl and O-methyl groups. The degradation of sugar units leads to the formation of heavy molecular weight ketones, including cyclopentenones and cyclopentanones. A comprehensive quantitative chemical analysis of biocrude is essential for optimizing pyrolysis processes and evaluating the techno-economic feasibility of its utilization as a feedstock for fuels and chemicals. Table 3 presents chemical compound profile present in biocrude. The chemical heterogeneity and instability of biocrude prevent its direct utilization in combustion applications. Many of these limitations can be attributed to the high oxygen content of biomass, which is carried over to the resulting biocrude.

4 Comparative Assessment of Hydrothermal and Pyrolysis Biocrudes

The production of biocrudes via HTL and pyrolysis represents two distinct thermochemical pathways for converting biomass into renewable fuels. These processes are central to bioenergy research, offering alternatives to conventional fossil fuels and contributing to a more sustainable energy future. While both methods convert biomass into biocrude, they differ significantly in process conditions, resulting in biocrudes with varying yields, quality, and properties. This section provides a detailed comparative assessment of biocrudes produced via HTL and pyrolysis, focusing on

Table 3 Physico-chemical profile of biocrude [62]

Properties	Value
C (wt%)	48–49
H (wt%)	5–6
O (wt%)	40–45
N (wt%)	0.5–0.9
HHV (MJ/kg)	15–25
Viscosity (cP)	100
Density (g/m^3)	1140–1150

Table 4 Comparison between HTL biocrude and pyrolysis biocrude

Property	HTL biocrude	Pyrolysis biocrude	References
Yield (% biomass input)	30–60%	40–75%	[19, 63]
Density (g/mL)	1.05–1.20	1.0–1.1	[66]
Viscosity (mPa s)	50–100	> 1000	[19, 67]
Acidity (pH)	6–8	2–3	[64, 66]
Carbon content (wt%)	75–80	60–70	[19, 63]
Oxygen content (wt%)	< 10	30–50	[19, 63]
Energy content (MJ/kg)	35–40	20–30	[57, 68]

yield, compositional analysis, physical and chemical properties, energy content, and potential applications (Table 4).

4.1 *Yield and Quality Comparison*

The yield of biocrude is a critical parameter, influencing the efficiency and feasibility of biofuel production. Both HTL and pyrolysis yield biocrudes, but under different conditions. HTL operates in subcritical or supercritical water (typically between 250 and 400 °C) under high pressure (10–25 MPa) [19]. This water-rich environment allows HTL to convert wet biomass without prior drying, making it energy-efficient for certain feedstocks. The biocrude yield from HTL typically ranges from 30 to 60% by weight of the biomass input, depending on the type of feedstock and process parameters [19], whereas pyrolysis is conducted at higher temperatures (400–600 °C) in the absence of oxygen. Fast pyrolysis, in particular, rapidly heats biomass, yielding biocrude in the range of 40–75%. However, the higher yields are often accompanied by greater production of char and gases [63]. In terms of quality, HTL biocrudes generally have lower oxygen content and higher energy density compared to pyrolysis biocrudes. HTL biocrudes exhibit lower viscosity, making them easier to handle in further refining processes. Pyrolysis biocrudes, on the other hand, are characterized by higher oxygen content (typically 30–50 wt%) [63], which reduces their heating value and increases the need for upgrading before they can be used as fuels.

4.2 *Compositional Analysis*

Compositional analysis provides insight into the molecular makeup of biocrudes, which is essential for determining their refining potential and suitability for different applications. Common techniques for compositional analysis include gas chromatography-mass spectrometry (GC–MS), Fourier-transform infrared spectroscopy (FT-IR), and nuclear magnetic resonance (NMR). GC–MS is used to identify volatile and semi-volatile organic compounds in biocrude. HTL biocrudes generally show a higher concentration of long-chain aliphatic hydrocarbons and fatty acids [64]. Pyrolysis biocrudes, in contrast, contain a greater proportion of phenolic compounds and oxygenates, which contribute to their higher acidity and instability during storage. Furthermore, FT-IR spectra reveal functional groups present in the biocrude. HTL biocrudes typically show peaks corresponding to alkanes, alkenes, and esters, indicating a lower oxygen content [65]. Pyrolysis biocrudes display more pronounced peaks for oxygenated compounds such as alcohols, aldehydes, and carboxylic acids, consistent with their higher oxygen content [64], whereas NMR analysis provides detailed information about the structural arrangement of carbon atoms within the biocrude. HTL biocrudes show signals corresponding to aliphatic and aromatic carbons, while pyrolysis biocrudes tend to have a higher concentration of oxygen-containing functional groups, as seen in the 13C-NMR spectra [64].

4.3 Physical and Chemical Property Analysis

The physical and chemical properties of biocrudes vary significantly between HTL and pyrolysis, affecting their handling, storage, and refining. HTL biocrudes typically have higher densities, ranging between 1.05 and 1.20 g/mL, compared to pyrolysis biocrudes, which range from 1.0 to 1.1 g/mL [66]. The higher density of HTL biocrudes reflects their more complex molecular structure, with a higher proportion of heavy hydrocarbons. Moreover, HTL biocrudes exhibit lower viscosities (ranging from 50 to 100 mPa s) than pyrolysis biocrudes, which can exceed 1000 mPa s at room temperature [67]. The high viscosity of pyrolysis biocrudes is primarily due to the presence of large amounts of high-molecular-weight oxygenates and polymers.

Pyrolysis biocrudes are generally more acidic, with a pH as low as 2–3, due to the presence of carboxylic acids and phenolic compounds [66]. In contrast, HTL biocrudes have a relatively neutral pH, around 6–8, making them less corrosive and easier to upgrade using conventional refining techniques [66]. The elemental analysis reveals HTL biocrudes contain higher levels of carbon (up to 75–80 wt%) and hydrogen (around 10 wt%) [63], with significantly lower oxygen content (typically less than 10 wt%). Pyrolysis biocrudes, on the other hand, have oxygen contents as high as 30–50 wt%, which reduces their energy content and necessitates extensive upgrading to remove the excess oxygen.

4.4 Energy Content and Potential Applications

The energy content of biocrudes is directly influenced by their chemical composition, particularly the carbon, hydrogen, and oxygen contents. HTL biocrudes, with their lower oxygen content and higher proportion of hydrocarbons, exhibit higher energy densities, typically in the range of 35–40 MJ/kg, which is closer to conventional petroleum fuels [68]. In contrast, pyrolysis biocrudes have lower energy contents, around 20–30 MJ/kg, primarily due to their higher oxygen content and the presence of oxygenated compounds, which reduce their calorific value [68].

Both types of biocrudes have potential applications in energy production, but they require upgrading to remove oxygen and improve stability. HTL biocrudes, due to their higher quality, are easier to refine and can be used as feedstocks for producing transportation fuels such as gasoline, diesel, and jet fuel [57]. Pyrolysis biocrudes, after substantial upgrading, can also be used in these applications, but they are more commonly considered for stationary power generation due to their lower energy density and higher upgrading costs.

5 Downstream Upgradation of Biocrude

Low heating value, presence of heteroatoms, moisture levels, high total acid number (TAN), and a reduced hydrogen-to-carbon (H/C) ratio are commonly viewed as unfavorable traits in fuels, limiting the direct application of biocrude oil as a transportation fuel [69–71]. Although a few studies indicate that certain amounts of water can actually enhance overall fuel quality by lowering viscosity and oxynitrogen emissions, boosting atomization potential, reducing acid value, and improving stability of biocrude oil [71, 72]. Yet, excess moisture can diminish the fuel's calorific value. Furthermore, biocrude oil with elevated oxygen content tends to be poorly compatible with conventional refinery fuels and displays low stability under ambient conditions, which can prompt unwanted reactions during storage and transport [69]. Consequently, efforts to develop more sustainable and efficient biofuel production have increasingly focused on upgrading processes that improve the quality and performance of biocrude.

5.1 Refining Processes

The pyrolysis and HTL processes yield similar types of bio-crude; therefore, both biocrude oils are generally upgraded using methods adapted from traditional petroleum refining technologies. Common methods explored for upgrading biocrude oil include both physical and chemical techniques, such as solvent addition, extraction, emulsification, hydrotreating, esterification, etc.

5.1.1 Separation

In HTL and pyrolysis processes, the products generally include a gas, an oily liquid, an aqueous liquid, and a solid residue. Research on biomass conversion via HTL and pyrolysis often uses physical separation techniques to isolate these fractions for analysis. These techniques help separate high-value products or prepare them for further processing aimed at fuel production and valuable product generation. Removing water from the oil fraction is especially important for upgrading, as residual water can deactivate catalysts during processing [73].

5.1.2 Solvent Addition

The primary purpose of using solvents is to enhance biocrude oil stability and reduce its viscosity. Typically, increased viscosity during storage is associated with water formation, which leads to lignin-rich sludge settling at the bottom. Both the biocrude oil and water contain oxygenated compounds and degraded lignin. Over time, the amount of water-insoluble material increases, contributing to the average molecular mass and viscosity of the biocrude oil. This increase is generally due to condensation and polymerization reactions [74].

Introducing a solvent to a two-phase mixture can improve the separation and extraction process. By decanting, the liquid product is split into aqueous and oil segments, yielding an oil portion with about 5% moisture content [75]. Selecting an appropriate solvent depends on its immiscibility with water to aid separation and its effectiveness in extracting organic compounds to maximize yield. Robbins' solute–solvent interaction chart, which examines how functional groups influence solubility via hydrogen bonding and donor–acceptor interactions, is one method for choosing solvents [76]. This chart is especially useful for targeting extractions based on specific functional groups. Due to the bio-crude's complex mixture, various solvents have been explored by researchers to find the optimal choice. The best solvent often varies with the bio-crude composition and feedstock type, with options like ethanol, isopropanol, methanol, acetone, and *N*,*N*-dimethylformamide (DMF) being effective for biocrude oil enhancement [77].

Solvent polarity is a crucial factor when selecting appropriate solvents, especially given the high number of polar compounds present in bio-crude and biocrude oil. For extraction in liquefaction studies, polar solvents like acetone, tetrahydrofuran (THF), ethanol, chloroform, and dichloromethane (DCM) are commonly used for characterizing oils [74]. Garcia-Perez et al., 2007 conducted successive extractions on bagasse pyrolysis biocrude oil using a series of solvents with increasing polarity, including pentane, benzene, dichloromethane, ethyl acetate, and methanol. They found the largest fractions were extracted with ethyl acetate and dichloromethane due to the high polarity of the biocrude oil's components. Conversely, bio-crude from microalgae, which contains a substantial amount of alkanes, may be more effectively extracted with nonpolar solvents [78, 79]. Valdez et al. [80] explored using nonpolar solvents like hexadecane, decane, hexane, and cyclohexane, alongside polar solvents such as methoxycyclopentane, dichloromethane, and chloroform, to separate bio-crude from algae liquefaction products. Nonpolar solvents yielded higher extraction rates, attributed to their structural similarity to bio-crude compounds, as supported by GC–MS analysis. Meanwhile, polar solvents extracted oil with a higher carbon content, likely recovering carbon-rich compounds similar to asphaltenes and resins.

Some research has also explored the use of multiple solvents in a mixture to fractionate oil products. Mixed solvent additives have shown benefits over single solvents for enhancing biocrude oil, due to their increased chemical interactions, which help to prevent polymerization and improve stability. For instance, a mix of 1 wt% methanol, 5.06 wt% DMF, and 1.94 wt% acetone was optimal for producing high-quality biocrude oil and enhancing storage properties [81]. Similarly, Zhu et al. [82] studied the impact of a compound additive (2.42 wt% n-butanol, 2.32 wt% dimethyl sulfoxide, and 3.25 wt% ethyl acetate) on biocrude oil during storage at 80 °C for up to 48 h.

5.1.3 Ultrasonication

Enhancing biocrude oil upgrading with solvents can be achieved through physical treatments like ultrasonication, which facilitates better molecular interaction by inducing frequent particle movement. Xu et al. [83] effectively applied ultrasonication to improve biocrude oil quality, using a 1:1 mixture of n-octanol and biocrude oil, with an ultrasonic power of 150 W for exposure times ranging from 2 to 20 min. Results showed that longer ultrasonication improved biocrude oil properties. For instance, untreated oil had a pH of 4.92 and HHV of 25.76 MJ/kg, while 20 min of ultrasonication raised the pH to 5.11 and HHV to 38.32 MJ/kg [84]. The pH increase was attributed to enhanced alcoholysis of organic compounds during ultrasonication. The process was also found to promote the breakdown of larger molecules, reducing moisture content and encouraging reactions like ring opening, which decreases ketone and alcohol levels while increasing alkanes and alkynes in the oil [85].

Using solvents is an effective and straightforward method to improve biocrude oil properties, like viscosity and calorific value. Solvents interact with components of biocrude oil, organic compounds, minerals, and alkali metals, helping to prevent polymerization and thus promoting stability. However, due to the complex nature of biocrude oil, there is limited understanding of the specific reaction mechanisms between solvents and oil constituents. Consequently, further research is needed to clarify these chemical interactions.

5.1.4 Emulsification

Emulsification, like solvent addition, can enhance the quality and energy content of biocrude oil, making the resulting mixture suitable for direct use in heat or power generation [86]. While polar solvents mix readily with biocrude, emulsification combines two immiscible liquids, suspending tiny droplets of one in the other. Successful emulsions typically create droplets ranging from 1 to 10 μm, whereas smaller droplets (1–100 nm) form microemulsions, which are more stable thermodynamically and resist phase separation across a wide temperature range [86–88]. Emulsifying biocrude oil with petroleum fuels such as diesel often requires a surfactant (like Span 80, Tween 80, or Atlox 4914), and co-surfactants (e.g., ethanol, methanol) to stabilize the emulsion [89]. Surfactants, composed of a hydrophilic "head" and lipophilic "tail," form a protective layer around droplets, stabilizing the mixture. Water-in-oil (W/O) emulsions arise when polar droplets are suspended in a nonpolar liquid, while oil-in-water (O/W) emulsions form when nonpolar droplets are in an aqueous solution. The hydrophilic–lipophilic balance (HLB) of surfactants, usually between 0 and 18, influences emulsion stability and must align with biocrude oil composition for optimal results. Oils high in aromatic hydrocarbons require surfactants with lower HLB values, while oxygen-rich biocrudes need higher HLBs for stable emulsions [77].

Emulsification without surfactants, using ultrasound and mechanical agitation, also shows promise for stable and cost-effective biocrude oil production, as shown by Zhang et al. [90], emulsifying biocrude with crude glycerol. These physical methods help to stabilize emulsions, reducing biocrude viscosity and improving calorific value and cetane number [87, 88]. However, the integration of techniques (ultrasonication or stirring) can add to overall upgrade costs of the process.

5.1.5 Hydrogenation

Hydrogenation, commonly used in petroleum refining, enhances hydrocarbon saturation and removes sulfur, nitrogen, and oxygen. This process helps prevent catalyst deactivation in subsequent steps, reduces coking, and improves fuel properties [91]. Biocrude, however, faces challenges like instability and viscosity increases due to polymerization, condensation, or component degradation through reactions like cracking, decarbonylation, decarboxylation, hydrocracking, hydrodeoxygenation, etc. [92]. Hydrogenation effectively addresses these reactions, helping to stabilize biocrude.

5.1.6 Esterification

Adding alcohols to bio-crude is a technique used to alter its chemical structure and enhance its physical properties. Alcohols react with organic acids in the bio-crude to form esters, compounds similar to those in biodiesel. Ethanol has been explored as a hydrogen alternative for upgrading pyrolysis biocrude oils. Zhang et al. [72] conducted a study using ethanol to treat rice husk pyrolysis biocrude oil with solid acid ($40SiO_2/TiO_2–SO_4^{2-}$) and solid base ($30K_2CO_3/Al_2O_3$–NaOH) catalysts. The esterified biocrude oils showed significant improvements in viscosity, density, and calorific value [74].

5.1.7 Hydroprocessing

Hydrotreating and hydrocracking are common methods used in hydroprocessing, described as follows:

Hydrotreatment

Hydrotreatment plays an essential role in refining various feedstocks, including fossil fuels and biocrude. This process involves treating the material with hydrogen in the presence of a metal-based catalyst at temperatures between 200 and 500 °C and pressures from 3 to 30 MPa [93, 94]. During hydrotreatment, oxygen is removed from biocrude oil's oxygenated compounds via hydrodeoxygenation (HDO), yielding hydrocarbons and water as by-products. Hydrotreatment achieves several goals: desulfurization, to lower sulfur content; denitrogenation, to reduce nitrogen-related emissions; deoxygenation, for better fuel stability and energy density; impurity removal, protecting equipment from metals; and hydrogenation, boosting energy content by saturating hydrocarbons. It is also valued for producing cleaner, higher-quality fuels. Hydrotreatment is seen as more efficient for biocrude upgrading than cracking, as it removes oxygen as H_2O instead of CO_2 and CO, thereby maintaining carbon yield [93, 95, 96].

Biocrude oils can be hydrotreated using various reactor types, such as batch, continuous, and down-flow reactors [97, 98]. Typically, hydrotreatment occurs in two stages. Initially, the biocrude is stabilized at a lower temperature of 100–300 °C with a catalyst, converting carboxyl and carbonyl groups into alcohols. In the second stage, the stabilized oil undergoes hydrodeoxygenation (HDO) and cracking reactions at a higher temperature of 350–400 °C in the presence of active catalysts [99, 100]. The main catalysts for HDO include transition metal and noble metals (e.g., Ru, Rh, Pd, Pt)-based catalysts, and metal sulfides (e.g., MoS_2, Ni–MoS_2) [101–103]. Noble metals are particularly effective for upgrading but come with high costs and significant hydrogen consumption, making them more expensive and vulnerable to sulfur poisoning [97, 101].

Numerous studies have documented the hydrotreatment of biocrudes obtained from pyrolysis and hydrothermal liquefaction (HTL), employing catalysts to convert biocrude into more valuable fuel or other products [100, 102, 104]. Hydrotreatment effectively upgrades biocrude by improving density, reducing heteroatoms, and enhancing properties like smoke points [105]. The process's economic scalability remains challenging, but hydrotreatment holds promise for renewable biofuel applications in industries such as aviation and automotive.

Hydrocracking

Hydrocracking combines cracking and hydrogenation using a bi-functional catalyst. It is generally carried out under more intense conditions following hydrotreatment, which prevents catalyst deactivation and aids in breaking down heavier molecules into lighter, saturated hydrocarbons [71].

5.1.8 Electrocatalytic Hydrogenation

Electrocatalytic hydrogenation, also known as electrochemical upgrading, is widely used to transform oxygenated or carbonyl compounds into valuable organic compounds or hydrocarbons [77]. Recently, it has been explored for biocrude oil, offering a potential alternative to conventional hydrogenation processes, which typically require high temperatures and hydrogen pressures up to 2000 MPa. This method still requires further development to optimize it for pilot-scale biocrude oil upgrading.

Typically, electrocatalytic hydrogenation takes place in an electrolytic membrane fuel cell, where biocrude oil hydrogenation uses protons derived from water oxidation, reducing the need for high-pressure hydrogen and high temperatures, thereby lowering costs. This setup involves an electrochemical cell with an anode and a cathode compartment, divided by a cation exchange membrane (CEM) or anion exchange membrane (AEM). At the anode, water oxidizes to produce O_2 and H^+, while at the cathode, oxygenated compounds in biocrude oil are hydrogenated as H^+ ions pass through the membrane, facilitating hydrogen evolution. Catalysts at the cathode reduce the overpotential for hydrogen evolution reaction (HER), improving conversion efficiency, while the anode may also be catalyst-enhanced for better performance. Although studies [106–108] show electrocatalytic hydrogenation of biocrude oil model compounds, only limited research has applied this method to actual biocrude oil.

This method could be effective for converting oxygenated compounds into hydrocarbons, yet it faces challenges, particularly the expense of membranes and hydrogen, along with electricity requirements. Renewable sources like microbial fuel cells (MFCs) might help meet the electricity needs, as MFCs use microorganisms to convert organic chemical energy into electricity [109, 110].

5.2 Production of Transportation Fuels

The primary aim of biocrude oil upgrading is to develop it as a viable alternative to heavy or light fuel oils for heating and transportation. Initial combustion tests of biocrude oil for industrial heating have shown promising results, suggesting it as a suitable replacement for heavy fuel oil. While biocrude oil applications at commercial scale are few, it has been utilized in combustion processes for boilers, turbines, and diesel engines [77]. For instance, in boiler systems, biocrude oil has shown acceptable emissions while producing heat efficiently [111]. However, some modifications to combustion systems may require to adapt to biocrude oil's unique properties. For example, biocrude oil was used in a small gas turbine (T216) with a dual-fuel setup, where diesel oil ignited the turbine before biocrude oil was introduced via a secondary nozzle. Although the power output of this mixed fuel system was lower than diesel alone (580 vs. 791 kWth), it resulted in reduced NO_x emissions [112]. Further tests using pure biocrude oil and biocrude oil/ethanol blends in micro gas turbines indicated higher emissions of CO and NO_x for biocrude oil, likely due to larger droplet formation from its viscosity and nitrogen content. Nonetheless, biocrude oil blends exhibited greater electrical efficiency than diesel fuel [113]. Diesel engines have also been explored for biocrude oil combustion. In a single-cylinder Ricardo diesel engine, biocrude oil achieved thermal efficiency comparable to diesel when preheated. However, longer ignition delays were observed, attributed to biocrude oil's lower cetane number and complex chemical makeup [114, 115].

A few examples of equipment adjustments for using biocrude oil as transportation fuel include blending bio-oil with solvents like ethanol or methanol to help mitigate ignition delays Pires et al. [116] demonstrated that blending biocrude oil with alcohols can enhance its fuel properties. They found that combining biocrude with methanol and tetra(ethylene glycol) dinitrate improves its cetane number, making it suitable for high-speed diesel engines. Additionally, blending the water-soluble fractions with glycerol was shown to reduce acidity and water content while increasing the heating value and density of water-soluble fuels, enhancing their potential as alternative fuel sources [116, 117]. The heavier fraction of biocrude oil, with a high heating value and a composition close to diesel, can be blended with diesel to enhance fuel value. Bertero et al. [118] proposed emulsifying pyrolyzed oil with diesel to power low-speed engines, achieving a 50% reduction in NO_x emissions. Additionally, heavy fractions can be used directly as fuel in heating or kettle systems, broadening their application potential [117].

Biocrude oil properties also depend on the feedstocks from which it is derived. Upgraded oils from specific biomass sources show significant promise. For instance, oil from animal waste primarily comprises alkanes with some nitriles, making it suitable for biofuel production. Spirulina yields hydrocarbons in the light diesel range with a higher aromatic content, while sewage sludge produces paraffinic hydrocarbons suitable for diesel and jet fuel applications. Biocrudes from dry and wet feedstocks contain valuable compounds like fatty acids and esters, such as oleic acid, which serve as key precursors in biodiesel production, highlighting their potential for renewable fuel applications. Miscanthus upgrades into an aromatic-rich oil ideal for gasoline. Aspen wood, when hydrotreated, generates a mix of hydrocarbons and valuable bio-blendstocks for low-sulfur diesel and jet fuel. Sewage sludge further aligns with conventional diesel and gasoline needs, underscoring its potential for biofuel production. Despite these advances, challenges such as managing heteroatom content in biocrudes remain critical for refining and fuel performance [119].

These findings highlight biocrude oil's versatile applications. However, as a transport fuel, biocrude oil requires further research due to its low heating value, instability, and poor physical properties. Effective upgrading to a drop-in transport fuel would involve complete oxygenate removal and adjustments to create naphthenes, paraffins, and aromatics to match conventional fuel properties. Producing biocrude oil affordably at large scales also remains challenging, which impacts its competitiveness as a practical fuel for diverse transportation and industrial uses [77].

5.3 Production of Chemicals and Materials

Biocrudes derived from both dry and wet feedstocks offer a diverse blend of chemical compounds, serving as essential components for various applications, especially in sustainable energy solutions. These compounds include fatty acids, esters, hydrocarbons, nitrogenous compounds, phenols, ketones, alcohols, and aldehydes, though their concentrations vary based on feedstock type [83, 84, 119–123]. For example, biocrudes from protein-rich biomass typically contain elevated nitrogen levels due to acid amides and heterocyclic aromatics like pyrroles, pyridines, and pyrazines, while algal feedstocks lead to higher acidity from free carboxylic acids [124].

Nitrogen-containing compounds are significant in biocrudes, with compounds like pyrrole offering potential applications in pharmaceuticals and fine chemicals [119]. Sulfur-containing compounds, such as thioethers like dimethyl sulfide, are also present and valuable for specialty chemicals and fuel additives. Various fractions in biocrude oil—such as acids, anhydrosugars, alcohols, ketones, aldehydes, esters, phenols, pyrolytic lignin, and furans—provide additional functionality [117]. Phenolic compounds, for instance, are essential in several industries, including chemicals, pharmaceuticals, food processing, and resin manufacturing. Research by Kim [125] highlights phenolic-rich biocrudes as effective replacements for fossil-based phenols, and Capunitan and Capareda [126] emphasize their value in adhesives and specialty polymers.

Biocrudes also contain platform chemicals like furfural, derived from pentose sugars, which are pivotal in producing resins and solvents, enhancing the biocrude's utility for sustainable chemical production and biofuel development. Heavy fractions of biocrude find uses in flavorings, agrichemicals, resins, fertilizers, and emission control agents; their mixability with other liquids makes them versatile for various applications [126]. Han et al. [127] demonstrate that reactions of carboxylic acids and phenols with lime to form calcium salts can aid in acid gas adsorption (e.g., SO_x). They also suggest that biocrude functional groups, such as carbonyl and hydroxyl groups, can produce slow-release nitrogen fertilizers that pollute groundwater less than mineral fertilizers, while lignin-rich humic substances condition soil.

The broad range of compounds in biocrudes enables their application across biofuel production and organic chemical synthesis. However, composition and structure are influenced by the feedstock and processing conditions. Understanding these variations and the underlying reaction mechanisms is critical to optimizing biocrude conversion processes for sustainable use.

5.4 Challenges and Opportunities in Downstream Upgradation of Biocrude Oil

Numerous studies on hydrothermal liquefaction (HTL) and pyrolysis have progressed at lab and pilot scales, but expanding to produce biocrude at large volumes efficiently is challenging [116]. Scaling up introduces obstacles in maintaining effective heat transfer, residence times, and consistent reaction conditions, especially in reactor design for large-scale operations [128]. Achieving steady temperature profiles and high pressures in large HTL reactors also proves complex, as temperature variations can hinder reaction kinetics [129–131]. Additionally, sustaining pressure safely and efficiently in large reactors adds to the complexity, with pressure control and safety measures becoming critical [132]. Energy efficiency is essential for cost-effectiveness, requiring a balance between input energy and output from biocrude and chemicals [133]. Thus, the reactor design, stable biomass supply, and detailed economic assessments across scales are vital for long-term feasibility and market positioning of biocrude upgrading processes. Future research should emphasize scaling thermochemical processes for commercial purposes, though the economic hurdles of capital and operational costs in large-scale production remain substantial [134–136].

Despite the challenges, downstream biocrude upgrading shows promise for enhancing fuel quality. However, challenges like heteroatoms (e.g., nitrogen and sulfur) in biocrude demand effective denitrogenation and desulfurization methods. Techniques such as solvent addition and emulsification can increase biocrude stability and energy density, though solvent addition may raise water content and lower pH, requiring emulsification with ultrasonication or stirring [77]. While ethanol and methanol reduce viscosity, emulsification for biofuel upgrade remains costly [88, 137]. Hydrotreatment is promising for high hydrocarbon yields, though its cost is impacted by the need for external hydrogen. Using hydrogen donors like ethanol for in-situ hydrogen production may cut costs, but feasibility studies on economic viability are still required. Catalyst challenges include coking issues with sulfided catalysts and sensitivity to water, which deactivates acidic supports like Al_2O_3 and zeolites [77, 100].

Electrocatalytic hydrogenation, though successful for model compounds, faces cost barriers due to pricey membranes and hydrogen. Advancing affordable, high-performance membranes is essential, while integrating electrochemical cells with hydrogen production (e.g., steam reforming) may further reduce costs, though it requires substantial research investment [77]. Incorporating renewable energy, like microbial fuel cells, could provide electricity sustainably. Techno-economic evaluations are needed to balance costs, hydrogen use, and emissions reduction.

Innovations in hydrogenation, deoxygenation, and catalyst development are key to improving biocrude upgrading efficiency. Research into sustainable catalysts, tailored for specific feedstocks, remains critical for enhancing selectivity and process efficiency. Additionally, assessing the environmental and health impacts of HTL, pyrolysis, and subsequent upgrading methods is essential for sustainable, eco-friendly processes [119]. Developing strategies to integrate HTL, pyrolysis, and downstream upgrades with existing refineries and fuel infrastructure would support biofuel adoption and distribution.

6 Economic Considerations

Due to this rise in demand for alternative and sustainable energy solutions, many different technologies have been invented or developed for the biomass conversion processes. Some are HTL and pyrolysis, and indeed each has different potential and advantages on its economic viability and its efficiency on energy as well as quality of the final products. The most widely practiced and known techniques of the conversions are hydrothermal as well as pyrolysis conversion techniques. Methods employed through hydrothermal methods use highly pressured and heated forms. They consume a great amount of energy and become pricey because of infrastructure costs, along with the overall requirements in operations. Hydrothermal process: HTA converts the biomass to a char product called hydrochar and other products of higher value. HTA involves the processing of biomaterials by applying extremely hot pressurized water, usually between 180 and 260 °C in which case pressure is involved during the said process. It results in converting the moist biomass without drying it. It thus enables wet agri-waste feed stocks together with food wastes suitability [138, 139]. It follows more at low pressure but is very demanding for very well-controlled temperatures like during pyrolysis. The other two methods include pyrolysis and hydrothermal liquefaction. Pyrolysis is different from HTL and requires an entirely different environment. Different properties are obtained with this process. There is the great advantage of no much drying required in hydrothermal liquefaction, giving it considerable opportunities to work on wet feedstocks. HTL represents an interesting alternative for the sustainable production of biofuel since the biocrude usually of a higher quality is produced in this process. For instance, pyrolysis is better for dry biomass, but it is relatively expensive since it takes further pretreatment stages.

6.1 Cost Analysis of Hydrothermal and Pyrolysis Processes

This section compares and contrasts the two processes they used to calculate the costs of producing biocrude oil: hydrothermal liquefaction and pyrolysis (Table 5). The important factors for the cost comparison analysis is illustrated in Fig. 3.

Table 5 Data-driven comparison table of pyrolysis and HTL

Aspect	HTL	Pyrolysis	References
Production cost	\$3.1–\$4.44/gge; \$2.6/gge(algae)	\$2.6–\$9.3/gge	[140, 141]
Energy requirements	Lower; utilizes wet feedstocks	Higher; necessitates drying biomass	[16, 140, 144, 145]
Minimum selling price	~ \$3.8/gge; upgraded ~ \$4.9/gge	Generally higher than HTL	[141]
Capital investments	Higher due to complex systems	High; specialized reactors needed	[142, 144]
Product yield/Quality	Higher yield; better energy density	Variable yields, requires upgrading	[140, 143]

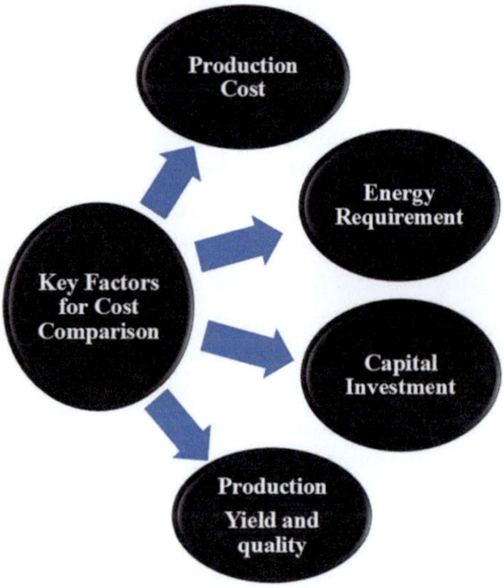

Fig. 3 Key factors for cost comparison of pyrolysis and HTL

6.1.1 Hydrothermal Liquefaction

Production Cost

Using second-generation feedstocks such as waste biomass and non-food crops, the expected feedstock costs for HTL biocrude, quoted in terms of gasoline-gallon equivalent (gge), are between \$3.1 and \$4.44 per gallon. The cost-effectiveness can be further improved for the third-generation feedstocks which include algae, and so on, which can be bought at around \$2.6/gge [140]. The minimum selling price of the biocrude produced is about \$3.8/gge with the upgraded gasoline cost close to \$4.9/gge [141].

Energy Requirement

Pressure range for the HTL process is as low as 5 MPa and can go up to 28 MPa. The temperature range is from 250 to 550 °C [16]. Drying biomass for pyrolysis would need at least 35% more energy than HTL technique; however, in this process, wet biomass can be used directly without drying [140].

Capital Investment

HTL reactors and pressure control systems are more complex, meaning that initial capital expenditures on these plants are more costly. However, wet feedstocks processing results in relatively lower operating costs that can cover these expenditures [142].

Production Yield and Quality

In most cases, HTL will have a higher yield of biocrude compared to pyrolysis, wherein the biocrude contains an energy density between 30 and 35 MJ/kg, which makes it of more value [143].

6.1.2 Pyrolysis

Production Cost

The production cost ranges, depending on the feedstock and technique used to acquire bio-oil. As estimated, this tends to be more expensive. Depending on the feedstock used for the production, production cost for pyrolysis-based bio-oils ranged from an estimated $2.6 to $9.3/gge in most cases [140].

Energy Requirements

In contrast, HTL can handle wet biomass, but pyrolysis uses more energy since it operates at a high temperature level of between usually (400–800 °C) with the absence of oxygen [144, 145].

Capital Investment

Due to the requirement of highly specialized reactors that must work at high temperatures and pressures, capital investments in pyrolysis facilities may also be large. They are more complex, which leads to a higher cost when setting them up [144].

Product Yield and Quality

The products of pyrolysis include biochar, syngas, and bio-oil. Bio-oil usually requires extensive refining because of its lower energy density and higher oxygen content compared to HTL biocrude [140, 143].

HTL is economically more viable for biocrude production than pyrolysis, mainly because it has the potential to handle wet biomass without significant pretreatment. HTL seems to be an emerging technology in renewable energy with increased product yields and better quality biocrude at lower operational costs. Pyrolysis, on the other hand, is a viable alternative but is always more expensive and involves rather complicated processing.

7 Environmental Impact Assessment

Interest in biocrude oil production through hydrothermal liquefaction and pyrolysis has been considerable, especially in view of the increasing need for renewable energy. Such technologies have to be evaluated on a critical scale in terms of their environmental implications, particularly water pollution, waste disposal, and greenhouse gas emissions.

The different feedstock and operational characteristics used in the production of biocrude oil by HTL and Pyrolysis processes have diverse implications for GHG emissions.

7.1 Air Pollution

HTL is an excellent method for carbon footprint reduction of biocrude as it digests wet biomass, which includes microalgae and sewage sludge. Pyrolysis results in much more greenhouse gas emissions compared to others because the process requires 400–800 °C for high temperature and energy during the process of drying up the biomass. General emissions from HTL have been relatively lower than those in pyrolysis. Due to the energy-intensive drying processes involved in the preparation of biomass, pyrolysis can produce as high as 210 g CO_2 eq/MJ. On the other hand, HTL can produce biocrude with GHG emissions as low as 11.4 g CO_2 eq/MJ of sustainable diesel [146]. The major cause of such a huge reduction in emissions is that HTL can directly handle wet biomass and not through the energy-intensive drying process like pyrolysis. In addition, the utilization of CO_2 collection technologies makes it possible for HTL processes to have a negative GHG footprint. This is particularly true in the case where algal feedstocks grown in wastewater are applied because they accumulate carbon throughout cultivation [147]. While it is possible to use bases of pyrolysis and hydrothermal liquefaction for renewable fuels, from which the latter would have smaller impacts on the environment mainly on greenhouse gases and utilizations of resources, on certain conditions, HTL recovered more efficiently with energies compared to pyrolysis and thus serves as the best method about the utilization of biocrude production and environmental amiability.

Besides improving its sustainability profile, HTL has the capability to accept wet feedstocks without significant preprocessing, thus reducing energy requirements [147]. Maximizing the environmental benefits of these technologies will require continuous evaluation and optimization as they advance.

7.2 Water Pollution

Water contamination is a major issue in both HTL and pyrolysis. HTL is a process carried out under high temperature and pressure, converting moist biomass into biocrude with the simultaneous production of an aqueous phase [147]. In this liquid state are nutrients, dissolved organic compounds, and even heavy metals and other potentially dangerous substances. Without proper management, the water phase formed during HTL may threaten aquatic ecosystems by the nutrients and dissolved organic molecules it comprises. The most notable advantage of HTL, however is that it can recycle nutrients and reduce pollution levels by combining it with wastewater treatment systems [148]. This by-product water can be converted into syngas by efficient treatment methods such as catalytic hydrothermal gasification, which reduces the potential for pollution. According to research, this means that it is now possible to have control of the quality of the aqueous phase and increase yield biocrude by finding the sweet spot for the HTL conditions. Pyrolysis, however, refers to the breakdown of dry biomass in an oxygenless heat process yielding bio-oil, solid char, and gaseous by-products [140]. The amount of liquid effluents yielded by pyrolysis is less than that of HTL; however, the oxygenated compounds present in high concentration in the bio-oil may enter into water sources if not disposed of properly and may give rise to leachate problems during transportation and storage. Hence, both these processes require proper water management measures to mitigate the consequences of pollution.

8 Life Cycle Analysis of Biocrude Production and Utilization

Life cycle analysis (LCA) refers to a structured approach taken in evaluating the environmental implications associated with every stage in the life of a product, from the extraction and processing of raw materials right up to usage and disposal. For biocrude manufacture, LCA is very fundamental for understanding whether such processes are sustainable and carry environmental costs in the way of those of hydrothermal liquefaction (HTL) or pyrolysis.

Studies on life cycle assessments indicate that HTL can present lower life cycle emissions of GHG when full production processes, starting with the cultivation and processing of feedstocks, are included in the assessment [149]. Advances in technologies in recent times include not only the co-liquefaction of mixed feedstocks but also catalysts' utilization, which increases yields of biocrude but also reduces energy input as well as emissions [150]. Life cycle analysis is useful in determining how sustainable a process such as hydrothermal liquefaction or pyrolysis could be in the production of biocrude. These processes involve feedstock, energy consumption patterns, and emissions profile, and they all play a role in overall environmental effect. Optimizing these processes will continue in research fields to make these processes more sustainable and toward renewable energy goals. LCA of the production and utilization of biocrude discusses feedstock selection, energy consumption, emissions, waste management, and overall sustainability.

8.1 Feedstock Selection

The feedstock is quite decisive in determining the environmentally sustainable production of biocrude. Algal biomass has recently emerged as promising feedstock due to short growth cycles and nutrient availability in wastewater. Studies suggested that the use of microalgae in hydrothermal liquefaction can yield adequate amount of biocrude without competing for agricultural ground or freshwater. Moreover, municipal solid waste (MSW) has been identified as a promising feedstock for biocrude production through HTL, which supports waste management strategies and generates energy at the same time [148]. The composition of the biomass determines the efficiency of HTL. Biomass feedstocks with higher percentages of hemicellulose and cellulose tend to produce more bio-oil yields because these compounds are prone to hydrothermal degradation.

Although the complex structure of lignin restricts its effective conversion, there is always a residual fraction left. Microalgae have high lipid content and low oxygen concentration, with bio-oil yields in the range of 15–55% and with similar HHV as regular diesel [148]. Co-liquefaction approaches where high-moisture feedstocks such as algae are combined with drier materials like wood pellets increase the overall yield and minimize water usage [150].

Pyrolysis, by contrast, demands dry feedstock, which is produced with energy-intensive drying pretreatment before conversion. That might increase the greenhouse gases associated with fossil fuel consumption in pretreatment [63]. Feedstocks for pyrolysis include typically agricultural residues, wood chip, and many types of lignocellulosic biomass. The chemical makeup of feedstocks largely determines the quality and production of bio-oil.

High levels of ash and other contaminants are often found in agricultural residues, including straw and corn stover, making them produce considerable quantities of bio-oil; nevertheless, they usually tend to lower the processing efficiency. Lignin present in woody biomass has relatively low yields compared to the lipid-rich feedstock sources such as microalgae. The size of particles also influences the pyrolysis efficiency: Finer particles enhance heat transfer rates and conversion rates.

HTL and pyrolysis show different prospects and challenges associated with the selection of feedstock. HTL ability to wet biomass without dry feedstocks increases its adaptability in particular organic waste streams which can be a challenge when processed. Pyrolysis requires feedstock having precise moisture and composition which would increase bio-oil yield. Adding different feedstocks improves the production of biocrude while promoting a circular economy through the conversion of waste into valuable resources [140].

8.2 Energy Consumption and Emission

The most important consideration for the sustainability of a process to produce biocrude is energy consumption. HTL operates at high temperatures and pressures with corresponding input of energy. Wet biomass can be converted into biocrude with minimal drying, thereby maximizing overall thermal efficiency [151]. Pyrolysis generally requires dry biomass, which means higher energy for drying and possibly further increasing the greenhouse gases associated with fossil fuel-based preprocessing steps [152].

8.3 Waste Management

Biocrude can be produced from biomass through the processes of pyrolysis and HTL. However, in order to minimize environmental impacts, the waste products generated by both methods are therefore of paramount importance and must be carefully managed.

8.3.1 Waste Management in Hydrothermal Liquefaction

Among the large by-product yield of HTL processes is an aqueous phase rich with organic compounds and solid residues, which are produced from wet biomass through biocrude conversion. In the absence of proper treatment, dissolved nutrients and organic materials in the liquid phase could contaminate bodies of water [140]. Such risks are mitigated with proper waste management. From HTL, fertilizers and inorganic materials are produced, which can either be recycled or used for energy recovery [153]. This process will further support a circular economy as it concerns the reuse and recycling of resources while minimizing waste. In addition, sewage sludge and other organic waste streams can be efficiently handled by integrating HTL with current waste management systems. By reducing landfill use and increasing the value

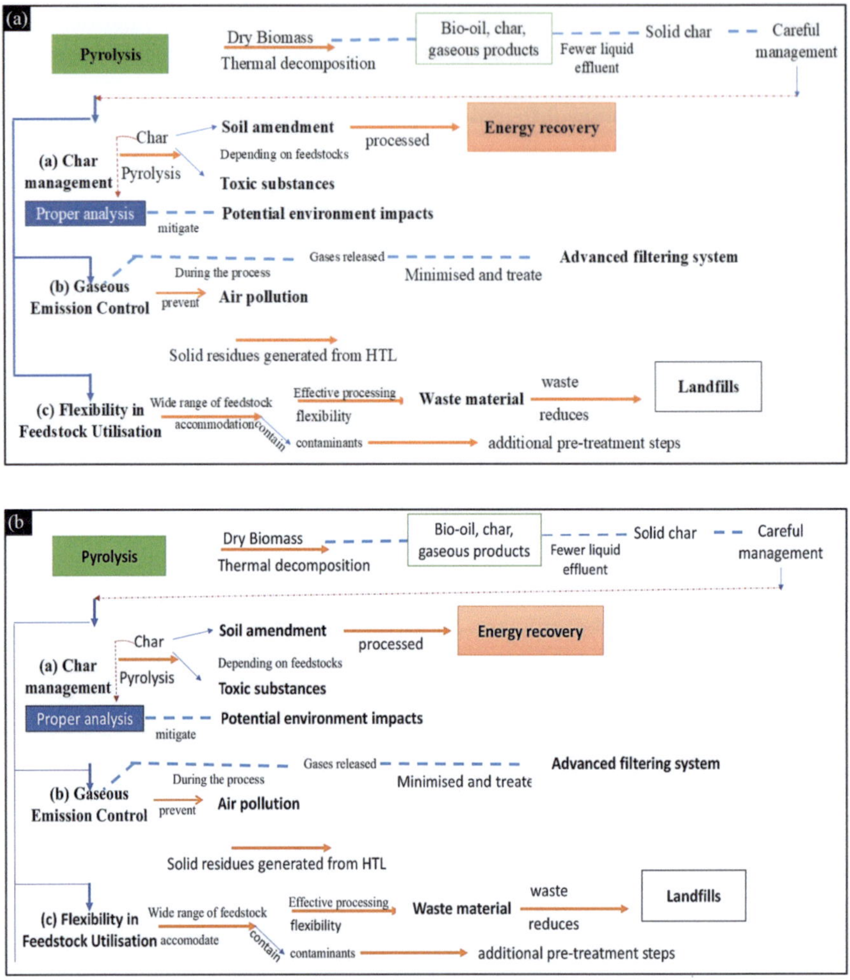

Fig. 4 Schematic representation of waste management in hydrothermal liquefaction (HTL) and pyrolysis **a** hydrothermal liquefaction (HTL). **b** Pyrolysis

of waste as biocrude, this integration improves sustainability [152]. By reducing the need for disposal, the ability to recycle the aqueous phase of the process further enhances waste management efficiency [148] (Fig. 4).

8.3.2 Waste Management in Pyrolysis

On the other hand, pyrolysis is the thermal decomposition of dried biomass in an oxygen-free environment, which produces bio-oil, char, and gaseous by-products. Although pyrolysis has fewer liquid effluents than HTL, it still produces solid char that needs careful handling. The char may contain toxic compounds depending on the feedstock used and thus requires thorough analysis and possible treatment before it can be used as a soil amendment or carbon sequestration material [153].

Moreover, there is gaseous waste from the pyrolysis itself that needs control to make the air pollution and in accordance with the environmental rule [149]. The primary step in preparation for a pyrolysis treatment of biomass includes drying them, which brings about unwanted waste such as nutrient-rich runoff possibly contaminating water sources [150]. Thus, as it reduces the solid waste residues produced during the processing of residual energy, this method shows challenges in controlling solid and also gaseous waste residue.

In conclusion, hydrothermal liquefaction and pyrolysis have special implications for waste management in the production of biocrude. HTL presents a more integrated process, where wet feedstocks are used and aqueous by-products are recycled with better efficiency. On the other hand, pyrolysis generates solid residues, which require careful handling to minimize potential environmental impacts. Sustained research and optimization are necessary to enhance the sustainability of both processes and minimize their waste management challenges.

9 Conclusions and Future Perspectives

Production of biocrude oil from HTL and pyrolysis processes has great potential to enhance energy security and sustainability. Biocrude is a renewable fuel derived from biomass, hence reducing reliance on imported energy sources and greenhouse gas emissions [154]. The benefit of HTL is that it can receive direct wet biomass, such as agricultural residue, municipal solid waste, and reduce the considerable energy consumption and water intensity of drying [151]. Pyrolysis instead takes dry feeds and produces a complex mixture of organic compounds which can be upgraded to valuable fuels and chemicals [155]. Both processes have advantages that are unique to themselves; HTL generally results in biocrude of higher energy density and lower in nitrogen content compared to pyrolysis-derived bio-oil. However, there are still challenges associated with the optimization of such processes to make them more economically viable, reduce emissions, and effectively manage wastes. Future research should target reactor designs, feedstock diversity, and integration into existing refining technologies to realize the maximum benefits of biocrude production for sustainable energy solutions.

There are large-scale implications for energy sustainability and security in the production of biocrude via HTL and pyrolysis. Biocrude is a renewable source of energy that can assist with reducing greenhouse gas emissions, mitigating climate change, and increasing energy independence—an important concern as the globe seeks alternatives to fossil fuel. Biocrude production can contribute to improved energy security. With the use of biomass resources already available in their locality, countries can strengthen energy resilience and thus reduce dependency on imported fossil fuels. For example, HTL can efficiently transform wet biomass, including agricultural residues and municipal solid waste (MSW), into biocrude with minimal pretreatment. This capability produces renewable energy and supports a circular economy by using waste materials that otherwise would end up in landfills. Furthermore, biocrude integrates easily into existing refining systems, which means it's easier to transition from fossil fuels and toward renewable energy. Biocrude derived from hydrothermal processing is upgradable to any given fuel grade or can be blended with fossil fuels due to its compatibility with traditional petroleum refining [155]. This offers a flexibility that enhances energy security through enhancing options for fuel production.

Primary factors influencing the sustainability of the product are feedstock and processes in biocrude production. Since HTL can directly process wet biomass, thus it bypasses the water intensively demanded drying processes normally used in pyrolysis. This feature reduces the carbon footprint of biomass preprocessing and also saves water resources. Studies have been carried out on life cycle assessment that revealed greenhouse gas emissions from HTL-produced biocrude are lower than traditional fossil fuels. Another important feature through which waste use of feedstock contributes to sustainability is minimizing landfill waste and maximizing resource recovery [150]. Advances in catalyst development for HTL and pyrolysis also enhance the production efficiency of biocrude while decreasing the emissions of processing. However, environmental impacts due to feedstock cultivation still persist. To make biocrude sustainable, responsible sourcing practices should avoid land-use change and deforestation [156]. To maximize the yield while at the same time minimizing the impact on ecological issues, intensive research into second-generation feedstock must be initiated. Among these are agricultural residues and dedicated energy crops grown on marginal lands.

Energy can be both produced and influenced by the product of biocrude production. However, this very much depends on the feasibility of producing at an economically viable level. The very high entrance requirements for the HTL and pyrolysis processes can thus be recuperated by sustained process improvement research and upscaling research and development. Governmental support through incentive policies or investments can enhance the economic feasibility of the production of biocrude.

Acknowledgements The authors would like to offer sincere thanks to the Department of Energy, Tezpur University, for providing all the necessary facilities required for conceptualization and manuscript preparation. The first author gratefully acknowledges the financial grant in the form of fellowship (Institutional fellowship) from Tezpur University, Assam, India.

Author Contributions The chapter was conceptualized and designed by Rupam Kataki (RK) and Priyanka Padhi (PP); Writing: The first draft was prepared by Priyanka Padhi (PP), Nilutpal Bhuyan (NB), Neelam Bora (NB), Mondita Athparia (MA), Bidisha Borah (BB), and Rupam Kataki (RK). The manuscript was reviewed and critically revised by Priyanka Padhi (PP) and Rupam Kataki (RK). The study was supervised by Rupam Kataki (RK). All authors have read and approved the final manuscript.

Funding: The authors declare that no funds, grants, or other support were received during the preparation of this manuscript.

Data Availability: All the relevant data is published solely within the paper.

Declarations

Disclosure Statement: The authors declare no conflicts of interest.

Financial interests: The authors declare they have no financial interests.

References

1. Usman M, Cheng S, Boonyubol S, Cross JS (2024) From biomass to biocrude: innovations in hydrothermal liquefaction and upgrading. Energy Convers Manage 302:118093. https://doi.org/10.1016/j.enconman.2024.118093
2. Datta D, Ghosal RK, Biswas S, Mandal E, Das B (2024) Biofuel production: an ecofriendly way to produce sustainable energy. In: Emerging sustainable technologies for biofuel production, 1st edn. Springer Nature Switzerland, Cham, pp 67–99. https://doi.org/10.1007/978-3-031-52167-6_4
3. Badoga S, Alvarez-Majmutov A, Xing T, Gieleciak R, Chen J (2020) Co-processing of hydrothermal liquefaction biocrude with vacuum gas oil through hydrotreating and hydrocracking to produce low-carbon fuels. Energy Fuels 34(6):7160–7169. https://doi.org/10.1021/acs.energyfuels.0c00937
4. Ilango V (2024) Bioenergy production using biomass wastes: challenges of circular economy. In: Valorization of biomass wastes for environmental sustainability: green practices for the rural circular economy. Springer Nature Switzerland, Cham, pp 171–188. https://doi.org/10.1007/978-3-031-52485-1_9
5. El-Araby R (2024) Biofuel production: exploring renewable energy solutions for a greener future. Biotechnol Biofuels Bioproducts 17(1):129. https://doi.org/10.1186/s13068-024-02571-9
6. Lababpour A (2018) Continuous hydrothermal liquefaction for biofuel and biocrude production from microalgal feedstock. ChemBioEng Rev 5(2):90–103. https://doi.org/10.1002/cben.201700017
7. Liu L, Zhang X, Zhao R, Song G, Tian L (2021) Pyrolysis of Phragmites hirsuta study on pyrolysis characteristics, kinetic and thermodynamic analyses. Int J Energy Res 45(10):15200–15216. https://doi.org/10.1002/er.6795
8. Cui Z, Cheng F, Jarvis JM, Jena U, Brewer CE (2021) Integrated extraction and catalytic upgrading of biocrude oil from co-hydrothermal liquefaction of crude glycerol and algae. Energy Fuels 35(15):12165–12174. https://doi.org/10.1021/acs.energyfuels.1c01336
9. Shahbeik H, Shafizadeh A, Gupta VK, Lam SS, Rastegari H, Peng W, Pan J, Tabatabaei M, Aghbashlo M (2023) Using nanocatalysts to upgrade pyrolysis bio-oil: a critical review. J Clean Prod 413:137473. https://doi.org/10.1016/j.jclepro.2023.137473
10. Cheng Z, Saha B, Vlachos DG (2018) Catalytic hydrotreatment of Humins to bio-oil in methanol over supported metal catalysts. Chemsuschem 11(20):3609–3617. https://doi.org/10.1002/cssc.201801535
11. Rasaq WA, Okpala COR, Igwegbe CA, Białowiec A (2024) Navigating pyrolysis implementation—a tutorial review on consideration factors and thermochemical operating methods for biomass conversion. Materials 17(3):725. https://doi.org/10.3390/ma17030725
12. Klemm M, Kröger M, Görsch K, Müller-Langer F, Majer S (2020) Fuel-driven biorefineries using hydrothermal processes. Chem Ing Tec 92(11):1653–1664. https://doi.org/10.1002/cite.202000093
13. Wang W, Bai J, Lu Q, Zhang G, Wang D, Jia J, Guan Y, Yu L (2021) Pyrolysis temperature and feedstock alter the functional groups and carbon sequestration potential of Phragmites

australis-and Spartina alterniflora-derived biochars. GCB Bioenergy 13(3):493–506. https://doi.org/10.1111/gcbb.12795
14. Oyedun AO, Patel M, Kumar M, Kumar A (2020) The upgrading of bio-oil via hydrodeoxygenation. Chem Catal Biomass Upgrading 35–60.https://doi.org/10.1002/9783527814794.ch2
15. Xia C, Pathy A, Paramasivan B, Ganeshan P, Dhamodharan K, Juneja A, Kumar D, Brindhadevi K, Kim SH, Rajendran K (2022) Comparative study of pyrolysis and hydrothermal liquefaction of microalgal species: analysis of product yields with reaction temperature. Fuel 311:121932. https://doi.org/10.1016/j.fuel.2021.121932
16. Mishra RK, Kumar P, Mohanty K (2022) Hydrothermal liquefaction of biomass for bio-crude production: a review on feedstocks, chemical compositions, operating parameters, reaction kinetics, techno-economic study, and life cycle assessment. Fuel 316:123377. https://doi.org/10.1016/j.fuel.2022.123377
17. Kohansal K, Sharma K, Haider MS, Toor SS, Castello D, Rosendahl LA, Zimmermann J, Pedersen TH (2022) Hydrotreating of bio-crude obtained from hydrothermal liquefaction of biopulp: effects of aqueous phase recirculation on the hydrotreated oil. Sustainable Energy Fuels 6(11):2805–2822. https://doi.org/10.1039/D2SE00399F
18. Gil A, Sancho-Sanz I, Korili SA (2024) Progress and perspectives in the catalytic hydrotreatment of bio-oils: effect of the nature of the metal catalyst. Ind Eng Chem Res 63(27):11759–11775. https://doi.org/10.1021/acs.iecr.4c00747
19. Elliott DC, Biller P, Ross AB, Schmidt AJ, Jones SB (2015) Hydrothermal liquefaction of biomass: developments from batch to continuous process. Biores Technol 178:147–156. https://doi.org/10.1016/j.biortech.2014.09.132
20. Dimitriadis A, Liakos D, Pfisterer U, Moustaka-Gouni M, Karonis D, Bezergianni S (2021) Impact of hydrogenation on miscibility of fast pyrolysis bio-oil with refinery fractions towards bio-oil refinery integration. Biomass Bioenerg 151:106171. https://doi.org/10.1016/j.biombioe.2021.106171
21. Castello D, Pedersen TH, Rosendahl LA (2018) Continuous hydrothermal liquefaction of biomass: a critical review. Energies 11:3165. https://doi.org/10.3390/en11113165
22. Heilmann SM, Jader LR, Harned LA, Sadowsky MJ, Schendel FJ, Lefebvre PA, von Keitz MG, Valentas KJ (2011) Hydrothermal carbonization of microalgae II. Fatty acid, char, and algal nutrient products. Appl Energy 88(10):3286. https://doi.org/10.1016/j.apenergy.2010.12.041
23. Brown TM, Duan P, Savage PE (2010) Hydrothermal liquefaction and gasification of Nannochloropsis sp. Energy Fuels 24(6):3639–3646. https://doi.org/10.1021/ef100203u
24. Bridgewater AV, Meier D, Radlein D (1999) An overview of fast pyrolysis of biomass. Org Geochem 30:1479. https://doi.org/10.1016/S0146-6380(99)00120-5
25. Pearce M, Shemfe M, Sansom C (2016) Techno-economic analysis of solar integrated hydrothermal liquefaction of microalgae. Appl Energy 166:19. https://doi.org/10.1016/j.apenergy.2016.01.005
26. Elliott DC (2011) Hydrothermal processing. In: Brown R (ed) Thermochemical processing of biomass. Wiley, Chichester, U.K, pp 200–231
27. Aysu T, Durak H, Güner S, Bengü AŞ, Esim N (2016) Bio-oil production via catalytic pyrolysis of Anchusa azurea: effects of operating conditions on product yields and chromatographic characterization. Biores Technol 205:7. https://doi.org/10.1016/j.biortech.2016.01.015
28. Biller P, Ross AA (2016) Production of biofuels via hydrothermal conversion. In: Luque R, Lin CSK, Wilson K, Clark J (eds) Handbook of biofuel production. Woodhead Publishing, pp 509–547
29. Rowbotham J, Dyer P, Greenwell H, Theodorou M (2012) Thermochemical processing of macroalgae: a late bloomer in the development of third-generation biofuels? Biofuels 3:441. https://doi.org/10.4155/bfs.12.29
30. Gollakota ARK, Kishore N, Gu S (2017) A review on hydrothermal liquefaction of biomass. Renew Sustain Energy Rev 81:1378. https://doi.org/10.1016/j.rser.2017.05.178

31. Bhuyan N, Bora N, Narzari R, Boruah K, Kataki R (2021) Thermo-catalytic conversion of non-edible seeds (extractive-rich biomass) to fuel oil. In: Shadangi KP (ed) Liquid biofuels: fundamentals, characterization, and applications. Wiley, pp 285–360. https://doi.org/10.1002/9781119793038.ch9
32. Peterson AA, Vogel F, Lachance RP, Fröling M, Antal MJ, Tester JW (2008) Thermochemical biofuel production in hydrothermal media: a review of sub- and supercritical water technologies. Energy Environ Sci 1(1):32–65. https://doi.org/10.1039/B810100K
33. Kumar S, Gupta R, Tyagi VV (2018) Hydrothermal treatment of biomass for bio-oil production: a review. Renew Sustain Energy Rev 81:1662–1689. https://doi.org/10.1201/9780429289798
34. Toor SS, Rosendahl L, Rudolf A (2011) Hydrothermal liquefaction of biomass: a review of subcritical water technologies. Energy 36(5):2328–2342. https://doi.org/10.1016/j.energy.2011.03.013
35. Elliott DC, Hart TR, Neuenschwander GG, Rotness LJ, Zacher AH (2009) Catalytic hydrothermal gasification of lignocellulosic biomass. Energy Fuels 23(7):4071–4078
36. Castello D, Haider MS, Rosendahl L (2020) Catalytic hydrothermal liquefaction of biomass into bio-crude: a review. Energy Convers Manage 209:112634
37. Tian C, Li B, Liu Z, Zhang Y, Lu H (2014) Hydrothermal liquefaction for algal biocrude production: a review. Renew Sustain Energy Rev 68(1):57–68
38. Teri G, Luo L, Savage PE (2014) Hydrothermal treatment of protein, polysaccharide, and lipids alone and in mixtures. Energy Fuels 28(12):7501–7509
39. Tamburini D, Ribechini E, Colombini MP (2016) New markers of natural and anthropogenic chemical alteration of archaeological lignin revealed by in situ pyrolysis/silylation-gas chromatography mass spectrometry. J Anal Appl Pyrol 118:249–258
40. Biller P, Sharma BK, Kunwar B, Ross AB (2015) Hydroprocessing of bio-crude from continuous hydrothermal liquefaction of microalgae. Fuel 159:197–205
41. Li K, Liu S, Liu X (2017) Chemical and physical properties of biocrude from hydrothermal liquefaction of microalgae. Renew Sustain Energy Rev 72:76–85
42. Zhou D, Zhang L, Zhang S, Fu H, Chen J (2010) Hydrothermal liquefaction of macroalgae Enteromorpha prolifera to bio-oil. Energy Fuels 24(7):4054–4061
43. Djandja OS, Wang Z, Chen L, Qin L, Wang F, Xu Y, Duan P (2020) Progress in hydrothermal liquefaction of algal biomass and hydrothermal upgrading of the subsequent crude bio-oil: a mini review. Energy Fuels 34(10):11723–11751
44. Lu J, Zhang Z, Zhang L, Fan G, Wu Y, Yang M (2021) Catalytic hydrothermal liquefaction of microalgae over different biochars. Catal Commun 149:106236
45. Velusamy K, Isabel JB, Periyasamy S, Thiruvenkadam A, Ravikumar H, Gupta SK, López-Maldonado EA (2024) Role of biochar as a greener catalyst in biofuel production: production, activation, and potential utilization–a review. J Taiwan Inst Chem Eng 105732
46. Roy M, Mohanty K (2023) Byproducts of a microalgal biorefinery as a resource for a circular bioeconomy. In: Biotic resources. CRC Press, pp 109–138
47. Mujtaba M, Fraceto LF, Fazeli M, Mukherjee S, Savassa SM, de Medeiros GA, Santo Pereira ADE, Mancini SD, Lipponen J, Vilaplana F (2023) Lignocellulosic biomass from agricultural waste to the circular economy: a review with focus on biofuels, biocomposites and bioplastics. J Clean Prod 402:136815. https://doi.org/10.1016/j.jclepro.2023.136815
48. Collard FX, Blin J (2014) A review on pyrolysis of biomass constituents: mechanisms and composition of the products obtained from the conversion of cellulose, hemicelluloses and lignin. Renew Sustain Energy Rev 38:594–608. https://doi.org/10.1016/j.rser.2014.06.013
49. Terashima N, Kitano K, Kojima M, Yoshida M, Yamamoto H, Westermark U (2009) Nanostructural assembly of cellulose, hemicellulose, and lignin in the middle layer of secondary wall of ginkgo tracheid. J Wood Sci 55:409–416. https://doi.org/10.1007/s10086-009-1049-x
50. Lourenço A, Pereira H (2018) Compositional variability of lignin in biomass. In: Lignin-trends and applications. pp 65–98.https://doi.org/10.5772/intechopen.71208
51. Dhyani V, Bhaskar T (2019) Pyrolysis of biomass. In: Biofuels: alternative feedstocks and conversion processes for the production of liquid and gaseous biofuels. Academic Press, pp 217–244. https://doi.org/10.1016/B978-0-12-816856-1.00009-9

52. Junior II, Do Nascimento MA, de Souza ROMA, Dufour A, Wojcieszak R (2020) Levoglucosan: a promising platform molecule? Green Chem 22(18):5859–5880. https://doi.org/10.1039/D0GC01490G
53. Xiong Z, Wang Y, Syed-Hassan SSA, Hu X, Han H, Su S et al (2018) Effects of heating rate on the evolution of bio-oil during its pyrolysis. Energy Conver Manage 163:420–427.https://doi.org/10.1016/j.enconman.2018.02.078
54. Luo S, Xiao B, Hu Z, Liu S, Guan Y, Cai L (2010) Influence of particle size on pyrolysis and gasification performance of municipal solid waste in a fixed bed reactor. Biores Technol 101(16):6517–6520. https://doi.org/10.1016/j.biortech.2010.03.060
55. Qureshi KM, Lup ANK, Khan S, Abnisa F, Daud WMAW (2021) Optimization of palm shell pyrolysis parameters in helical screw fluidized bed reactor: effect of particle size, pyrolysis time and vapor residence time. Cleaner Eng Technol 4:100174. https://doi.org/10.1016/j.clet.2021.100174
56. Lu Q, Li WZ, Zhu XF (2009) Overview of fuel properties of biomass fast pyrolysis oils. Energy Convers Manage 50(5):1376–1383. https://doi.org/10.1016/j.enconman.2009.01.001
57. Demirbas A (2009) Biofuels securing the planet's future energy needs. Energy Convers Manage 50(9):2239–2249. https://doi.org/10.1016/j.enconman.2009.05.010
58. Wildschut J (2009) Pyrolysis oil upgrading to transportation fuels by catalytic hydrotreatment
59. Zhang Q, Chang J, Wang T, Xu Y (2007) Review of biomass pyrolysis oil properties and upgrading research. Energy Convers Manage 48(1):87–92. https://doi.org/10.1016/j.enconman.2006.05.010
60. Oasmaa A, Czernik S (1999) Fuel oil quality of biomass pyrolysis oils state of the art for the end users. Energy Fuels 13(4):914–921. https://doi.org/10.1021/ef980272b
61. Xu Y, Hu X, Li W, Shi Y (2011) Preparation and characterization of bio-oil from biomass. In: Progress in biomass and bioenergy production. pp 197–222.https://doi.org/10.5772/16466
62. Chukwuneke JL, Ewulonu MC, Chukwujike IC, Okolie PC (2019) Physico-chemical analysis of pyrolyzed bio-oil from swietenia macrophylla (mahogany) wood. Heliyon 5(6). https://doi.org/10.1016/j.heliyon.2019.e01790
63. Bridgwater AV (2012) Review of fast pyrolysis of biomass and product upgrading. Biomass Bioenerg 38:68–94. https://doi.org/10.1016/j.biombioe.2011.01.048
64. Mullen CA, Boateng AA (2010) Chemical composition of bio-oils produced by fast pyrolysis of two energy crops. Energy Fuels 24(1):210–218. https://doi.org/10.1021/ef900799c
65. Gai C, Zhang Y, Chen W-T, Liu T, Dong Y (2015) An investigation of reaction pathways of hydrothermal liquefaction by deuterium labeling. Green Chem 17(5):3815–3824. https://doi.org/10.1039/c5gc00600k
66. Zhou L, Cai J, Olofsson G (2013) Physical properties of biocrude from the hydrothermal liquefaction of swine manure. Fuel 109:292–299. https://doi.org/10.1016/j.fuel.2013.02.060
67. Westerhof RJM, Kuipers NJM, Kersten SRA, van Swaaij WPM (2009) Viscosity measurements of pyrolysis liquids: their meaning and fate. Energy Fuels 23(1):705–715. https://doi.org/10.1021/ef800843h
68. Jones SB, Zhu Y, Anderson DB, Hallen RT, Elliott DC, Schmidt AJ, Albrecht KO (2013) Process design and economics for the conversion of algal biomass to hydrocarbons: whole algae hydrothermal liquefaction and upgrading (PNNL-23227). Pacific Northwest National Lab (PNNL), Richland, WA (United States). https://doi.org/10.2172/1126297
69. Lian X, Xue Y, Zhao Z, Xu G, Han S, Yu H (2017) Progress on upgrading methods of bio-oil: a review. Int J Energy Res 41(13):1798–1816
70. Veses A, Aznar M, López JM, Callén MS, Murillo R, García T (2015) Production of upgraded bio-oils by biomass catalytic pyrolysis in an auger reactor using low cost materials. Fuel 141:17–22
71. Ghadge R, Nagwani N, Saxena N, Dasgupta S, Sapre A (2022) Design and scale-up challenges in hydrothermal liquefaction process for biocrude production and its upgradation. Energy Convers Manag: X 14:100223
72. Zhang S, Yan Y, Li T, Ren Z (2005) Upgrading of liquid fuel from the pyrolysis of biomass. Biores Technol 96(5):545–550

73. Venderbosch RH, Ardiyanti AR, Wildschut J, Oasmaa A, Heeres HJ (2010) Stabilization of biomass-derived pyrolysis oils. J Chem Technol Biotechnol 85(5):674–686
74. Ramirez JA, Brown RJ, Rainey TJ (2015) A review of hydrothermal liquefaction bio-crude properties and prospects for upgrading to transportation fuels. Energies 8(7):6765–6794
75. Peterson AA, Vogel F, Lachance RP, Fröling M, Antal MJ Jr, Tester JW (2008) Thermo-chemical biofuel production in hydrothermal media: a review of sub-and supercritical water technologies. Energy Environ Sci 1(1):32–65
76. Frank TC, Dahuron L, Holden BS, Prince WD, Seibert AF, Wilson LC Liquid-liquid extraction and other liquid-liquid operations and equipment. In: Green DW, Perry RH (eds) Perry's chemical engineering handbook, 8th ed. McGraw-Hill, New York, NY
77. Kumar R, Strezov V (2021) Thermochemical production of bio-oil: a review of downstream processing technologies for bio-oil upgrading, production of hydrogen and high value-added products. Renew Sustain Energy Rev 135:110152
78. Hossain FM, Kosinkova J, Brown RJ, Ristovski Z, Stephens E, Hankamer B et al (2015) The chemical-physical properties of bio-crude derived from the hydrothermal liquefaction of microalgae. Algal Res
79. Jena U, Das KC (2011) Comparative evaluation of thermochemical liquefaction and pyrolysis for bio-oil production from microalgae. Energy Fuels 25(11):5472–5482
80. Valdez PJ, Dickinson JG, Savage PE (2011) Characterization of product fractions from hydrothermal liquefaction of Nannochloropsis sp. and the influence of solvents. Energy Fuels 25(7):3235–3243
81. Zhu L, Li K, Zhang Y, Zhu X (2017) Upgrading the storage properties of bio-oil by adding a compound additive. Energy Fuels 31(6):6221–6227
82. Zhu L, Li K, Ding H, Zhu X (2018) Studying on properties of bio-oil by adding blended additive during aging. Fuel 211:704–711
83. Xu X, Li Z, Sun Y, Jiang E, Huang L (2018) High-quality fuel from the upgrading of heavy bio-oil by the combination of ultrasonic treatment and mutual solvent. Energy Fuels 32(3):3477–3487
84. Xu D, Lin G, Liu L, Wang Y, Jing Z, Wang S (2018) Comprehensive evaluation on product characteristics of fast hydrothermal liquefaction of sewage sludge at different temperatures. Energy 159:686–695
85. Mohapatra H, Kleiman M, Esser-Kahn AP (2017) Mechanically controlled radical polymerization initiated by ultrasound. Nat Chem 9(2):135–139
86. Leng L, Li H, Yuan X, Zhou W, Huang H (2018) Bio-oil upgrading by emulsification/microemulsification: a review. Energy 161:214–232
87. Jiang X, Ellis N (2010) Upgrading bio-oil through emulsification with biodiesel: mixture production. Energy Fuels 24(2):1358–1364
88. Jiang X, Ellis N (2010) Upgrading bio-oil through emulsification with biodiesel: thermal stability. Energy Fuels 24(4):2699–2706
89. de Luna MDG, Cruz LAD, Chen WH, Lin BJ, Hsieh TH (2017) Improving the stability of diesel emulsions with high pyrolysis bio-oil content by alcohol co-surfactants and high shear mixing strategies. Energy 141:1416–1428
90. Zhang M, We MYSLS, Wu H (2018) Direct emulsification of crude glycerol and bio-oil without addition of surfactant via ultrasound and mechanical agitation. Fuel 227:183–189
91. Alfke G, Irion WW, Neuwirth OS (2008) Oil refining. In: Elvers B (ed) Handbook of Fuels: Energy Sources for Transportation. Wiley-VCH, Weinheim, Germany
92. Mortensen PM, Grunwaldt JD, Jensen PA, Knudsen KG, Jensen AD (2011) A review of catalytic upgrading of bio-oil to engine fuels. Appl Catal A 407(1–2):1–19
93. Auersvald M, Shumeiko B, Vrtiška D, Straka P, Staš M, Šimáček P et al (2019) Hydrotreatment of straw bio-oil from ablative fast pyrolysis to produce suitable refinery intermediates. Fuel 238:98–110
94. Zhang X, Wang T, Ma L, Zhang Q, Jiang T (2013) Hydrotreatment of bio-oil over Ni-based catalyst. Biores Technol 127:306–311

95. Mohan D, Pittman CU Jr, Steele PH (2006) Pyrolysis of wood/biomass for bio-oil: a critical review. Energy Fuels 20(3):848–889
96. Chen B, Shi ZM, Jiang SJ, Tian H (2016) Catalytic cracking mechanisms of tar model compounds. J Cent South Univ 23(12):3100–3107
97. Cheng S, Wei L, Alsowij MR, Corbin F, Julson J, Boakye E, Raynie D (2018) In situ hydrodeoxygenation upgrading of pine sawdust bio-oil to hydrocarbon biofuel using Pd/C catalyst. J Energy Inst 91(2):163–171
98. Lau PL, Allen RW, Styring P (2013) Continuous-flow Heck synthesis of 4-methoxybiphenyl and methyl 4-methoxycinnamate in supercritical carbon dioxide expanded solvent solutions. Beilstein J Org Chem 9(1):2886–2897
99. Capunitan JA, Capareda SC (2014) Hydrotreatment of corn stover bio-oil using noble metal catalysts. Fuel Process Technol 125:190–199
100. Han Y, Gholizadeh M, Tran C-C, Kaliaguine S, Li C-Z, Olarte M et al (2019) Hydrotreatment of pyrolysis bio-oil: a review. Fuel Process Technol 195:106140
101. Sanna A, Vispute TP, Huber GW (2015) Hydrodeoxygenation of the aqueous fraction of bio-oil with Ru/C and Pt/C catalysts. Appl Catal B 165:446–456
102. Jahromi H, Agblevor FA (2018) Hydrodeoxygenation of aqueous-phase catalytic pyrolysis oil to liquid hydrocarbons using multifunctional nickel catalyst. Ind Eng Chem Res 57(39):13257–13268
103. Mu W, Ben H, Du X, Zhang X, Hu F, Liu W et al (2014) Noble metal catalyzed aqueous phase hydrogenation and hydrodeoxygenation of lignin-derived pyrolysis oil and related model compounds. Bioresource Technol 173:6–10
104. Reyhanitash E, Tymchyshyn M, Yuan Z, Albion K, Van Rossum G, Xu CC (2016) Hydrotreatment of fast pyrolysis oil: effects of esterification pretreatment of the oil using alcohol at a small loading. Fuel 179:45–51
105. Cheremisinoff NP, Rosenfeld PE (2009) The petroleum industry. In: Cheremisinoff NP, Rosenfeld PE (eds) Handbook of pollution prevention and cleaner production: best practices in the wood and paper industries. William Andrew, Massachusetts, USA, pp 1–97
106. Elangovan S, Larsen D, Bay I, Mitchell E, Hartvigsen J, Millett B et al (2017) Electrochemical upgrading of bio-oil. ECS Trans 78:3149–3158
107. Lister TE, Diaz LA, Lilga MA, Padmaperuma AB, Lin Y, Palakkal VM, Arges CG (2018) Low-temperature electrochemical upgrading of bio-oils using polymer electrolyte membranes. Energy Fuels 32(5):5944–5950
108. Li Z, Garedew M, Lam CH, Jackson JE, Miller DJ, Saffron CM (2012) Mild electrocatalytic hydrogenation and hydrodeoxygenation of bio-oil derived phenolic compounds using ruthenium supported on activated carbon cloth. Green Chem 14(9):2540–2549
109. Kumar R, Singh L, Zularisam AW (2016) Exoelectrogens: recent advances in molecular drivers involved in extracellular electron transfer and strategies used to improve it for microbial fuel cell applications. Renew Sustain Energy Rev 56:1322–1336
110. Kumar R, Singh L, Zularisam AW, Hai FI (2018) Microbial fuel cell is emerging as a versatile technology: a review on its possible applications, challenges and strategies to improve the performances. Int J Energy Res 42(2):369–394
111. Oasmaa A, Kyt M, Sipil K (2001) Pyrolysis oil combustion tests in an industrial boiler. In: Bridgwater AV (ed) Progress in thermochemical biomass conversion. Blackwell Science Ltd., Oxford, UK, pp 1468–1481
112. Strenziok R, Hansen U, Knstner H (2001) Combustion of bio-oil in a gas turbine. In: Bridgwater AV (ed) Progress in thermochemical biomass conversion. Blackwell Science Ltd., Oxford, UK, pp 1452–1458
113. Buffi M, Cappelletti A, Rizzo AM, Martelli F, Chiaramonti D (2018) Combustion of fast pyrolysis bio-oil and blends in a micro gas turbine. Biomass Bioenerg 115:174–185
114. Shihadeh A, Hochgreb S (2000) Diesel engine combustion of biomass pyrolysis oils. Energy Fuels 14(2):260–274
115. Laesecke J, Ellis N, Kirchen P (2017) Production, analysis and combustion characterization of biomass fast pyrolysis oil–biodiesel blends for use in diesel engines. Fuel 199:346–357

116. Pinheiro Pires AP, Arauzo J, Fonts I, Domine ME, Fernandez Arroyo A, Garcia-Perez ME, et al (2019) Challenges and opportunities for bio-oil refining: a review. Energy Fuels 33(6):4683–4720
117. Drugkar K, Rathod W, Sharma T, Sharma A, Joshi J, Pareek VK et al (2022) Advanced separation strategies for up-gradation of bio-oil into value-added chemicals: a comprehensive review. Sep Purif Technol 283:120149
118. Bertero M, de la Puente G, Sedran U (2012) Fuels from bio-oils: bio-oil production from different residual sources, characterization and thermal conditioning. Fuel 95:263–271
119. Usman M, Cheng S, Boonyubol S, Cross JS (2024) From biomass to biocrude: Innovations in hydrothermal liquefaction and upgrading. Energy Convers Manage 302:118093
120. Tian Y, Wang F, Djandja JO, Zhang SL, Xu YP, Duan PG (2020) Hydrothermal liquefaction of crop straws: effect of feedstock composition. Fuel 265:116946
121. Conti F, Toor SS, Pedersen TH, Seehar TH, Nielsen AH, Rosendahl LA (2020) Valorization of animal and human wastes through hydrothermal liquefaction for biocrude production and simultaneous recovery of nutrients. Energy Convers Manage 216:112925
122. Chen F, Wang Y, Zheng L, Wu L, Ding X (2023) Hydrothermal liquefaction of lignocellulosic biomass with potassium phosphate and iron and their binary mixture: a comprehensive investigation on the yields and compositions of biocrude and solid residue. Biores Technol 386:129532
123. Adedeji OM, Russack JS, Molnar LA, Bauer SK (2022) Co-hydrothermal liquefaction of sewage sludge and beverage waste for high-quality bio-energy production. Fuel 324:124757
124. Matayeva A, Bianchi D, Chiaberge S, Cavani F, Basile F (2019) Elucidation of reaction pathways of nitrogenous species by hydrothermal liquefaction process of model compounds. Fuel 240:169–178
125. Kim JS (2015) Production, separation and applications of phenolic-rich bio-oil–a review. Biores Technol 178:90–98
126. Capunitan JA, Capareda SC (2013) Characterization and separation of corn stover bio-oil by fractional distillation. Fuel 112:60–73
127. Han Y, Paiva Pinheiro Pires A, Denson M, McDonald AG, Garcia-Perez M (2020) Ternary phase diagram of water/bio-oil/organic solvent for bio-oil fractionation. Energy Fuels 34(12):16250–16264
128. Wagner JL, Le CD, Ting VP, Chuck CJ (2017) Design and operation of an inexpensive, laboratory-scale, continuous hydrothermal liquefaction reactor for the conversion of microalgae produced during wastewater treatment. Fuel Process Technol 165:102–111
129. Deepika C, Mrinal, Pohrmen CB, Jaiswal KS, Sangmesh B, Jaiswal KK et al (2024) Hydrothermal liquefaction of wet microalgal biomass for biofuels and platform chemicals: advances and future prospects. Discov Appl Sci 6(5):245
130. Sharma N, Jaiswal KK, Kumar V, Vlaskin MS, Nanda M, Rautela I et al (2021) Effect of catalyst and temperature on the quality and productivity of HTL bio-oil from microalgae: a review. Renewable Energy 174:810–822
131. Motavaf B, Savage PE (2021) Effect of process variables on food waste valorization via hydrothermal liquefaction. ACS EST Eng 1(3):363–374
132. Johannsen I, Kilsgaard B, Milkevych V, Moore D (2021) Design, modelling, and experimental validation of a scalable continuous-flow hydrothermal liquefaction pilot plant. Processes 9(2):234
133. Subhash GV, Rajvanshi M, Kumar GRK, Sagaram US, Prasad V, Govindachary S, Dasgupta S (2022) Challenges in microalgal biofuel production: a perspective on techno economic feasibility under biorefinery stratagem. Biores Technol 343:126155
134. Ranjbar S, Malcata FX (2023) Hydrothermal liquefaction: how the holistic approach by nature will help solve the environmental conundrum. Molecules 28(24):8127
135. Torres-Mayanga PC, Lachos-Perez D, Mudhoo A, Kumar S, Brown AB, Tyufekchiev M et al (2019) Production of biofuel precursors and value-added chemicals from hydrolysates resulting from hydrothermal processing of biomass: a review. Biomass Bioenergy 130:105397

136. Liu R, Fei W, Shen C (2014) Influence of acetone addition on the physicochemical properties of bio-oils. J Energy Inst 87(2):127–133
137. Garcia-Perez M, Chaala A, Pakdel H, Kretschmer D, Roy C (2007) Characterization of bio-oils in chemical families. Biomass Bioenergy 31:222–242
138. Zhang B, Biswal BK, Zhang J, Balasubramanian R (2023) Hydrothermal treatment of biomass feedstocks for sustainable production of chemicals, fuels, and materials: progress and perspectives. Chem Rev 123(11):7193–7294. https://doi.org/10.1021/acs.chemrev.2c00673
139. Reza T (2022) Thematic issue: hydrothermal conversion of biomass. Biomass Convers Biorefinery 1–2.https://doi.org/10.1007/s13399-021-01927-7
140. Ranjbar S, Malcata FX (2023) Hydrothermal liquefaction: how the holistic approach by nature will help solve the environmental conundrum. Molecules 28(24):8127. https://doi.org/10.3390/molecules28248127
141. Snowden-Swan LJ, Zhu Y, Jones SB, Elliott DC, Schmidt AJ, Hallen RT, Maupin GD (2016) Hydrothermal liquefaction and upgrading of municipal wastewater treatment plant sludge: a preliminary techno-economic analysis, rev. 1 (No. PNNL-25464-Rev. 1). Pacific Northwest National Lab (PNNL), Richland, WA (United States). https://doi.org/10.2172/1327165
142. None N (2020) 2019 project peer review report (No. DOE/EE-2045). EERE Publication and Product Library, Washington, DC (United States)
143. Haarlemmer G, Briand M, Roubaud A, Roussely J, Déniel M (2018) Economic evaluation of a hydrothermal liquefaction process. Detritus 3(September):84–92. https://doi.org/10.31025/2611-4135/2018.13695
144. Pahnila M, Koskela A, Sulasalmi P, Fabritius T (2023) A review of pyrolysis technologies and the effect of process parameters on biocarbon properties. Energies 16(19):6936. https://doi.org/10.3390/en16196936
145. Clifford CB (2021) EGEE 439: alternative fuels from biomass sources. Lesson 7.2. Energy Institute, The Pennsylvania State University
146. Zhang B, Wu J, Deng Z, Yang C, Cui C, Ding Y (2017) A comparison of energy consumption in hydrothermal liquefaction and pyrolysis of microalgae. Trends Renewable Energy 3(1):76–85. https://doi.org/10.17737/tre.2017.3.1.0013
147. Garcia-Garcia G, Martín-Lara MÁ, Calero M, Blázquez G (2024) Environmental impact of different scenarios for the pyrolysis of contaminated mixed plastic waste. Green Chem 26(7):3853–3862. https://doi.org/10.1039/D3GC04396G
148. Shah SIH, Seehar TH, Raashid M, Nawaz R, Masood Z, Mukhtar S, Kalam MA (2024) Biocrude from hydrothermal liquefaction of indigenous municipal solid waste for green energy generation and contribution towards circular economy: a case study of urban Pakistan. Heliyon 10(17). https://doi.org/10.1016/j.heliyon.2024.e36758
149. Penke C, Moser L, Batteiger V (2021) Modeling of cost optimized process integration of HTL fuel production. Biomass Bioenerg 151:106123. https://doi.org/10.1016/j.biombioe.2021.106123
150. Li S (2024) Reviewing air pollutants generated during the pyrolysis of solid waste for biofuel and biochar production: toward cleaner production practices. Sustainability 16(3):1169. https://doi.org/10.3390/su16031169
151. Watanabe MM, Isdepsky A (2021) Biocrude oil production by integrating microalgae polyculture and wastewater treatment: Novel proposal on the use of deep water-depth polyculture of mixotrophic microalgae. Energies 14(21):6992. https://doi.org/10.3390/en14216992
152. Grande L, Pedroarena I, Korili SA, Gil A (2021) Hydrothermal liquefaction of biomass as one of the most promising alternatives for the synthesis of advanced liquid biofuels: a review. Materials 14(18):5286. https://doi.org/10.3390/ma14185286
153. Karka P, Johannsen I, Papadokonstantakis S (2024) Hydrothermal liquefaction integrated with wastewater treatment plants—life cycle assessment and technoeconomic analysis of process system options. Sustainable Energy Fuels. https://doi.org/10.1039/D3SE01211E
154. Shah AA, Sharma K, Haider MS, Toor SS, Rosendahl LA, Pedersen TH, Castello D (2022) The role of catalysts in biomass hydrothermal liquefaction and biocrude upgrading. Processes 10(2):207. https://doi.org/10.3390/pr10020207

155. Yu J, Biller P, Mamahkel A, Klemmer M, Becker J, Glasius M, Iversen BB (2017) Catalytic hydrotreatment of bio-crude produced from the hydrothermal liquefaction of aspen wood: a catalyst screening and parameter optimization study. Sustainable Energy Fuels 1(4):832–841. https://doi.org/10.1039/C7SE00090A
156. Ladu L, Morone P (2024) Sustainability assessments of bio-based products: from research to practice (and standards). Societal Impacts 3:100041. https://doi.org/10.1016/j.socimp.2024.100041

Chapter 11
Biocrude Oil by Fischer–Tropsch Process

Tengku Sharifah Marliza, Nurul Asikin Mijan, Wan Nor Adira Wan Khalit, and Mohd Razali Shamsuddin

Abstract The current chapter presents a detailed of Fisher-Tropsch (FT) process to produce biocrude oil. FT process is a catalytic chemical reaction that convert the syngas, the mixture gas consists of hydrogen and carbon monoxide into liquid hydrocarbons. Biocrude oil, sometimes called bio-oil, is a synthetic fuel typically characterised by its dark colour, viscous consistency, and composition of numerous oxygenated organic compounds along with water. This chapter thoroughly explores the reaction mechanism of the FT process, including intermediate products and elemental reactions. Additionally, the operational conditions for both low- and high-temperature FT processes are discussed, together with the design considerations for FT reactors, catalysts, and the distribution of products. Biocrude oil produced from FT process has many applications in different areas include road, air, and marine transport sectors. In this regard, the commercial use of biocrude oil, its efficiency and performance also described in this chapter.

Keywords Fisher-Tropsch · Biocrude oil · Catalyst · FT product · Syngas

T. S. Marliza (✉)
Department of Science and Technology, Universiti Putra Malaysia, Bintulu, Sarawak, Malaysia
e-mail: t_marliza@upm.edu.my

N. A. Mijan
Department of Chemical Sciences, Faculty of Science and Technology, Universiti Kebangsaan Malaysia, Bangi, Selangor, Malaysia
e-mail: nurul.asikin@ukm.edu.my

W. N. A. W. Khalit
Centre for Foundation Studies in Science, Universiti Putra Malaysia, Serdang, Selangor, Malaysia
e-mail: noradira@upm.edu.my

M. R. Shamsuddin
Preparatory Centre for Science and Technology, Universiti Malaysia Sabah, Kota Kinabalu, Sabah, Malaysia
e-mail: razalishamsuddin@ums.edu.my

M. Aslam et al. (eds.), *Production and Biorefining of Biocrude Oil: Current Status and Future Developments*, Advances in Sustainability Science and Technology,
https://doi.org/10.1007/978-981-96-5198-6_11

1 Introduction

Fisher-Tropsch (FT) process is a catalytic method that converts synthetic gas, which is carbon monoxide, CO, and hydrogen, H_2 into a petroleum-like products. FT process may produce from a range of carbonaceous feedstocks like natural gas, coal, and biomass with the presence of metal catalysts, reaction occurred around 150–300 °C at a certain pressure.

Biocrude oil or bio-oil, or sometimes also known as pyrolysis oil, is one of biomass feedstocks produced from liquefaction reaction using high-temperature and high-pressure liquid phase thermal processing technique. Biocrude oil is a synthetic fuel contain of complex mixture of oxygenated organic molecules with a large molecular weight range. In order to obtained biocrude oil, dried biomass went through a heating process with the absence of oxygen in a reactor and temperature of reaction at 500 °C followed by cooling process, then separation the oil from aqueous phase and other processes. The obtained biocrude oil resulted in non-volatile characteristics, less in thermal stability, corrosiveness, limited solubility with fossil fuels, and propensity to undergo polymerisation upon exposure to air. Figures 1 and 2 show an overview of the FT process and a general process flow for pyrolysis process for the biocrude oil production, respectively.

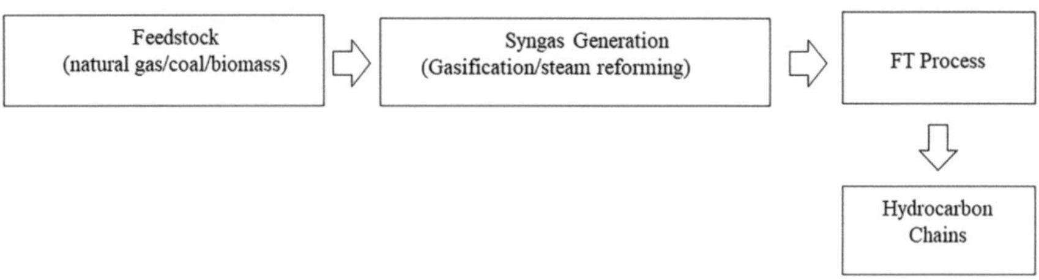

Fig. 1 An overview of FT process

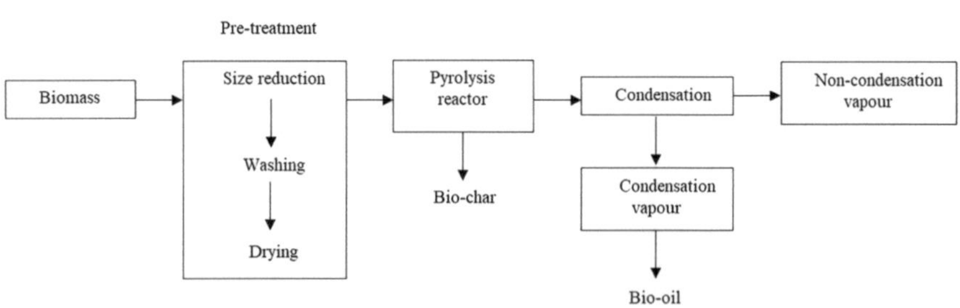

Fig. 2 A general process flow for pyrolysis to produce bio-oil [1]

2 Fischer–Tropsch History

The Fischer–Tropsch (FT) process has significantly contributed to the development of synthetic fuels and has a fascinating history that spans over a century. The FT process is named after its inventors, Franz Fischer and Hans Tropsch (Fig. 3). In the year of 1923, these German chemists developed the process at the Kaiser Wilhelm Institute for Coal Research in Mülheim, Germany. Their aim was to find a way to convert coal, which was abundant in Germany but limited in petroleum resources. The first FT industrial factory opened in Germany in 1936, and by the 1940s, FT liquids production exceeded one million tonnes per year.

During World War II, Germany utilised the FT process on a massive scale due to its limited access to oil reserves. This led to the synthetic fuels production from coal and other feedstocks, which significantly supported the German war effort. After World War II, the FT process was largely forgotten in the Western world, while it continued to be researched and applied in the Soviet Union and South Africa [2].

In the 1952, South Africa began to develop its FT industry, known as Sasol (South African Coal, Oil, and Gas Corporation) which is located in Sasolburg. This highly complicated facility combined the two versions of the FT process. Sasol used this process to manufacture liquid fuels from coal and subsequently a natural gas, helping the country overcome oil embargoes and limited access to conventional petroleum resources during the apartheid era. In the late twentieth century and early twenty-first century, rising concerns about energy security and environmental sustainability rekindled interest in the FT process.

Meanwhile, in 1993, the first commercial gas-to-liquid plant using cobalt-catalysed FT process began the production in Bintulu, Sarawak Malaysia by Shell. This facility produces a range of hydrocarbon products that includes naphtha, kerosene, and distillates. However, the manufacturing of paraffin and wax is optional [3]. Since then, most of the researchers and companies worldwide started investigating its potential to produce clean synthetic fuels from many feedstocks including municipal waste and biomass.

The FT process, when applied to natural gas or biomass feedstocks, it can produce synthetic fuels with lower emissions compared to conventional petroleum-based

Fig. 3 Inventors of Fischer–Tropsch synthesis (FTS) process, Franz Fischer (left) and Hans Tropsch (right) [4]

fuels. It remains a crucial technology in the quest for sustainable and alternative sources of transportation fuels. Nevertheless, this technology has made it an appealing choice for lowering greenhouse gas emissions and diminishing climate change.

3 Fischer–Tropsch Mechanism

The FT process is a catalytic chemical reaction that converts carbon monoxide (CO) and hydrogen (H_2) into hydrocarbons with varying molecular weights. Most favourable catalysts used in FT reaction are iron- or cobalt-based. Several chemical reactions take place simultaneously in the FT process that produces both desired and unwanted products. The reactions below outline the general types of reactions that occur in the FT reaction of hydrocarbons (paraffins, olefins, and alcohols), where the term -CH_2- refers to the methylene group of normal paraffins [4].

$$\text{Alkane: } n\mathrm{CO} + (2n+1)\mathrm{H_2} \rightarrow \mathrm{H(CH_2)}n\mathrm{H} + n\mathrm{H_2O}$$

$$\text{Alkene: } n\mathrm{CO} + 2n\mathrm{H_2} \rightarrow (\mathrm{CH_2})n + n\mathrm{H_2O}$$

$$\text{Alcohol: } n\mathrm{CO} + 2n\mathrm{H_2} \rightarrow \mathrm{H(CH_2)}n\mathrm{OH} + (n-1)\mathrm{H_2O}$$

$$\text{Carbonyl: } n\mathrm{CO} + (2n-1)\mathrm{H_2} \rightarrow (\mathrm{CH_2})n\mathrm{O} + (n-1)\mathrm{H_2O}$$

However, FT process conditions are typically selected to optimise the production of grater molecular weight hydrocarbon liquid fuels, as these are higher-value products. The reaction is extremely exothermic, with a typical standard reaction enthalpy of $-$ 165 $\mathrm{kJ/mol_{CO}}$ along with CO feed concentrations exceeding 25% and high CO conversions [5]. The production of methane ($n = 1$) is undesirable. The majority of the alkanes produced are typically straight-chain, making them suitable for diesel fuel. In addition to alkane production, competing reactions generate small amounts of alkenes, alcohols, and other oxygenated hydrocarbons [6]. Other side reactions occur during the process, with the water–gas-shift reaction being the most prominent.

$$\mathrm{CO} + \mathrm{H_2O} \rightarrow \mathrm{H_2} + \mathrm{CO_2}$$

Depending on the catalyst, temperature and type of method used, hydrocarbons ranging from methane to higher molecular paraffins and olefins can be produced. Small quantities of low-molecular weight oxygenate such as alcohol and organic acids are also produced.

4 Fischer–Tropsch Product and Elemental Reaction

Converting a mixture of H_2 and CO into aliphatic products is a multi-step reaction with several intermediate compounds. In general, FT process involving a sequence of elementary polymerisation reaction steps, namely chain initiation, chain propagation, and chain termination [7]. CO is converted to alkanes through the hydrogenation process, followed by hydrogenolysis of C–O bonds which involved the cleavage with H_2 and finally the formation of C–C bonds. Figure 4 depicts the dissociative adsorption of CO on metal atoms and the formation of carbide species. The insertion of adsorbed dissociated hydrogen into these carbide species generates the active CH_2 intermediate, which initiates the propagation step. The alkyl chain then desorbs from the metal after hydrogenation and β-scission, resulting in olefins or paraffin [8].

There have been several proposed mechanistic pathways of FT reaction, however, all reaction stages still involving with initiation, propagation, and termination steps. A reviewed by Qi et al. [9], proposed three main mechanisms of FT reaction involved a carbide mechanism, hydroxycarbene (enol) mechanism, and CO insertion mechanism. In carbide mechanism, authors indicating that the dissociative of CO adsorption serves as chain initiation step. The methylene group (CH_2) produced by the surface carbon hydrogenation is supposed to be the chain propagation monomer. Hydroxycarbene intermediates (–CHOH) facilitate C–C coupling in hydroxycarbene mechanism, whereas the CO insertion mechanism indicates chain growth by inserting CO into alkyl intermediates [9]. Chen et al. [10] had proposed the mechanism of C–O dissociation on cobalt FT catalyst. Direct CO dissociation on a Co catalyst was investigating using $^{12}C^{16}O/^{13}C^{18}O$ scrambling, both in the presence and absence of H_2. The initial scrambling rate without H_2 was notably higher than the rate of CO consumption under CO hydrogenation conditions, suggesting that the surface had reactive sites that capable to dissociate CO without the present of H atoms.

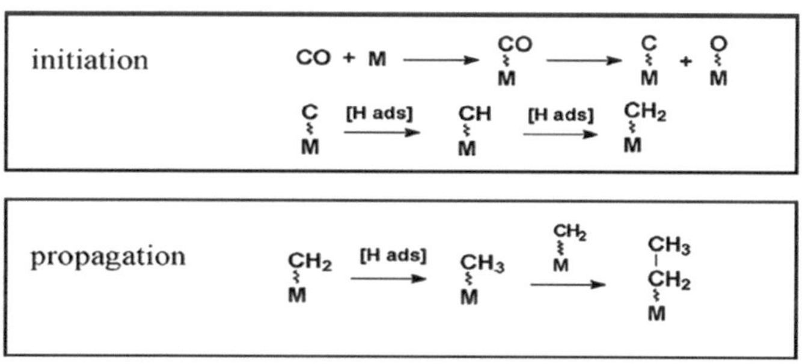

Fig. 4 Mechanism pathway of FT reaction [8]

5 Fischer–Tropsch Reactor

As previously noted in the preceding sections, the Fischer–Tropsch synthesis is a catalytic process utilised for the artificial generation of hydrocarbon chemical compounds and fuels derived from syngas feed, typically conducted using iron- or cobalt-based catalysts. Catalysts are integral to the functioning of the FT reactor. It is essential to take into account that FT reactors can be classified based on the thermodynamic approach utilised. Iron-based catalysts employed in a process with elevated temperatures (300–350 °C) facilitate the production of gasoline and linear, low-molecular weight olefins. This is also regarded as high-temperature FT (HTFT). Iron or cobalt catalysts are utilised in lower temperature procedures (200–240 °C) to produce diesel fuels and high molecular weight linear waxes, known as low temperature FT (LTFT) [11]. It is significant to highlight that a primary distinguishing characteristic between the HTFT and LTFT reactors lies in the lack of a fluid stage enveloping catalyst fragments present in HTFT reactors.

Throughout the years, there have been fundamentally four commercially utilised FT reactor designs. The multi-tubular fixed bed, the slurry reactor, and the fluidised-bed reactor, which includes both a fixed bed and a circulating fluidised bed reactor (CFB) (Fig. 5), are presented here [12, 13]. The fixed-bed reactor consists of multiple modest tubes, where the catalyst operates as an agent with surface activity. The moisture encases the passageways, precisely controlling temperature by expertly managing vaporisation pressure. The slurry type reactor is widely employed and consists of both liquid and robust elements, where the catalyst is not stationary but moves as small particles in conjunction with the reaction materials. The diversity of fluidised bed reactors is significant, but they are essentially characterised by the perpetual motion of the catalyst in the system. The CFB reactor utilises a fluidised bed of catalyst particles, which are kept in suspension through a flow of reactants. A fluidised bed reactor is a type of reactor that utilises a bed of catalyst particles, which are fluidised by the reactants.

5.1 High-Temperature Fischer–Tropsch Process

Figure 5a illustrates the inaugural commercial high-temperature Fischer–Tropsch fixed fluidised-bed reactor employed for Fischer–Tropsch synthesis at the Brownsville, Texas facility during the mid-1950s. It is employed to transform synthesis gas into hydrocarbons under designated circumstances. The reactors functioned at around 2 MPa and a temperature range of 300–350 °C. This temperature range exceeds the standard working temperature range for reactors, which is around 220–250 °C, by 100 °C. The FFB reactor is an optimal reactor for the HTFT process. Nonetheless, a significant issue was the deterioration of the catalyst, which resulted in separation difficulties and a swift loss of catalyst from the reactor zone. Furthermore, the FT reactions are significantly exothermic, releasing energy in the form of

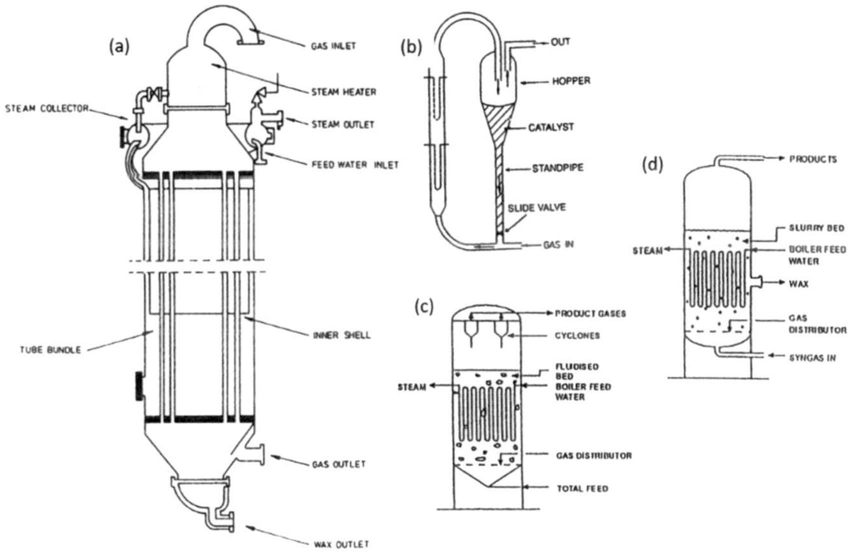

Fig. 5 The diagram of the different commercial FT reactor designs (**a**) Multi-tubular fixed-bed FT reactor, (**b**) circulating fluidised beds (CFBs), (**c**) FFB reactor, and (**d**) slurry phase bubbling bed reactor

heat from the system to its environment. Consequently, it is imperative to promptly dissipate the reaction heat from the catalyst particles to avert overheating, which could lead to the catalyst's deactivation from sintering and fouling, along with an unwanted rise in methane production. Rauch et al. [11] proposed that elevated rates of heat exchange can be attained by driving syngas at high linear velocities through long, narrow tubes containing catalyst particles to create turbulent flow, or more effectively, by employing a fluidised catalyst bed reactor known as CFB.

The CFB reactors, referred to as Synthol reactors, typically function at 350 °C with 375 bar pressure. In the FT process, the amalgamated gas supply (both fresh and recycled) will be introduced into the reactor from the bottom. The catalyst will descend the standpipe and pass through the sliding valve. The catalyst will be conveyed by the elevated gas velocity into the process region, where thermal energy is dissipated by heat transfer mechanisms. Subsequently, the by-product vapours and catalyst will be transferred into a large-diameter catalyst hopper, whereupon the resulting catalyst precipitates and the output fumes are discharged through a rotating chamber. While CBD is regarded as an optimal reactor for HTFT, it possesses certain limitations. CBD reactors are intricate systems that necessitate the circulation of substantial quantities of catalyst, resulting in significant erosion in specific areas of the reactor [14]. Furthermore, increasing catalyst loading in the CBD reactor is unfeasible. This is due to CFB utilising two phases of fluidised catalyst. The catalyst moves downward via the pipeline in a saturated portion while being transported upward through the "reaction" zone (Fig. 5b). To inhibit the ingress of feed gas into the standpipe, the variation in pressure across the standpipe must surpass the pressure in the reaction zone. The bulk density of iron-based catalysts declines with increased working temperatures due to carbon build-up, which simultaneously reduces the differential pressure across the standpipe. Therefore, augmenting the catalyst concentration in the reaction cycle is impractical to counteract the usual decline in catalyst activity with time. Here is the list of the beneficial characteristics of FFB versus CFB;

i. The cost of constructing the FFB is significantly cheaper. The FFB reactor is significantly lower in size with identical capacity.
ii. An expanded reaction section allows for the installation of numerous cooling coils, hence facilitating the introduction of additional reactants or gases through increased volumetric flow or elevated operating pressure.
iii. The entire catalyst charge is going to engage in the reaction; but, in the scenario of CFB, only a small fraction will do so.
iv. Carbon deposition is negligible in the FFB.

The catalyst selectivity pertinent to the spatial distribution of paraffins in the HTFT process, as indicated by the Schultz-Flory [SF] factor, is:

$$a = Cn + 1/Cn$$

Cn represents the quantity of molecules of paraffin generated with n carbon atoms, typically approximately 0.7. This leads to the generation of over 50% of C1–C4 gases on a weight-per-weight basis, while yielding fewer than twenty percent of fuel and larger-molecule hydrocarbon derivatives, respectively.

5.2 *Low-Temperature Fischer–Tropsch Process*

The fixed bed and slurry reactor (Fig. 5a, b) are employed in the LTFT process. A cobalt catalyst is frequently employed in fixed-bed reactors [12]. This process generally produces the majority of middle-distilled substances (kerosene, diesel), naphtha, and waxes. In a fixed-bed reactor, the tubes are often filled with catalyst material. The synthesis gas traverses the tubes, and at the reactor's base, the vapour-liquid products are collected. Slurry reactors have attracted interest considering to the distribution of the slurry, which enables more effective management of the process heat in a slurry bed reactor compared to a fixed-bed reactor. During this procedure, the catalyst will be disseminated across the liquid and gaseous reactants via a gas distributor, producing small gas bubbles. Reactants penetrate into the medium and disseminated to the catalyst fragments when gas bubbles rise through the reactor. Depending on the catalyst system, here is generally where they are transformed into volatile and fluid-like substances. The gaseous substances produced will rise into the gas bubbles and then accumulate at the top of the reactor. Liquid products may be recovered from the residual fluid utilising filtration, sedimentation, or hydrocyclone techniques [14].

Despite the slurry reactor demonstrating superior heat transfer rates that facilitated effective thermal regulation of the reaction, the recovery of liquid products remains relatively cumbersome due to their amalgamation with the suspending liquid. The challenge is exacerbated by the propensity of catalyst particles to degrade in the slurry, resulting in the formation of micro catalyst fragments that are also somewhat challenging to distinguish from the liquid products. Fixed-bed reactors lack the phase mixing that facilitates effective heat management in slurry bubble column reactors; however, they circumvent the challenges associated with liquid and catalyst separation. The following enumerates the advantages of a fixed-bed reactor in comparison with a slurry reactor:

i. The process of expanding to massive manufacturing reactors is conceptually straightforward, involving merely an increase in the number of tubes.
ii. Generally, there are no complications regarding liquid–solid segregation, since the catalyst is securely immobilised across the passageways.

Nonetheless, the fixed-bed reactor has various limitations, such as constraints in the variation in temperature along the tubes, restrictions in the depth of the primary tube sheets, challenges in the transfer between gas and liquid phases, and a limited capacity for solid catalysts. Utilising a fixed-bed reactor renders the continuation of LHFT impractical, as continuous catalyst replenishment is unachievable. The reactor must be deactivated prior to the introduction of new catalyst for the ensuing reaction. The design of an FT reactor is a complex process that necessitates meticulous consideration of multiple parameters, as elucidated in the aforementioned explanation and prior work [15–17]. Essential design concerns for an FT reactor include:

i. Heat Removal: The Fischer–Tropsch reaction is exothermic, indicating that heat is produced during the process. The evacuation of heat is essential to prevent catalyst deactivation and to sustain the required reaction temperature.
ii. Temperature Regulation: Temperature regulation is crucial to guarantee that the reaction occurs at the intended rate and to avert undesirable side effects.
iii. Catalyst Selection: The selection of a catalyst is contingent upon multiple criteria, including the intended product, the feedstock, and the operational parameters.
iv. Reactant Feed Rate and Composition: The feed rate and composition of the reactants can influence the selectivity and yield of the intended products.
v. Mass and Heat Transfer: The reactor design must provide effective mass and heat transmission to optimise the conversion of reactants into products.
vi. Pressure Regulation: Pressure regulation is crucial to guarantee that the reaction occurs at the intended rate and to avert undesirable side effects.
vii. Safety Considerations: The reactor's design must incorporate safety factors, including the potential for explosion or fire hazards.

Each reactor type possesses distinct advantages and problems, making the selection of the suitable reactor essential for the success of FT synthesis in a specific application.

6 Fischer–Tropsch Product Distribution

The FT process constitutes a chemical transformation that converts syngas into a range of hydrocarbons, encompassing paraffin (alkanes), olefin (alkenes), and oxygenated hydrocarbons. The distribution of products is impacted by various factors, including temperature, pressure, gas ratio, and catalyst. The discussion explores into the categories of products and their carbon chain distribution during the formation of hydrocarbons, including alkanes, alkenes, and oxygenated products. Predominantly, the FT process generates hydrocarbons with the formula (C_nH_{2n+2}), where n typically ranges from 10 to 20 [18]. Most produced alkanes are straight chains, making them suitable for use as diesel fuel. However, concurrent reactions also produce minor amounts of alkenes, along with alcohols and other oxygenated hydrocarbons. The products of the Fischer–Tropsch process exhibit a characteristic molecular weight distribution that can be described by the Shultz-Flory distribution [19]. This distribution is affected by various factors, including the feed composition, temperature, and pressure. The distribution of carbon numbers in the products depends on the operational conditions, with higher temperatures and pressures promoting the formation of heavier hydrocarbons [20].

The production of paraffins is a fundamental aspect of the FT process. The reactions yielding alkanes follow the formula $(2n + 1)H_2 + n\ CO \rightarrow C_nH_{2n+2} + n\ H_2O$, where the value of n typically ranges from 10 to 20 [21]. The majority of the alkanes generated are straight-chain in structure, making them well-suited for use as diesel fuel. Alongside the formation of alkanes, competing reactions also lead to the generation of minor quantities of alkenes. The formation of alkenes is less favourable compared to alkanes, as it may result in the undesirable production of by-products. Furthermore, the FT process gives rise to small amounts of oxygenated hydrocarbons, such as alcohols and organic acids. These oxygenated products can undergo additional reactions, contributing to a more complex distribution of hydrocarbons. The FT process yields a diverse range of hydrocarbons, including alkanes, alkenes, and oxygenated compounds. The distribution of carbon chain lengths in these hydrocarbons is influenced by factors such as the feedstock composition, temperature, and pressure. A comprehensive understanding and optimisation of these factors are crucial for tailoring the FT process to produce hydrocarbons with specific properties for a variety of applications.

6.1 Factors Affecting Product Distribution

Researchers have extensively investigated the FT process, which involves converting a synthesis gas mixture of carbon monoxide and hydrogen into liquid hydrocarbons. Numerous studies have examined the influence of process parameters on the distribution of FT products. Extensive laboratory experiments conducted over prolonged periods (TOS: 14,350 h) have provided insights into the FT synthesis process, which is representative of industrial-scale operations. The thermodynamic equilibrium distribution of FT process products has been analysed through the lens of minimising Gibbs free energy [22]. A comprehensive examination of the entire FT product distribution has been undertaken to determine the selectivity of catalysts employed in the process.

Thermodynamic equilibrium analysis is crucial for understanding FT process products, and the significance of such a distribution should not be underestimated. This method provides flexibility for complex problems where FT reaction pathways are unclear. Additionally, this equilibrium distribution may have significant implications for enhancing selectivity towards desired products [23]. However, additional research is required to determine FT product distribution in thermodynamic equilibrium.

At each temperature in the thermodynamic equilibrium system, there exists a linear relationship between the number of carbon atoms and the concentration of C4 plus products. An increase in reaction temperature leads to a shift in selectivity towards lower carbon number hydrocarbons, aligning with the relative stability of the products. Hence, low carbon number hydrocarbons are thermodynamically favoured at high temperatures. The chain growth probability is somewhat sensitive to reaction temperature, decreasing with an increase, consistent with available experimental literature. Moreover, the ratio of C3 product to C2 product increases with temperature, also consistent with experimental literature, indicating the extent of C2 product reincorporation.

The hydrocarbon product distribution during the FT process adheres to an Anderson-Schulz-Flory (ASF) distribution, expressed as $Wn/n = (1 - \alpha)2\alpha n - 1$, where Wn is the weight fraction of hydrocarbons with n carbon atoms, and α is the chain growth probability. This distribution stems from the reaction mechanism, encompassing side reactions and operating conditions [24]. High-pressure conditions favour the thermodynamic preference for hydrocarbons with elevated carbon numbers, and the likelihood of chain growth exhibits sensitivity to pressure. Regardless of the specific temperature and pressure values applied, products formed under thermodynamic equilibrium adhere to the ASF general polymerisation distribution when carbon numbers surpass approximately three. The FT synthesis can be conceptualised as a straightforward polymerisation reaction, wherein carbon monoxide or a C1 species derived from it functions as the monomer [25]. Schulz's equation, derived for free radical polymerisation, applies if there is a constant probability of chain growth. Flory and Anderson extended efforts to account for products formed by chain branching and to develop chain growth mechanisms. The polymerisation process aligns with the ASF distribution when assuming chain growth probability is independent of chain length.

The product distribution in FT synthesis is influenced by several key factors:

i. Catalyst Type: The choice of catalyst, such as iron or cobalt, can impact the diversity of the hydrocarbon products generated.
ii. Catalyst Particle Size: The size of the catalyst particles can affect the product distribution, with smaller particles often leading to the formation of longer-chain

hydrocarbons, and larger particles tending to yield shorter-chain hydrocarbons [18].

iii. Temperature: The temperature of the FT process can influence the distribution of the products. Higher temperatures typically favour the production of shorter-chain hydrocarbons, while lower temperatures generally result in the formation of longer-chain hydrocarbons.

iv. Pressure: The pressure under which the FT process is conducted can also impact the product distribution. Increased pressures tend to promote the formation of longer-chain hydrocarbons, whereas lower pressures are more likely to yield shorter-chain hydrocarbons [26].

v. Syngas Composition: The composition of the syngas feed, particularly the ratio of hydrogen-to-carbon monoxide, can affect the product distribution. A higher hydrogen-to-carbon monoxide ratio typically leads to the production of longer-chain hydrocarbons, while a lower ratio favours the formation of shorter-chain hydrocarbons [27].

6.2 Types of Products

The FT process, which converts synthesis gas into liquid hydrocarbons, can be conducted in two principal operating modes: Low-temperature Fischer–Tropsch synthesis (LTFT) and high-temperature Fischer–Tropsch synthesis (HTFT). LTFT synthesis typically occurs within the temperature range of 200 to 250°C and focuses on optimising the production of long-chain hydrocarbon waxes. Conversely, HTFT, operated at temperatures between 320 and 350 °C, aims to generate lower molecular weight alkenes and liquid hydrocarbon products suitable for use as gasoline and diesel fuels. Generally, the FT process operates within the temperature range of 150–300 °C (302–572 °F) [28]. Elevated temperatures result in faster reactions and increased conversion rates but tend to favour the production of lighter hydrocarbons, such as alkenes and liquid products within the gasoline and diesel fuel range. LTFT processes primarily target maximum wax production, while HTFT operations focus on generating low-molecular mass alkenes and liquid products within the gasoline (mainly) and diesel fuel range.

Figure 6 illustrates product distribution for high and low-temperature FT synthesis involving various chemical reactions. Typically, these chemical reactions are driven by the desired product profile. LTFT is preferred for maximising wax production, whereas HTFT is favoured for producing lighter hydrocarbons, including alkenes and liquid products within the gasoline and diesel fuel range.

The FT process involves a polymerisation-type reaction, yielding a diverse spectrum of hydrocarbons, predominantly linear paraffins and linear olefins, as well as oxygenates, primarily alcohols. The carbon numbers in the product range from C1 to even exceeding C120, encompassing a broad spectrum. Depending on the catalyst, reactor, and operating conditions, FT synthesis can generate an extensive array of hydrocarbons, including light hydrocarbons, gasoline, diesel fuel, and wax [29].

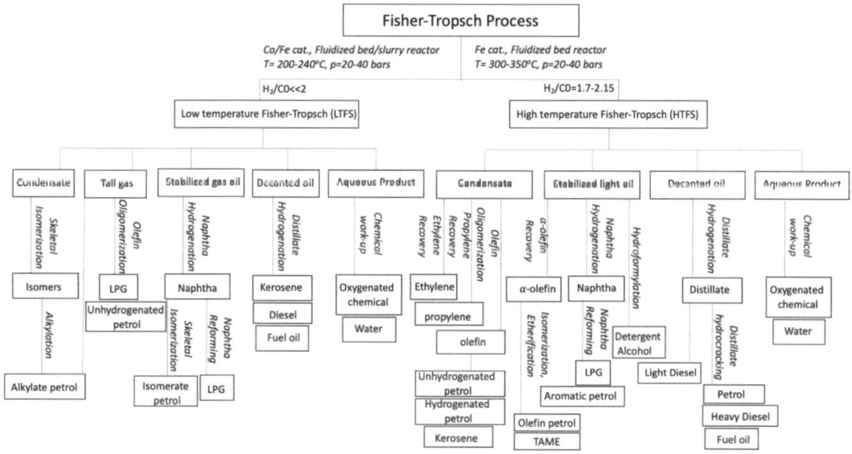

Fig. 6 Flow chart of product distribution by LTFT and HTFT process reactions

Olefin Products

Olefins, is a key component in the chemical industry, can be generated through the FT process. The FT to olefin (FTO) process is designed to efficiently convert syngas into light olefins, emphasising high selectivity using a suitable catalyst, minimising methane formation, and preventing excess CO_2 production [30]. Iron-based catalysts, widely available, are commonly employed in the FTO process. Olefins produced in the FT process primarily consist of propylene. The FT process allows for the direct production of light olefins from sources such as natural gas, biomass, solid waste, and CO_2, facilitated by established gasification/reforming technology. A significant challenge remains in achieving desirable olefin selectivity, especially for the C5+ range, while minimising undesired C1 by-products. Recent studies have concentrated on developing highly selective catalysts for light olefin production through the FT process, including photothermal FT synthesis driven by solar light under low pressure [31].

FT products consistently contain olefins in varying amounts, contingent on the conditions employed in all cases investigated herein. The low carbon number hydrocarbons formed in the FT process are olefinic. Product compositions exhibit fluctuating olefin amounts based on the temperature applied. Elevated reaction temperatures result in a shift towards greater selectivity for hydrogenated products, with low concentrations of C2 olefinic products, high concentrations of C3 olefinic products, and essentially constant concentrations of high-molecular weight olefinic products in the C4–C10 range [32]. Thus, under specified conditions, olefins produced in the Fischer–Tropsch process are predominantly primary, essentially comprising propylene. Conversely, olefin amounts decrease with reduced temperature, indicating that higher temperature operation favours olefinic product formation.

Paraffin Product

The production of paraffin hydrocarbons is a crucial aspect of the FT process. These straight-chain, saturated hydrocarbons have extensive industrial applications, including in the manufacturing of fuels, lubricants, and solvents. Numerous scholarly investigations have been conducted to explore the synthesis and characteristics of paraffin hydrocarbons within the FT process. For instance, a study led by Muleja et al. [29] scrutinised the impacts of reaction conditions on the selectivity and yield of paraffin hydrocarbons. The researchers observed that elevated reaction temperatures and prolonged residence times were correlated with heightened paraffin hydrocarbon selectivity. Similarly, an investigation by Patzlaff et al. [33] explored the influence of diverse feedstocks on paraffin hydrocarbon production in the FT process, revealing that feedstock composition significantly affected both the yield and composition of the paraffin hydrocarbons.

The FT process can produce a variety of paraffins, from light gases like methane to heavy compounds in diesel and waxes. The paraffin formed depend on reaction conditions and how the products are separated. Temperature is key—higher temperatures favour lighter paraffins like gasoline, while lower temperatures favour heavier paraffins for diesel and waxes [24]. Advances in FT technology focus on catalysts to optimise the production of specific paraffins for applications like transportation fuels. Furthermore, research has explored the hydrocracking of FT paraffin mixtures to acquire engine fuels, showcasing continual efforts to refine the production and utilisation of paraffins derived from the FT process [23].

The specificity of methane production in the FT process is contingent upon the operational conditions, particularly the hydrogen-to-carbon monoxide feed ratio. An elevation in this ratio prompts a shift towards hydrocarbons with lower carbon numbers, primarily methane [22]. In specified conditions, the selectivity for methane remains relatively subdued due to the utilisation of low feed ratios of hydrogen-to-carbon monoxide. This has noteworthy implications for refining the selectivity of desired products. However, higher feed ratios tend to favour the unwanted formation of methane, known as methanation. Notwithstanding this limitation, a variety of feed compositions can be utilised in the practical implementation of the FT process.

6.3 Oxygenates Products

The FT process gives rise to various oxygenated by-products, such as alcohols and organic acids, which are formed due to side reactions, including the water–gas-shift reaction, occurring during the synthesis process [34]. The distribution of these oxygenated compounds is significantly influenced by the reaction conditions, particularly temperature and catalyst type. Specifically, higher temperatures tend to favour the production of lighter oxygenates, while lower temperatures promote the formation of heavier oxygenates.

Comprehending the distribution of oxygenated by-products in the FT process is crucial for optimising the production of desired products. The oxygenated compounds encompass alcohols, which are straight-chain or branched hydrocarbons containing hydroxyl groups, and organic acids, which are hydrocarbons with carboxylic acid groups [31]. These by-products exhibit distinct properties, including molecular structure, polarity, reactivity, physical characteristics, and diverse applications.

Oxygenated compounds can exhibit a diverse array of molecular structures, ranging from straight-chain to branched hydrocarbons. For instance, the alcohol family encompasses compounds with varying carbon chain lengths, including methanol, ethanol, and butanol. The presence of polar functional groups, such as the hydroxyl (–OH) group in alcohols, imparts polarity, enabling these compounds to engage in hydrogen bonding interactions with other polar molecules and enhancing their solubility in polar media. Furthermore, the reactivity of these oxygenated compounds varies, as alcohols can undergo oxidation reactions to form aldehydes or ketones, while organic acids can undergo dehydration reactions to produce alkenes or alkanes. Physical properties like density and viscosity depend on the compound type and molecular structure, e.g. straight-chain alcohols have lower densities than branched alcohols [30].

The oxygenated compounds from the FT process find applications in diverse areas, including transportation fuels, chemical feedstocks, and intermediate products for chemical production. Methanol and ethanol, for instance, can serve as oxygenates in gasoline, while butyric acid and acetic acid contribute to the production of butyl alcohol.

Recent research has concentrated on developing new catalysts and process technologies to enhance FT reaction selectivity, aiming to maximise the production of specific oxygenated compounds suitable for various applications. Catalysts based on ruthenium, nickel, cobalt, and iron have been explored to minimise methane formation and optimise the production of desired oxygenated products [35].

Iron and cobalt are the most used active metal catalysts in FTS, as these provide technical and economic advantages. The world's largest-scale implementation of FT technology is a series of plants operated by Sasol in South Africa, a country with large coal reserves, but little oil. With a capacity of 165,000 Bpd at its Secunda plant, Sasol uses coal and natural gas as feedstocks and produces various synthetic petroleum products, including most of the country's diesel fuel [36].

7 Fischer–Tropsch Catalysis

Catalyst deactivation poses a challenge in the FT process, potentially leading to a decline in reaction rate and product yield. Strategies to counteract catalyst deactivation involve the use of promoters, which can enhance the mechanical properties of the catalyst and stabilise a high metal surface area.

The selection of reactor design also influences catalyst performance, with conditions favouring FT reactions varying based on factors like Fe catalysts, high feed rates, and high pressure. Diverse reactor designs, including fixed-bed reactors, fluidised-bed reactors, and trickle-bed reactors, have been explored to optimise catalyst performance. In some instances, the catalysts utilised in the Fischer–Tropsch process can be recovered and reused, contributing to reduced overall costs. Techniques for catalyst recovery include methods such as hydrogen recovery units, which extract the necessary hydrogen from the tail gas or feed syngas stream [37]. Cobalt-based catalysts are frequently preferred due to their extended activity, diminished water–gas shift activity, and lower modified product yield. Ongoing research efforts focus on developing innovative catalysts and support materials to enhance the FT process [38]. Certain promoters, like rhenium, platinum, and ruthenium, are believed to improve the activity of cobalt-based catalysts. The selection of catalyst, support materials, reactor configuration, and recovery methods significantly influences the efficiency and effectiveness of the Fischer–Tropsch process. Continuing research aims to engineer innovative catalysts and process improvements to optimise the Fischer–Tropsch process, enabling the production of a diverse range of hydrocarbons for various applications.

7.1 Type of Catalysts

The catalyst implemented has a significant impact on the degree of selectivity of reactions and distribution of products. Iron (Fe), cobalt (Co), nickel (Ni), and ruthenium (Ru) operate as active catalysts in the FT process [4]. Fe and cobalt are the most often used because of their ability to transform inexpensive precursors into complicated hydrocarbon mixtures, demanding additional refining [39]. Co-based catalysts have a stronger preference for methane, an undesirable hydrocarbon, and short-chain hydrocarbons. In contrast, Fe catalysts are known for their ability to enhance the formation of long-chain hydrocarbons, which are excellent for diesel fuel [40].

The catalyst used influences the formation of oxygenated molecules, resulting in different quantities of alcohols and organic acids as by-products. The FT process normally operates at temperatures that vary between 150 to 300 °C and pressures ranging from one-to-many tens of atmospheric pressure. FT reactions are highly influenced by operational factors such as catalyst type, syngas flow rates, and pressure. Fe catalysts show favourable to reactions with high gas flow rates, and high-pressure conditions. Current studies have focused on discovering catalysts and process innovations to control the selectivity of FT reactions, to optimise the production of specific hydrocarbon products suitable for a variety of applications, such as fuels for transportation and chemical feedstocks [37]. The choice of catalyst is a pivotal determinant of the efficiency and selectivity of the FT process, exerting influence over the type and quality of hydrocarbon products obtained.

Cobalt Catalyst

In the FT process, catalysts derived from Co have several both benefits and drawbacks. One of their key advantages is their excellent intrinsic activity at lower temperatures, which makes them ideal for FT synthesis at low temperatures [41]. Furthermore, they show a high likelihood of chain elongation, resulting in the synthesis of desired long-chain hydrocarbons suited for diesel fuel. Cobalt-based catalysts, while exhibiting enhanced selectivity towards the production of desirable long-chain hydrocarbons suitable for diesel fuel, also demonstrate some drawbacks. These catalysts display increased selectivity towards undesirable methane and short-chain hydrocarbons, which negatively impacts the product distribution and leads to a greater proportion of lower-value compounds [42]. Additionally, certain cobalt-based catalysts may have significant interactions with the support material, resulting in the formation of undesired mixed metal oxides. These factors present challenges that must be addressed to optimise the Fischer–Tropsch process [42].

Recognising the drawbacks of cobalt-based catalysts, recent scholarly investigations have focused on mitigating these challenges and enhancing the performance of these FT catalysts. While cobalt-based catalysts exhibit commendable activity and selectivity for FT synthesis, it is essential to account for their propensity towards methane production and potential interactions with the catalyst support during their implementation.

Iron Catalyst

The utilisation of iron as a catalyst in the FT process offers several distinct advantages. Iron is widely available and cost-effective, making it an appealing option for large-scale commercial applications [43]. Additionally, iron-based catalysts exhibit a high degree of resistance to deactivation caused by sulphur poisoning, which is a significant problem in the FT synthesis. Owing to their pronounced selectivity towards the production of long-chain hydrocarbons, iron catalysts are capable of selectively generating compounds suitable for diesel fuel [44]. Furthermore, ongoing investigations have revealed that the active centres within iron-based catalysts are readily accessible, contributing to their overall effectiveness in the FT process [45–47].

Despite the advantages of employing Fe catalysts, such as their widespread availability and resistance to sulphur poisoning, they also exhibit notable drawbacks in comparison with Co-based catalysts. Specifically, Fe catalysts demonstrate lower catalytic activity and selectivity in FT reactions [39]. Additionally, Fe-based catalysts are more prone to oxidation and coke formation, leading to a gradual decline in their activity and selectivity over time. This challenge is particularly pronounced in fixed-bed reactors, where sulphur contamination and carbon build-up can potentially deactivate the catalyst. Furthermore, the accumulation of significant quantities of carbon in fluidised bed reactors may also influence catalyst efficiency. These limitations have necessitated ongoing research and development efforts aimed at mitigating the drawbacks associated with the use of Fe-based catalysts in the FT process [35, 46].

Nickel Catalyst

The utilisation of nickel as a catalyst in the FT process presents both advantages and disadvantages. Nickel exhibits exceptional catalytic activity, making it an intriguing candidate for FT hydrocarbon synthesis from syngas. However, nickel catalysts possess a strong hydrogenation capacity, which results in the preferred production of methane at high temperatures and the formation of undesired light hydrocarbons, such as C4 gases, thereby lowering the average liquid hydrocarbon output [30, 41, 48]. The tendency of nickel to enhance methane generation is a significant drawback, as methane is not the targeted end product of the FT process. Furthermore, nickel catalysts demonstrate lower selectivity in creating higher molecular weight hydrocarbons, which are useful for the production of diesel fuels and other liquid hydrocarbon products [30]. In contrast, cobalt and iron catalysts are more widely employed in the FT process due to their stronger selectivity towards the target hydrocarbons and a lesser propensity to induce methane generation [33]. Consequently, while nickel has high catalytic activity, its ability to produce light hydrocarbons, particularly methane, and its reduced selectivity for desirable liquid hydrocarbon compounds are significant drawbacks in the context of the FT process.

Ruthenium Catalyst

Ruthenium has emerged as a viable Fischer–Tropsch catalyst, demonstrating greater efficiency at lower temperatures compared to other transition metals such as nickel, cobalt, and iron [38]. However, its high cost and limited catalytic activity restrict its widespread industrial deployment [23]. The use of ruthenium as an FT catalyst offers several advantages, including its excellent catalytic activity, selectivity, and stability under reaction conditions. Notably, ruthenium-based catalysts comprise solely the metal without any additional promoters, resulting in a straightforward system that is amenable to mechanistic investigation [36].

The disadvantages of using ruthenium as an FT catalyst are its expensive price and limited availability. The scarce supply of Ru makes scaling up the fabrication of Ru-based catalysts for applications in the industry challenging [49]. In addition, the employing of Ru catalysts may need greater temperatures of reaction to achieve the appropriate product distribution, resulting in higher energy consumption and operational expenses [26, 29].

7.2 Support Catalyst

Oxides (e.g., alumina, silica), carbon-based materials, and zeolites are typical support materials that have a major influence on catalyst performance [30]. For example, research on Co catalysts supported by carbon-based materials for FT synthesis has shown a major disadvantage such as high Co-support interaction, which leads to a generation of undesired mixed metal oxides [31]. Research on Fe-based catalysts indicates that operating conditions, such as elevated input flow rates and pressure, may enhance FT reaction [44]. Additionally, the catalyst's selectivity towards specific hydrocarbons, including methane, is influenced by the choice of support material. For instance, low-temperature cobalt-based catalysts tend to exhibit greater selectivity towards methane, an undesirable product [32]. Consequently, the nature of the support material has a significant impact on catalytic activity, selectivity, and stability in the FT process. Ongoing studies are focused on optimising support material selection to improve FT catalytic performance and regulate product distribution to meet specific industrial requirements [18].

Extensive research has been undertaken to explore the impact of various oxide supports on the catalytic activity and selectivity of FT catalysts. In-situ investigations of cobalt catalysts incorporating metal oxides such as alumina (Al_2O_3) and magnesia (MgO) have revealed the influence of the support material on the structure and behaviour of the active cobalt clusters. A significant issue with oxide supports is the potential for substantial interactions between the cobalt and the support, leading to the formation of undesirable mixed metal oxides that can alter the catalytic characteristics [42]. The current perspective emphasises the importance of metal-support interactions in shaping catalytic behaviour, particularly the relationship between active site availability and particle size [32]. Furthermore, the choice of oxide support has been demonstrated to modify the selectivity of the FT process, with lower temperature cobalt-based catalysts exhibiting increased selectivity towards methane, an undesirable product. Consequently, the selection of the appropriate oxide support is crucial for tailoring FT catalysts to achieve the desired activity, selectivity, and stability for the targeted hydrocarbon product outputs [50]. Continued research in this area is necessary to develop new and more effective FT catalysts.

Carbon-based materials have gained prominence as catalyst supports due to their exceptional thermal and chemical stability, superior thermal conductivity, and expansive surface area [42]. The carbon support structure can influence the selectivity and distribution of produced hydrocarbons, as well as the rate of CO turnover. Additionally, the quantity of surface oxygen groups on carbon supports has been shown to impact the FT synthesis process [51]. While it remains challenging to synthesise highly dispersed and stable catalysts on carbon-based supports, ongoing research focuses on modifying the surface chemical characteristics of these materials to enhance their performance as catalyst supports.

Carbon nanotubes, carbon nanofibers, and graphene have been investigated as catalyst supports due to their special characteristics, such as large surface area and good mechanical strength [51]. The utilisation of carbon-based supports, nonetheless, is not without drawbacks, such as the unwanted strong cobalt-support interactions that result in mixed metal oxide production. Furthermore, carbon accumulation on cobalt catalysts can affect the FT process by forming amorphous carbon, which can block active sites and limit catalytic activity [18].

Considering their distinctive microporous structure and alterable acidity, zeolites can impact mass transfer and product selectivity in FT synthesis [40]. Several research has been conducted using zeolite-supported Co catalysts for FT synthesis, to tailor the molecular weight distribution of hydrocarbons. Research has shown that the zeolite support material can influence the composition of FT products, enabling the selective generation of isoparaffins and diesel fuel compounds [35, 40]. The acidity of the zeolite support appears to be an important factor affecting the distribution of products, particularly the production of branched paraffins. Furthermore, the utilisation of zeolite nanosheets has been observed to increase the synthesis of olefins during the Fischer–Tropsch process, demonstrating the ability of well-constructed zeolite promoters to influence reaction pathways and product yields [34]. Overall, the specific type of zeolite support employed has a substantial impact on the catalytic performance and product selectivity in the Fischer–Tropsch process, allowing for the tailoring of certain hydrocarbon products. These findings emphasise the importance of understanding the influence of zeolite characteristics on the Fischer–Tropsch synthesis when developing efficient catalysts that generate the desired hydrocarbon distributions.

7.3 Promoter

Incorporating promoters, substances that enhance the catalyst's activity and selectivity, can boost the efficiency of the FT process. These promoters can be added to the catalyst to improve its performance by increasing surface area, enhancing the dispersion of active metals, and altering electrical properties [50]. The specific promoter used in the FT process significantly impacts the reaction kinetics and product distribution. For instance, combining promoters like potassium, copper, and zinc with cobalt-based catalysts has shown enhanced selectivity towards higher molecular weight hydrocarbons, while rhenium and platinum have been found to improve the activity of iron-based catalysts [49, 50].

Emerging research has concentrated on unravelling the fundamental mechanisms driving the impacts of promoter additives in the FT process. Recent investigations of Co/Al_2O_3 catalysts have found that the incorporation of promoters such as potassium and copper can enhance the catalyst's surface area, leading to increased activity and selectivity towards higher molecular weight hydrocarbons [52]. The utilisation of promoters appears to be an effective approach for enhancing catalyst performance and optimising product yields in the FT process. The selection of a specific promoter is influenced by the unique characteristics of the catalyst and the prevailing reaction conditions, and ongoing research endeavours are focused on developing innovative and more effective promoter systems to further improve the efficiency of the FT process [18, 36, 49, 52].

7.4 Catalyst's Recovery and Reuse

The economic and environmental consequences associated with the recovery and recycling of catalysts in the FT process are significant. Over time, these catalysts can become deactivated due to phenomena such as sintering, carbon deposition, and impurity poisoning, including from sulphur [53]. Consequently, the development of effective procedures for catalyst reactivation and regeneration is critical. Research efforts have focused on understanding the mechanisms of deactivation and devising regeneration methods, including oxidation, reduction, and washing, to remove carbon and other deactivating compounds [44, 48]. Furthermore, exploration of promoters and novel catalyst compositions has been conducted to enhance stability and activity, thereby prolonging operational lifespans [26]. By recovering and recycling catalysts, the FT approach becomes more sustainable and cost-effective, increasing its feasibility for large-scale fuel and chemical production. Current research endeavours aim to optimise catalyst management strategies and improve the overall efficiency and environmental performance of the FT process.

The catalyst employed in the FT process can gradually lose its effectiveness due to various factors, such as the presence of impurities or sintering. A redox-based method is utilised to recover and reuse the catalyst, involving the reduction of the deactivated catalyst with hydrogen to restore its activity. Importantly, the catalyst must be passivated after the reduction step to prevent oxidation of the metal particles when exposed to air [54]. Ex-situ studies have been conducted to investigate the reduction in catalytic activity caused by sulphur poisoning of a cobalt-based catalyst [53]. Recent research has focused on the development of novel catalysts and process methods to manage the selectivity of the Fischer–Tropsch reaction, with the aim of enhancing the production of specific oxygenated compounds suitable for applications such as transportation fuels and chemical feedstocks [45].

The incorporation of promoters such as Rh, Pt, and Ru is a common approach to enhance the physicochemical properties of the catalyst and facilitate the reduction of cobalt oxide to the active metallic phase [36]. The redox process is crucial for the recovery and recycling of the catalyst in the Fischer–Tropsch process, as it enables the optimisation of product yields and the minimisation of undesirable by-products. Passivation of the catalyst after reduction is necessary to prevent metal particle oxidation upon exposure to air. Additionally, the addition of promoters can improve the mechanical characteristics of the catalyst as well as the reduction of cobalt oxide to the active metal phase, resulting in enhanced catalytic performance [18, 24, 29, 36].

8 Commercialisation

The commercialisation of FT in biocrude oil involves the manufacture of liquid hydrocarbons from organic matter using the FT process. Biocrude oil can be used as a raw material for the FT process to produce liquid hydrocarbons and synthetic fuels. The FT process can provide a sustainable approach to fuel production by converting waste materials into liquid hydrocarbons and synthetic fuels, reducing dependence on carbon-rich fossil fuels [55]. The commercialisation of FT in biocrude oil can provide a cost-effective approach to fuels production, reduce carbon emissions, and provide surrogate source of liquid fuels.

The commercialisation of FT in Ras Laffan, Qatar involves the production of gas-to-liquids (GTL) products using the FT process. The Pearl GTL factory in Ras Laffan Industrial City, Qatar, is a strategic partnership between Qatar Petroleum and Shell, and it is the world's largest supplier of GTL products, providing around ten times the total German FT capacity in 1944. The plant uses natural gas as a raw material for the FT process to generate liquid hydrocarbons and synthetic fuels. The FT process can generate varied length hydrocarbons from any carbonaceous feedstock, such as coal, natural gas, or biomass. The FT process can provide a sustainable approach to fuels production by converting waste materials into liquid hydrocarbons and synthetic fuels, reducing dependence on carbon-rich fossil fuels.

Sasol operates two FT synthesis-based fuels production plants in Secunda, South Africa, known as Sasol II and Sasol III. The two facilities comprise 80 Sasol-Lurgi Fixed-Bed Dry Bottom (FBDB) gasifiers, with a combined production of roughly 150,000 barrels per day (bpd) from both units, peaking at 160,000 bpd. Sasol's feedstocks include coal and natural gas, and it manufactures a wide range of synthetic petroleum products, together with the majority of the country's diesel fuel. South Africa is currently a global leader in the area of FT synthesis, and Sasol has more experience than any other company in the transformation of hydrocarbon gas-to-liquid fuels.

In addition, PetroSA which is a South African oil and gas company also has been involved in the transformation of synthetic fuels using FT technology. PetroSA operates a GTL plant in Mossel Bay, South Africa, which uses natural gas as a raw material for the FT process to manufacture liquid hydrocarbons and synthetic fuels. The plant has a mass-production capacity of 36,000 barrels per day. The FT process comprises a catalytic chemical reaction in which the synthetic gas's carbon monoxide (CO) and hydrogen (H_2) are transformed into liquid hydrocarbons. The FT process is used to transform liquid hydrocarbons in both coal liquefaction and gas-to-liquids technology. The products of the FT process can be refined into numerous fuel types, like gasoline, diesel, and jet fuel.

Nonetheless, Shell Middle Distillate Synthesis (SMDS) process is a gas-to-liquids (GTL) technology developed by Shell which converts natural gas into high-quality middle distillate products like diesel and jet fuel. The SMDS process involves a two-stage process in which a catalyst is first reduced using GTL off-gases to produce CO_2 and steam that is used for the next stage. In the second stage, the reduced catalyst is used to change syngas (a mixture of CO and H_2) into middle distillate products. The SMDS process is a type of FT synthesis that uses a cobalt-based catalyst. The SMDS process is applied in Malaysian and South African GTL plants operated by Shell. The SMDS approach combines the chain-length-independent FT chain-growth reaction with a chain-length-dependent cracking mechanism in two stages. The SMDS process uses a cobalt-based catalyst and can produce naphtha, kerosene, and gas oil with varying yield production.

Meanwhile, Velocys is a company that specialises in the development of small-scale GTL technologies that use FT synthesis to transform natural gas and other feedstocks into high-end products like diesel and jet fuel. Velocys' FT technology is based on microchannel reactors that offer high heat and mass transfer rates, resulting in high conversion rates and selectivity. The FT process incorporates a catalytic chemical reaction in which the synthetic gas's CO and H_2 are transformed into liquid hydrocarbons. Commercial FT facilities utilise various technologies, and the syncrudes generated in the various plants are not the same. Velocys' technology adds significant intellectual property and optimisation to the long-established FT process, enabling the production of sustainable aviation fuels (SAF) safely and effectively in a reactor at a large commercial scale. Velocys also produces the world's most compact FT reactor for the synthesis of hydrocarbons and allows the transformation of advanced biofuels from sustainable carbon sources such as household waste and agricultural residues.

Rentech, Inc. has developed a proprietary FT process for the transformation of synthetic fuels from a variety of raw materials, like coal, natural gas, and biomass. The company has built its first commercial synthetic fuels plant using the FT process near Natchez, Mississippi. Rentech has also partnered with ClearFuels Technology Inc. to build a pilot-scale biorefinery in Florida that will use gasification and FT processes to produce second-generation biodiesel from energy wood. Additionally, Rentech has licensed its FT process to Donyia Polo Petrochemicals Ltd. in Mumbai, India, for the transformation of natural gas to synthetic hydrocarbons. The FT process has also been studied for its potential applications in the co-production of FT fuels and electricity with CO_2 capture.

Other than that, the FT process can also be used to provide a diverse range of chemicals, including methanol and olefins [56]. Methanol is a critical component in the manufacture of a broad petrochemical products. Methanol may be used to produce higher-value-added products through processes like methanol to olefin, methanol to propylene, and methanol to petrol [57]. In the methanol-to-olefins (MTO) process, methanol is preheated to gaseous phase prior to be injected into the MTO reactor for olefin synthesis [58]. The FT process can also be adjusted to produce high selectivity catalysts for the transformation of olefins and higher hydrocarbons. A methanol-mediated method or a modified FT synthesis (FTS) can be used to convert CO_2 into liquid fuel-grade hydrocarbons [59]. In the first approach, CO_2 is converted to methanol, which is subsequently turned into hydrocarbons using the commercial zeolite-catalysed methanol-to-gasoline (MTG) process. The FT process can produce light olefins using iron-based catalysts [39]. The process aims to transform syngas to light olefins with higher selectivity over a suitable catalyst, lower methane transformation, and minimise carbon deposition. The FT to olefin (FTO) process is based on a one-step process using synthesis gas without alteration of H_2/CO ratio.

Apart from that, FT process also has been commercialised in the production of lubricants and waxes. FT wax is a by-product wax produced in the manufacture of liquid fuels and other products originating from coal and natural gas. FT wax has good compatibility with polyvinyl chloride (PVC) and can be used as both internal and external lubricants in PVC processing and production. FT synthetic wax is mainly utilised as an outermost lubricant in PVC processing and production because it has small molecular weight, low melt viscosity, high melting point, and narrow melting range. FT wax can replace imported FT wax and polyethylene wax in the fields of PVC processing, ink coating, and rubber protection. FT wax is abundant in saturated straight-chain hydrocarbons, which enable for a narrower carbon number distribution, reduce oil content, higher melting point, and greater hardness [60]. FT wax is characterised by low steric hindrance and very low viscosity when melted. FT wax is used in processed paper, hot melts, polish, ink, lubricants, asphalt, and rubber. There are a few industries that use the FT process to produce lubricants and waxes such as Myers, TER Chemicals, Sasol, Exxon Mobile, and Shell.

9 Research Development

9.1 FT Development Effort to Industrial Scale

Here are several significant advancements in the large-scale implementation of the FT method:

Process Development Strategy

The advancement of the FT technique entails conducting upward process investigations to select the ideal procedure system, reactor innovation, and operation circumstances, while downstream efforts concentrate on compiling every development endeavour into an extensive procedure direction [61]. The aim of the reactor model is to forecast the efficiency of a commercial FT method, incorporating all the scaling up principles for hydraulics and transference. Considering procedure effectiveness, the commercial manufacturing of renewable fuels in fluid reactions (Fig. 7) is favoured because of a consistent heat profile, adjustable hydraulics, versatile catalyst loading, and reduced costs contrasted with rigid reactors.

Catalyst Performance

The advancement of catalysts is essential for attaining optimal preference and operation in the FT procedure. Innovators have worked on establishing innovative catalysts for FT production that boost product selection while improving both the reaction speed and output of the desired end products. Table 1 presents significant advancements in the fabrication of FT catalysts.

The focus of progress in FT catalysts has been on the construction of the catalyst, the effectiveness of the end products, the capability of the catalyst, and the support for the catalyst. These developments have arisen from thorough foundational studies and a profound understanding of the critical elements that affect FT chemistry.

Scale-Up

The FT approach must be escalated to achieve economic viability. The most comprehensive application of FT technology worldwide is found in numerous Sasol-operated facilities in South Africa, a nation abundant in coal reserves while lacking in oil [17]. However, the efficacy of the FT process is limited by the number of trains and requires

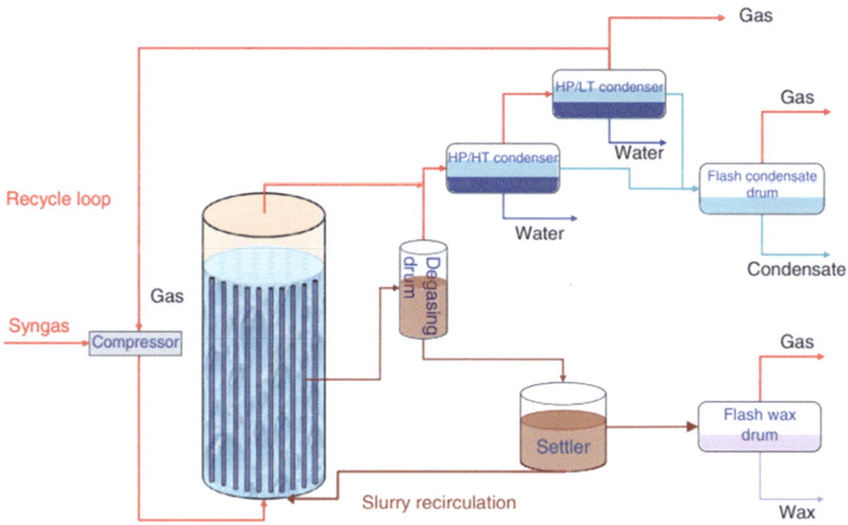

Fig. 7 Flow chart illustrating the reactor model of large-scale FT process [61]

Table 1 Key developments in the research and development of FT catalysts

Key development of catalyst	Justifications
Catalyst design	The architecture of catalysts is essential for attaining elevated activity and selectivity in the Fischer–Tropsch method. Experts have been investigating the application of innovative bimetallic catalysts, primarily those related with cobalt-based catalysts. Enhancing the effectiveness of the FT procedure for fuel containing hydrocarbons is crucial [8, 63]
Product selectivity	The FT technique parameters are often selected to optimise the production of greater molecular mass hydrocarbon liquefied fuel which are more valuable products. Scientists have been optimising the product selectivity of the Fischer–Tropsch technique to generate long-chain hydrocarbons and clean energy
Catalyst performance	The achievement of optimal selection and efficiency in the Fischer–Tropsch system depends on catalyst efficacy. Researchers have been examining the catalyst's unpredictable actions over time in fixed-bed and slurry reactors to obtain insights into its evolution in realistic process conditions
Catalyst support	The support for catalysts is essential in Fischer–Tropsch manufacturing, since its surface region and pore arrangement markedly improve catalytic efficacy by affecting the shape, variation, and reducibility of the reactive component. Current studies on the impact of support in Fischer–Tropsch catalysis have primarily concentrated on evaluating the catalytic characteristics of cobalt catalysts using mesoporous and zeolite supports, including SiO_2, Al_2O_3, and TiO_2 [8, 64]

many rows of tubes for scalability, suggesting that catalyst replacement upon deactivation will be labour-intensive [7]. This facilitates the implementation of FT across both minor and major dimensions.

The industrial-scale development of FT necessitates the fulfilment of several phases. Figure 8 depicts recent developments in Fischer–Tropsch (FT) facilities utilising an iron-based FT catalyst within fluid vessels measuring an inner diameter of 5.3 m [62]. Laboratory measurements for FT have been undertaken since the 1980s. In 2009, a capacity of 80–90% was achieved, followed by peak performance in 2010 with minor modifications, and then escalated to 120% efficiency in 2011 and 2012. Since 2015, the deployment of Fischer–Tropsch synthesis using a suspension reactor in manufacturing facilities has begun. The FT catalysts are crucial for the effective advancement of sustainable FT process developments. The initial phases of commercial plant development begin with catalyst formulation to tackle the active phase, followed by an analysis of the system, microkinetics, kinetics, catalysis, and hydrodynamics or thermodynamics, employing a pilot plant. The completion of the experimental facility will facilitate the next phase, wherein the FT may be regarded as implemented in the demonstration plant. At this stage, the FT process will focus on mathematical models of the reactor, its configuration, or simulations. Upon completion, the focus transitions to method integration and optimisation at the production operational scale.

To thoroughly evaluate the devised technique and catalyst, ensuring confidence in their dependable scale-up prior to industrialisation, two distinct instruments were established and carried out by IFPEN/ENI/Axens.

i. According to previous research, a Fischer–Tropsch prototype operation with an occupancy of 20 barrels per day has been established at the ENI refinery site in Sannazzaro, Italy (SNZ) [61]. Evaluation of catalyst's efficiency (activity and selectivity) in actual slurry bubble circumstances over an extended period, as well as the assessment of various liquid/solid segregation strategies and slurry catalyst

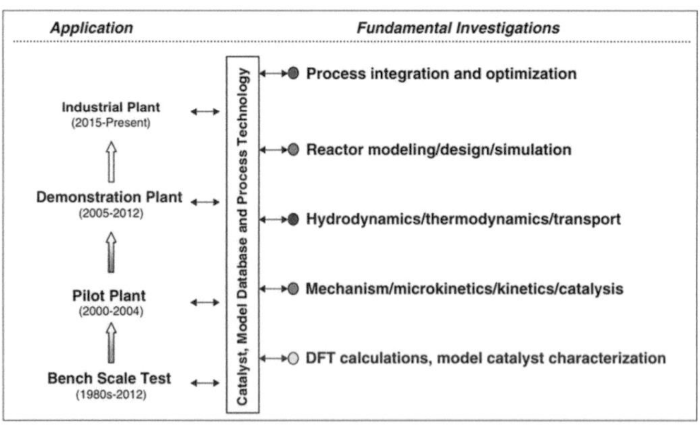

Fig. 8 Schematic for knowledge integration in the MTFT process development [62]

management, were examined. Since 2001, the trial operation has functioned a slurry bubble column under Fischer–Tropsch conditions for more than 20,000 h.

ii. The application of a Large Validation Tool (LVT) aims to measure granular crystallisation on a commercial-scale catalysts and to validate the mechanical endurance under a full-sized commercial reactor. This apparatus has a high-capacity Continuous Stirring Tank Reactor (CSTR) precisely engineered to replicate physical strain encountered by the catalyst at a commercial basis (up to 15,000 bpd) and the substance environment during the FT production cycle concurrently.

10 Conclusion

The current chapter has been discussed the detail of FT process including history, reaction mechanism, FT intermediate product and elemental reaction, type of reactor used in FT reaction, product distribution and catalysis, FT commercialisation and research development of FT process. As conclusion, FT process has centred on process of development strategy in industrial-scale advancement, catalyst performance, and the scaling up of operations. These achievements have evolved through extensive fundamental research and a profound comprehension of the key factors influencing FT chemistry.

References

1. Gahane D, Biswal D, Mandavgane SA (2022) Life cycle assessment of biomass pyrolysis. Bioenergy Res 15:1387–1406. https://doi.org/10.1007/s12155-022-10390-9
2. Demirbas A (2009) Biorefineries: current activities and future developments. Energy Convers Manag 50(11):2782–2801. https://doi.org/10.1016/j.enconman.2009.06.035
3. Leckel D (2009) Diesel production from fischer–Tropsch: the past, the present, and new concepts. Energy Fuels 23(5):2342–2358. https://doi.org/10.1021/ef900064c
4. Mahmoudi H, Mahmoudi M, Doustdar O, Jahangiri H, Tsolakis A, Gu S, Wyszynski ML (2017) A review of Fischer Tropsch synthesis process, mechanism, surface chemistry and catalyst formulation. Biofules Eng. 2:11–31. https://doi.org/10.1515/bfuel-2017-0002
5. Fratalocchi L, Visconti CG, Groppi G, Lietti L, Tronconi E (2018) Intensifying heat transfer in Fischer-Tropsch tubular reactors through the adoption of conductive packed foams. J Chem Eng 349:829–837. https://doi.org/10.1016/j.cej.2018.05.108
6. Takao K, Frank D, Eiichiro M, David G, Masaaki T (2001) Coal liquefaction Ullmann's encyclopedia of industrial chemistry. Wiley-VCH, Weinheim. https://doi.org/10.1002/14356007.a07_197
7. Chai J, Jiang J, Gong Y, Wu P, Wang A, Zhang X, Wang T, Meng X, Lin Q, Lv Y, Men Z, Wang P (2023) Recent mechanistic understanding of Fischer-Tropsch synthesis on Fe-carbide. Catalysts 13:1052. https://doi.org/10.3390/catal13071052
8. Jahangiri H, Bennett J, Mahjoubi P, Wilson K, Gu S (2014) A review of advanced catalyst development for Fischer-Tropsch synthesis of hydrocarbons from biomass derived syn-gas. Catal Sci Technol 4:2210–2229. https://doi.org/10.1039/C4CY00327F

9. Qi Z, Chen L, Zhang S, Su J, Somorjai GA (2020) A mini review of cobalt-based nanocatalyst in Fischer-Tropsch synthesis. Appl Catal A Gen 602:117701. https://doi.org/10.1016/j.apcata.2020.117701
10. Chen W, Zijlstra B, Filot IAW, Pestman R, Hensen EJM (2018) Mechanism of carbon monoxide dissociation on a cobalt Fischer-Tropsch catalyst. ChemCatChem 10:136–140. https://doi.org/10.1002/cctc.201701203
11. Rauch R, Hofbauer H, Sacareanu S, Chiru A (2011) From gasification to synthetic fuels via Fischer-Tropsch synthesis. Bull Transsilvania Univ Brasov Series I Eng Sci 3:33–40. http://hdl.handle.net/20.500.12708/161934
12. Gavrilović L, Jørgensen EA, Pandey U, Putta KR, Rout KR, Rytter E, Hillestad M, Blekkan EA (2021) Fischer-Tropsch synthesis over an alumina-supported cobalt catalyst in a fixed bed reactor—effect of process parameters. Catal Today 369:150–157. https://doi.org/10.1016/j.cattod.2020.07.055
13. Steynberg AP, Dry ME, Davis BH, Breman BB (2004) Chapter 2—Fischer-Tropsch reactors. In: Steynberg A, Dry M (eds) Studies in surface science and catalysis, vol 152. Elsevier, pp 64–195. https://doi.org/10.1016/S0167-2991(04)80459-2
14. Speight JG (2020) Chapter 8—hydrocarbons from synthesis gas. In: Speight JG (eds) Handbook of industrial hydrocarbon processes, 2nd edn. Gulf Professional Publishing, pp 343–386. https://doi.org/10.1016/B978-0-12-809923-0.00008-4
15. Geertsema A (n.d.) Considerations in the selection of a FT reactor for synfuels production. https://www.fischer-tropsch.org/DOE/_conf_proc/USDOE-INDLIQ/91010533/de91010533_geertsema.pdf
16. Khumalo N, Hausberger B, Glasser D, Hildebrandt D, Holland S (2007) The challenge of designing a Fischer-Tropsch reactor. In: 5th International conference heat transfer fluid mechanical thermodynamics South Africa, 1–4 July 2007. http://hdl.handle.net/2263/42823
17. Lee HJ, Choi JH, Garforth A, Hwang S (2015) Conceptual design of a Fischer-Tropsch reactor in a gas-to-liquid process. Ind Eng Chem Res 54:6749–6760. https://doi.org/10.1021/acs.iecr.5b00556
18. Zhang Q, Deng W, Wang Y (2013) Recent advances in understanding the key catalyst factors for Fischer-Tropsch synthesis. J Energy Chem 22:27–38. https://doi.org/10.1016/S2095-4956(13)60003-0
19. Förtsch D, Pabst K, Groß-Hardt E (2015) The product distribution in Fischer-Tropsch synthesis: an extension of the ASF model to describe common deviations. Chem Eng Sci 138:333–346. https://doi.org/10.1016/j.ces.2015.07.005
20. Kuipers EW, Scheper C, Wilson JH, Vinkenburg IH, Oosterbeek H (1996) Non-ASF product distributions due to secondary reactions during Fischer-Tropsch synthesis. J Catal 158:288–300. https://doi.org/10.1006/jcat.1996.0028
21. Schulz H, Claeys M (1999) Kinetic modelling of Fischer-Tropsch product distributions. Appl Catal A-Gen 186:91–107. https://doi.org/10.1016/S0926-860X(99)00166-0
22. Hillestad M (2015) Modeling the Fischer-Tropsch product distribution and model implementation. Chem Prod Process Model 10:147–159. https://doi.org/10.1515/cppm-2014-0031
23. Dry ME (1982) Catalytic aspects of industrial Fischer-Tropsch synthesis. J Mol Catal 17:133–144. https://doi.org/10.1016/0304-5102(82)85025-6
24. Van Der Laan GP, Beenackers AACM (1999) Kinetics and selectivity of the Fischer-Tropsch synthesis: a literature review. Catal Rev Sci Eng 41:255–318. https://doi.org/10.1081/CR-100101170
25. Yang R, Zhou L, Gao J, Hao X, Wu B, Yang Y, Li Y (2017) Effects of experimental operations on the Fischer-Tropsch product distribution. Catal Today 298:77–88. https://doi.org/10.1016/j.cattod.2017.05.056
26. Chen J, Yang C (2019) Thermodynamic equilibrium analysis of product distribution in the Fischer-Tropsch process under different operating conditions. ACS Omega 4:22237–22244. https://doi.org/10.1021/acsomega.9b03707
27. Masuku CM, Hildebrandt D, Glasser D (2011) The role of vapour-liquid equilibrium in Fischer-Tropsch product distribution. Chem Eng Sci 66:6254–6263. https://doi.org/10.1016/j.ces.2011.09.005

28. Filip L, Zámostný P, Rauch R (2019) Mathematical model of Fischer-Tropsch synthesis using variable alpha-parameter to predict product distribution. Fuel 243:603–609. https://doi.org/10.1016/j.fuel.2019.01.121
29. Muleja AA, Yao Y, Glasser D, Hildebrandt D (2016) A study of Fischer-Tropsch synthesis: product distribution of the light hydrocarbons. Appl Catal A Gen 517:217–226. https://doi.org/10.1016/j.apcata.2016.03.015
30. Liu Y, Kamata H, Ohara H, Izumi Y, Ong DSW, Chang J, Poh CK, Chen L, Borgna A (2020) Low-olefin production process based on Fischer-Tropsch synthesis: process synthesis, optimization, and techno-economic analysis. Ind Eng Chem Res 59:8728–8739. https://doi.org/10.1021/acs.iecr.0c00542
31. Lin T, Gong K, Wang C, An Y, Wang X, Qi X, Li S, Lu Y, Zhong L, Sun Y (2019) Fischer-Tropsch synthesis to olefins: catalytic performance and structure evolution of Co2C-based catalysts under a CO_2 environment. ACS Catal 9:9554–9567. https://doi.org/10.1021/acscatal.9b02513
32. Xiong H, Motchelaho MAM, Moyo M, Jewell LL, Coville NJ (2011) Correlating the preparation and performance of cobalt catalysts supported on carbon nanotubes and carbon spheres in the Fischer-Tropsch synthesis. J Catal 278:26–40. https://doi.org/10.1016/j.jcat.2010.11.010
33. Patzlaff J, Liu Y, Graffmann C, Gaube J (1999) Studies on product distributions of iron and cobalt catalyzed Fischer-Tropsch synthesis. Appl Catal A-Gen 186:109–119. https://doi.org/10.1016/S0926-860X(99)00167-2
34. Tomasek S, Lonyi F, Valyon J, Wollman A, Hancsok J (2020) Hydrocracking of Fischer-Tropsch paraffin mixtures over strong acid bifunctional catalysts to engine fuels. ACS Omega 5:26413–26420. https://doi.org/10.1021/acsomega.0c02711
35. Corrao E, Salomone F, Giglio E, Castellino M, Ronchetti SM, Armandi M, Pirone R, Bensaid S (2023) CO_2 conversion into hydrocarbons via modified Fischer-Tropsch synthesis by using bulk iron catalysts combined with zeolites. Chem Eng Res Des 197:449–465. https://doi.org/10.1016/j.cherd.2023.07.052
36. Zhang Q, Kang J, Wang Y (2010) Development of novel catalysts for Fischer-Tropsch synthesis: tuning the product selectivity. ChemCatChem 2:1030–1058. https://doi.org/10.1002/cctc.201000071
37. Makhura E, Rakereng J, Rapoo O, Danha G (2019) Effect of the operation parameters on the Fischer Tropsch synthesis process using different reactors. In: Procedia manufacturing. Elsevier B.V., pp 349–355. https://doi.org/10.1016/j.promfg.2019.05.051
38. Yahyazadeh A, Dalai AK, Ma W, Zhang L (2021) Fischer-Tropsch synthesis for light olefins from syngas: a review of catalyst development. Reactions 2:227–257. https://doi.org/10.3390/reactions2030015
39. Gholami Z, Gholami F, Tišler Z, Hubacek J, Tomas M, Baciak M, Vakili M (2022) Production of light olefins via Fischer-Tropsch process using iron-based catalysts: a review. Catalysts 12:174. https://doi.org/10.3390/catal12020174
40. Lapidus AL, Eliseev OL, Davydov PE, Volkov AS (2006) Catalytic performance of Co/zeolite in Fischer-Tropsch synthesis. In: DGMK/SCI-conference synthesis gas chemistry, pp 185–189. https://www.osti.gov/etdeweb/servlets/purl/20840770
41. Sineva L, Mordkovich V, Asalieva E, Smirnova V (2023) Zeolite-containing Co catalysts for Fischer-Tropsch synthesis with tailor-made molecular-weight distribution of hydrocarbons. Reactions 4:359–380. https://doi.org/10.3390/reactions4030022
42. Fu T, Jiang Y, Lv J, Li Z (2013) Effect of carbon support on Fischer-Tropsch synthesis activity and product distribution over Co-based catalysts. Fuel Process Technol 110:141–149. https://doi.org/10.1016/j.fuproc.2012.12.006
43. Shultz JF, Hall WK, Seligman B, Anderson RB (1955) Hagg iron carbide in Fischer-Tropsch Synthesis. J Am Chem Soc 77:213–221. https://doi.org/10.1021/ja01606a079
44. Janbroers S, Louwen JN, Zandbergen HW, Kooyman PJ (2009) Insights into the nature of iron-based Fischer-Tropsch catalysts from quasi in situ TEM-EELS and XRD. J Catal 268:235–242. https://doi.org/10.1016/j.jcat.2009.09.021

45. Dry ME (1990) Fischer-Tropsch synthesis over iron catalysts. Catal Letters 7:241–252. https://doi.org/10.1007/BF00764506
46. Gao Y, Shao L, Yang S, Hu J, Zhao S, Dang J, Wang W, Yan X, Yang P (2023) Recent advances in iron-based catalysts for Fischer–Tropsch to olefins reaction. Catal Commun 181. https://doi.org/10.1016/j.catcom.2023.106720
47. Okoye-Chine CG, Mubenesha S (2022) The use of iron ore as a catalyst in Fischer–Tropsch synthesis—a review. Crystals (Basel) 12. https://doi.org/10.3390/cryst12101349
48. Borugadda VB, Kamath G, Dalai AK (2020) Techno-economic and life-cycle assessment of integrated Fischer-Tropsch process in ethanol industry for bio-diesel and bio-gasoline production. Energy 195. https://doi.org/10.1016/j.energy.2020.116985
49. Hammache S, Goodwin JG, Oukaci R (2002) Passivation of a Co-Ru/-Al2O3 Fischer-Tropsch catalyst. Catal Today 71:361–367. https://doi.org/10.1016/S0920-5861(01)00463-1
50. Soled SL, Iglesia E, Fiato RA, Baumgartner JE, Vroman H, Miseo S (2003) Control of metal dispersion and structure by changes in the solid-state chemistry of supported cobalt Fischer-Tropsch catalysts. Top Catal 26:101–109. https://doi.org/10.1023/B:TOCA.0000012990.83630.f9
51. Chen Y, Wei J, Duyar MS, Ordomsky DD, Khodakov AY, Liu J (2021) Carbon-based catalysts for Fischer-Tropsch synthesis. Chem Soc Rev 50:2337–2366. https://doi.org/10.1039/D0CS00905A
52. Zhang CH, Yang Y, Teng BT, Li TZ, Zheng HY, Xiang HW, Li YW (2006) Study of an iron-manganese Fischer-Tropsch synthesis catalyst promoted with copper. J Catal 237:405–415. https://doi.org/10.1016/j.jcat.2005.11.004
53. Khodakov AY (2009) Fischer-Tropsch synthesis: Relations between structure of cobalt catalysts and their catalytic performance. Catal Today 144:251–257. https://doi.org/10.1016/j.cattod.2008.10.036
54. Samidin S, Ahmad KN, Shamsuddin MR, Tahari AMN, Lahuri AH, Salleh F, Mijan NA, Jamil MSM, Isahak WNRW, Yarmo MA, Yusop MR (2023) Role of Cr-doped NiO in reduction under a low concentration of H2 and CO. Surf Interfaces 41. https://doi.org/10.1016/j.surfin.2023.103106
55. Pearson R, Coe A, Paterson J (2021) Innovation in fischer-Tropsch: a sustainable approach to fuels production a cost-effective method of converting any carbon source into high-quality liquid hydrocarbon fuels. Johnson Matthey Technol Rev 65(3):395–403. https://doi.org/10.1595/205651321X16143384043486
56. Abdellatief TMM, Ershov MA, Kapustin VM, Chernysheva EA, Savelenko VD, Makhmudova AE, Potanin DA, Salameh T, Abdelkareem MA, Olabi AG (2022) Innovative conceptional approach to quantify the potential benefits of gasoline-methanol blends and their conceptualization on fuzzy modeling. Int J Hydrogen Energy 47(82):35096–35111. https://doi.org/10.1016/j.ijhydene.2022.08.076
57. Khalili-Garakani A, Nezhadfard M, Iravaninia M (2022) Enviro-economic investigation of various flare gas recovery and utilization technologies in upstream and downstream of oil and gas industries. J Clean Prod 346:131218. https://doi.org/10.1016/j.jclepro.2022.131218
58. Gogate MR (2019) Methanol-to-olefins process technology: current status and future prospects. In: Petroleum science and technology, vol 37. Taylor and Francis Inc., pp 559–565. https://doi.org/10.1080/10916466.2018.1555589
59. Tavares M, Westphalen G, Ribeiro de Almeida JMA, Romano PN, Sousa-Aguiar EF (2022) Modified fischer-tropsch synthesis: A review of highly selective catalysts for yielding olefins and higher hydrocarbons. In: Frontiers in nanotechnology, vol 4. Frontiers Media S.A. https://doi.org/10.3389/fnano.2022.978358
60. Nippon Seiro (2023) FT Wax. https://www.seiro.co.jp/en/product_en/list/detail?id=539
61. Boyer C, Gazarian J, Lecocq V, Maury S, Forret A, Schweitzer JM, Souchon V (2016) Development of the Fischer-Tropsch process: from the reaction concept to the process book. Oil Gas Sci Technol 71:44. https://doi.org/10.2516/ogst/2015032
62. Xu J, Yang Y, Li YW (2013) Fischer-Tropsch synthesis process development: Steps from fundamentals to industrial practices. Curr Opin Chem Eng 2:354–362. https://doi.org/10.1016/j.coche.2013.05.002

63. Mazurova K, Miyassarova A, Eliseev O, Stytsenko V, Glotov A, Stavitskaya A (2023) Fischer-Tropsch synthesis catalysts for selective production of diesel fraction. Catalysts 13:1215. https://doi.org/10.3390/catal13081215
64. Tavasoli A, Sadagiani K, Khorashe F, Seifkordi AA, Rohani AA, Nakhaeipour A (2008) Cobalt supported on carbon nanotubes—A promising novel Fischer-Tropsch synthesis catalyst. Fuel Process Technol 89:491–498. https://doi.org/10.1016/j.fuproc.2007.09.008

Chapter 12
Life Cycle and Techno-Economic Assessment of the Conversion of Biocrude Oil from Pyrolysis to Chemicals

Elias Martinez-Hernandez and Myriam A. Amezcua-Allieri

Abstract Many countries whose main source of wealth relies on fossil fuels are facing a critical challenge in crafting a sustainable path to incorporate low carbon products into energy portfolio. Low carbon technology products produce less pollution than their traditional energy counterparts and will play a vital role in the transition to a low carbon economy. Low carbon technologies include biomass systems to produce biofuels. The biocrude oil from pyrolysis is a complex mixture of decomposition products from biomass components. Although some direct uses of this oil are proposed, higher value and practical applications require converting the biocrude oil into other products. In this work, the conversion of biocrude oil into value-added aromatic products is evaluated. The process consists of the pyrolysis of lignin from a bioethanol process, the separation of the oil and its hydrodeoxygenation into a mixture of benzene, toluene, and xylene (BTX), and a mixture of cycloalkanes which are reformed to generate the hydrogen for the process. The BTX mixture is separated and sold as value-added products. This chapter describes the economic and life cycle environmental sustainability assessment of the whole process is carried out and results are compared against fossil-based alternatives.

Keywords Pyrolysis · Biocrude oil · Techno-economic analysis · Life cycle assessment

1 Introduction

Biocrude oil in general is a liquid emulsion of oxygenated organic compounds, polymers, and water which contains up to 40% of oxygen. Biomass pyrolysis oil contains a complex and diverse mixture of chemicals ranging from small molecules, such as formaldehyde and acetic acid to polyphenols, anhydro-sugars, and other

E. Martinez-Hernandez · M. A. Amezcua-Allieri (✉)
Energy Efficiency and Biofuels Division, Mexican Petroleum Institute, Mexico City, Mexico
e-mail: mamezcua@imp.mx

M. Aslam et al. (eds.), *Production and Biorefining of Biocrude Oil: Current Status and Future Developments*, Advances in Sustainability Science and Technology,
https://doi.org/10.1007/978-981-96-5198-6_12

Table 1 Range of values reported for biocrude oil yield from different types of pyrolysis [2]

Pyrolysis type	Min yield (%)	Max yield (%)
Slow pyrolysis	14	75
Intermediate pyrolysis	11	53
Flash pyrolysis	27	85
Catalytic pyrolysis	12	56
Microwave pyrolysis	27	75

oligosaccharides. Due to such compositional complexity, the utilization of biocrude oil from pyrolysis and the establishment of application standards is challenging. One of the firs potential uses is as a substitute for fossil fuel oils, for which there is the ASTM D7544 standard that specifies pyrolysis oil quality parameters when used as a liquid biofuel. Other potential ways to use biocrude oil are in the coprocessing in crude oil refineries to generate hydrocarbons, or its conversion into chemicals and materials.

As a result of a wide variation in chemical compositions of biomass, pyrolysis conditions, and reactor types, there is a great variation in biocrude oil yields reported. For example, a relatively high yield of 85% has been obtained for pure cellulose using flash pyrolysis (Ansari et al. 2019), while the lowest yield reported is for municipal solid waste using intermediate pyrolysis (500 °C, heating rate of 25 °C/min) [1]. A summary of the ranges reported in the literature is provided in Table 1.

The valorization of biocrude oil depends on the type of products contained. The major product types can be grouped into anhydro-sugars, furans, phenols, phenolic aldehydes, and ketones and other oxygen containing compounds. The abundance or presence of each group depends on the lignocellulosic composition of biomass. However, the biocrude oil product distribution from separated biomass fractions has been investigated, as shown in Fig. 1 for cellulose, hemicellulose, and lignin. Clear differences can be observed, such as the presence of phenols occurring in lignin and hemicellulose, while anhydro-sugars appear only in the case of hemicellulose and cellulose. As expected, lignin generates more aromatic compounds, such as phenols and phenolic aldehydes and ketones, while cellulose produces more anhydro-sugars. Hemicellulose yields the wider range of product types with majority being furans and other oxygen containing compounds.

2 Conversion of Lignin Biocrude Oil into Value-Added Products

The lignin fraction of biomass, produced from bioethanol plants, pulp and paper mills, and other processes, is rich in aromatic oxygenated structures. This has attracted attention of researchers to generate value-added products, such as aromatic and cyclic paraffins. As a result, the pyrolysis of lignin for bio-oil production is being intensively

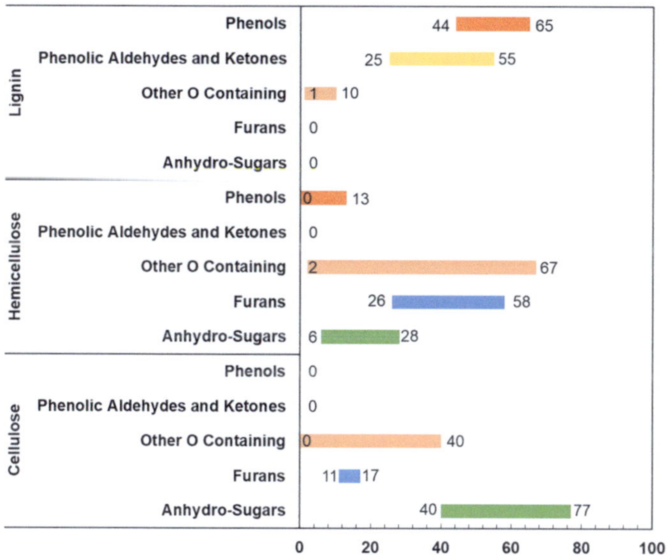

Fig. 1 Range of products in biocrude oils obtained from pyrolysis of cellulose, hemicellulose, and lignin fractions of biomass. *Data source* [2]

studied. The major reactions during lignin pyrolysis can be summarized as follows [3]:

1. Demethoxylation, demethylation, decarboxylation, and decarbonylation parallel reactions of side chains present in lignin form light gases, such as methane, H_2, CO, and CO_2.
.2 Cracking of lignin polymeric structures generates monophenols and oligomeric phenolic compounds which will form the biocrude oil product.
3. Intermediates and phenolic compounds can repolymerize and yield char, while the carbonization of polymeric structures also leads to char.

Typical bio-oil yields from lignin range between 15 and 35% depending on the type of lignin which can be from the Kraft process (Kraft lignin), alkali process (alkali lignin), or organosolv extraction process (organosolv lignin), or more recently from ionic liquid extraction (ionic liquid lignin) [3]. A recent review reported biocrude oil yields of up to 55% from lignin [4]. The major components identified in lignin pyrolysis oil include guaiacols, phenols, eugenol, syringols, cresols, and catechols. The product distribution obtained depends on temperature and lignin source. For example, monophenols are the major products with pyrolysis at 450–525 °C [3]. The presence of methoxy and hydroxyl functionalities in the lignin affects the formation of char and may reduce conversion toward value-added aromatics [5]. The formation of such aromatic compounds from lignin has generated interest in lignin pyrolytic biocrude oil for chemicals production, such as benzene, toluene, and xylenes by means of catalytic upgrading or conversion for oxygen removal and improving stability and

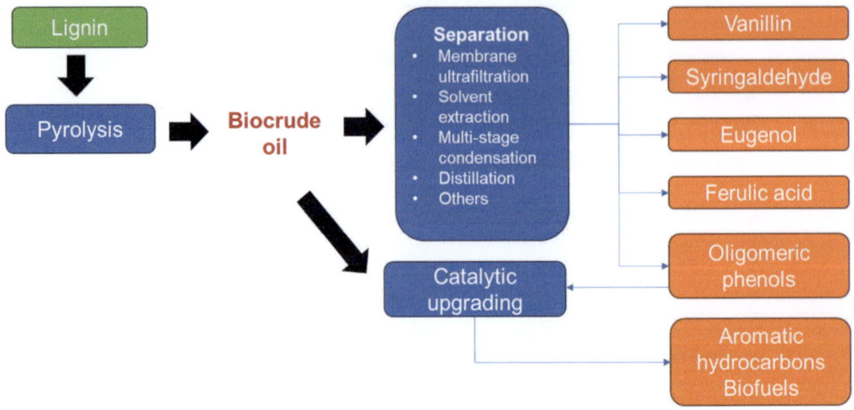

Fig. 2 Valorization routes for biocrude oil from lignin pyrolysis

reducing acidity and corrosiveness of the biocrude oil. Figure 2 shows alternative routes for valorization of biocrude oil from lignin pyrolysis. The valorization can be trough separation of monomeric phenols using a range of methods, such as membrane ultrafiltration, solvent extraction, multi-stage condensation, distillation, and combinations thereof. Vanillin is a typical product industrially obtained from lignin, while other products of interest include syringaldehyde, eugenol, and ferulic acid [6].

The production of aromatics and hydrocarbons require catalytic upgrading of biocrude oil. Biocrude oil upgrading by catalytic hydrotreatment or hydrodeoxygenation (HDO) is the most widely studied process for removing oxygen. Removal of oxygen from phenolic compounds present in the biocrude oil occurs by deoxygenation (DDO) to form aromatics, or by hydrogenation (HYD) followed by deoxygenation [7]. Catalyst should be designed to avoid hydrogenation of aromatic compounds in bio-oils to obtain higher aromatic yield and reduce hydrogen consumption. The integration of the pyrolysis of lignin with biocrude oil processing through biorefinery schemes can be a key element that determines the economic viability of the process, but the environmental impact of the process is an important aspect that must also be analyzed. This chapter presents the techno-economic and life cycle analysis of a process for the conversion of lignin to biocrude which is processed to obtain aromatics and cyclic paraffins.

3 Techno-economic Analysis of Lignin Bio-Oil Conversion

The techno-economic analysis comprises the process simulation and estimation of capital and operating costs to determine unit cost of production. These are presented in the following sections.

3.1 Process Simulation

The process can be simulated using process simulators, such as Aspen Plus®, SuperPro Designer®, and others. The objective is to establish the overall mass and energy balances which are the basis for costing and economic assessment. In this work, SuperPro Designer was used. The biorefinery process involves pyrolysis of lignin into bio-oil which is then processed by hydrodeoxygenation to obtain aromatics (benzene, toluene, and xylenes, a product known as BTX) and cyclic paraffins as a fuel blend stock product. The process also includes on-site hydrogen production by steam reforming of part of the cyclic paraffins, as well as on-site heat and power generation from char and off-gases generated from the process. The process was simulated in SuperPro Designer® as shown in Fig. 3.

The lignin is fed into a drying step to remove water before entering the pyrolyzer. The gaseous and solid products are then separated in a cyclone. The solid are recovered as char and is sent to energy production for the process. The gaseous product is then passed through a cooler and then flashed to separate a liquid product and separate non-condensable gases. These off-gases are sent also to energy production. The liquid fraction passes through a decanter to separate water and recover the bio-oil. Due to the complexity of bio-oil composition, it was considered that the bio-oil is made up by hydroxy, methoxy, and methyl phenols which are model compounds for lignin pyrolysis products. These compounds can be hydrodeoxygenated into benzene, toluene, and xylene and cycloparaffins using a MoO_3–NiO–Al_2O_3 catalyst [8].

The bio-oil is sent to the hydrodeoxygenation (HDO) section where is heated and reacted with hydrogen in the HDO reactor. The reactor effluent is cooled down and flashed to separate the liquid from the gas fraction. The gas fraction containing CO_2, CO, H_2, and water is sent to the water–gas shift reactor (WGSR). The liquid stream contains the hydrocarbons generated which are sent to the product purification section

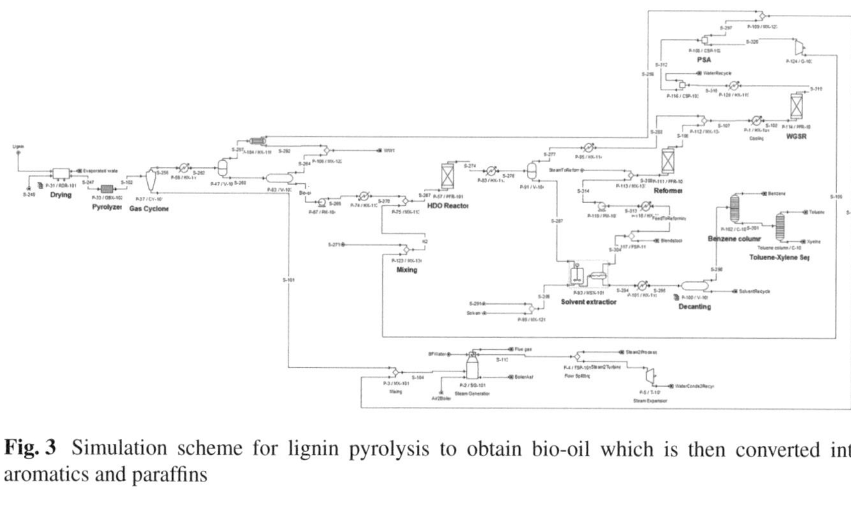

Fig. 3 Simulation scheme for lignin pyrolysis to obtain bio-oil which is then converted into aromatics and paraffins

Table 2 HDO reactor product yield used in simulation [8]

Species	Reactor yield (% mol, bio-oil basis)	Price (US$/kg)
Benzene	13	0.91
Toluene	44	0.738
p-Xylene	26	0.935
Methyl cyclohexane	12	0.65
Cyclohexane	6	0.65

where a solvent extraction unit separates the aromatics (BTX) from the paraffins using a methanol/water mixture. The solvent is recovered by decanting, and the BTX fraction is then separated using a series of two conventional distillation columns. The cyclic paraffins separated from the solvent extraction is sold as blend stock but part of it is diverted toward the hydrogen production section which uses a steam reformer. The effluent from the reformer passes through the WGSR to increase the hydrogen generated. Finally, the hydrogen is recovered and purified using pressure-swing absorption (PSA) and returned to the HDO section. Off-gases from hydrogen production are also sent to energy production.

The main HDO reactor product yield used in the simulation and the prices used in the economic analysis are shown in Table 2.

3.2 *Economic Assessment*

The economic assessment was carried out to establish the investment required, the capital and operating costs and the unit cost of production, return on investment, and payback period. The calculations are based on costing factors for facility-dependent costs (investment, maintenance, engineering, instrumentation and control, construction, etc.). Labor costs, utility costs, and other operating costs are also considered based on the mass and energy balances. The assessment is assumed for 2023, considering 4% inflation rate, a 9% interest rate, a 30% tax rate and 15 USD/h of labor costs, 15-year project lifetime, 10-year depreciation time, and 5% savage value. A sensitivity analysis to economy of scale was carried out to establish a processing scale for the biorefinery based on the unit cost of production.

4 Life Cycle Assessment of Lignin Bio-Oil Conversion

Life cycle assessment (LCA) is a well-known methodology to evaluate environmental impacts of products, processes, services, or technologies. LCA methodology was carried out in this study to evaluate the environmental impacts of producing BTX from fuel oil (base scenario, BS) and comparing these impacts obtaining BTX form bio crude oil from lignin using pyrolysis (alternative scenario, AS).

4.1 Method

An attributional LCA was carried out based on the methodology established in the ISO 14040 and 14044 [9, 10]. An LCA model was used to analyze the environmental impacts of bio crude oil production from cradle to grave. The mass and energy balances obtained, together with environmental releases, included the input and output parameters in the SimaPro model. The environmental impacts of producing and using bio crude oil from lignin were assessed relative to a fuel crude oil. Comparison was performed on a per megajoule of fuel equivalent basis.

In each scenario, calculation and evaluation of the emissions and impacts to the different media: water, air, and soil as well as impacts derived from the elaboration of main products throughout the value chain in order to identify and quantified hot spots. Using the results of environmental impacts and indicators of emissions, impacts to environment were established, considering all the stages in the productive process.

4.1.1 Life Cycle Impact Categories and Functional Unit

The CML Baseline 2000 Method [11] was used to carry out the impact evaluation considering the following categories: Global warming potential (GWP), human toxicity potential (HTP), Photochemical Ozone Creation Potential (POCP), abiotic resource depletion (AD), acidification (AP), eutrophication (EP), ozone layer depletion (ODP), freshwater aquatic ecotoxicity (FAETP), sea water aquatic ecotoxicity (SAETP), and terrestrial ecotoxicity (TEP) potentials. Mitigation is considered as the elimination, reduction, or control of the adverse environmental impact. Therefore, mitigation was represented as a positive percentage if it has a reduction in the amounts of pollutants reflected in the indexes of the environmental impact categories.

4.1.2 System Boundary

For both base and alternative scenarios, the system includes the impacts associated with the corresponding technological route. In the case of BS, it includes the extraction of crude oil, its transport to the refinery, the refining to be transformed into fuel oil, and its transport to a regional storage center. For the AS case, the system includes the production of biomass (lignin) in the field, its transportation to the biorefinery where the process takes place and involves pyrolysis of lignin into bio-oil which is then processed by hydrodeoxygenation to obtain aromatics (BTX) and cyclic paraffins as a fuel blend stock product. On-site hydrogen production as well as on-site heat and power generation from char and off-gases generated from the process were

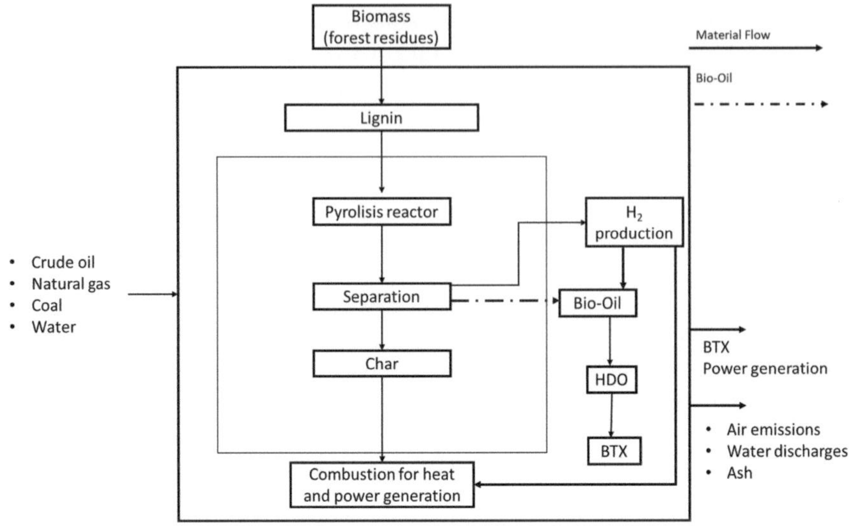

Fig. 4 System boundaries (alternative scenario, AS) for lignin biocrude oil conversion into aromatics

also considered. The impacts associated with transport to the consumer or the final use of crude oil produced is not considered. Figure 4 shows the system boundaries limits for the AS.

4.1.3 Description of Technological Routes and Assumptions

Different life cycle stages of bio-oil production, the production of bio-oil was further divided into five-unit processes, consisting of feedstock preparation and drying, a pyrolysis reactor, hydrodeoxygenation, char recovery, and a furnace process (heat/ energy generation). On-site hydrogen production as well as on-site heat and power generation from char and off-gases generated from the process were also considered.

The functional unit was 1 MJ of energy to compare the process for producing BTX from fuel oil (base scenario, BS) and comparing these impacts obtaining BTX form bio crude oil from lignin using pyrolysis (alternative scenario, AS).

4.1.4 Life Cycle Inventory

Data for fuel, transportation, electricity use, and other were obtained from Ecoinvent database version 2.1. All feedstock was obtained from forest residues [12], reported weights were oven dry, and bio-oil density was 1.2 kg/L.

The CML Baseline 2000 Method [11] is used to carry out the impact evaluation considering the following main categories: global warming potential (GWP, kg CO_2

eq), human toxicity potential (HTP, kg 1,4-DB eq), Photochemical Ozone Creation Potential (POCP, kg C_2H_4 eq), abiotic resource depletion (AD, kg Sb eq), acidification (AP, kg SO_2 eq), eutrophication (EP, kg PO_4 eq), ozone layer depletion (ODP, kg CFC-11 eq), freshwater aquatic ecotoxicity (FAETP, kg 1,4-DB eq), seawater aquatic ecotoxicity (SAETP, kg 1,4-DB eq), and terrestrial ecotoxicity (TEP, kg 1,4-DB eq) potentials.

5 Results

5.1 Techno-economic Analysis

The process was simulated for various capacities to find an appropriate lignin input scale. It is well known that biomass processing requires large scales to be competitive and feasible. Thus, the results of unit production cost per kg of product mix were plotted against the lignin input as shown in Fig. 5.

It can be observed that the curve becomes practically flat after 800 t/d capacity which means there is no further economy of scale advantage beyond this processing capacity. Thus, the capacity of 800 t/d was selected for further analysis.

Figure 6 shows the overall mass balance for an 800 t/d lignin biorefinery to produce BTX. The lignin can be supplied from ethanol biorefineries.

Table 3 shows the mass balance of the main streams. From a total lignin input of 800 t/d, 200 t/d of bio-oil was produced. Table 4 shows the mass balance of the rest of the streams in the process, such as solvent, recycles, water, air, steam, and waste and emission flows. The overall product yield on a bio-oil basis is 8.6% benzene, 31.5% toluene, 19.6% xylene, and 6.8% fuel blend stock. Thus, 66.5% of the bio-oil

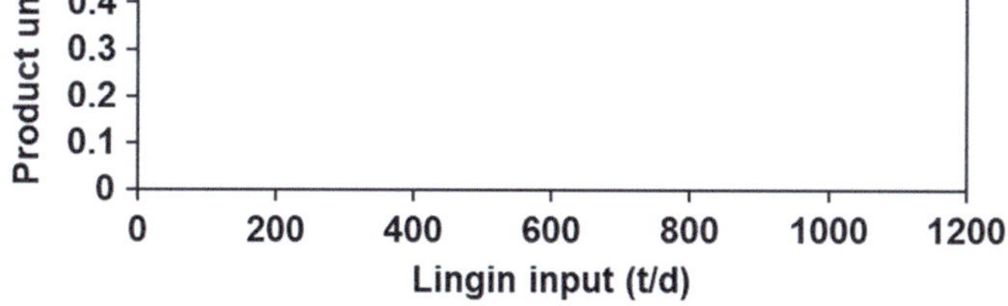

Fig. 5 Simulation scheme for lignin processing into bio-oil which is then converted into aromatics and paraffins

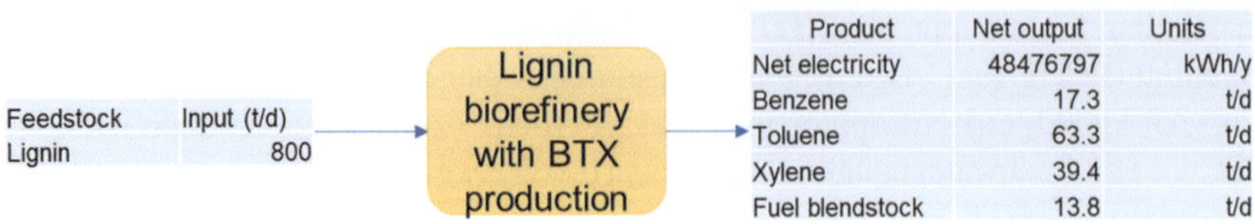

Fig. 6 Overall mass balance of bioethanol biorefinery with BTX production from lignin bio-oil

is recovered as products. 60% of the cyclic paraffins produced from the HDO of bio-oil was used for hydrogen production, which decreased the net product output. The char and off-gases are used as fuels and provided the whole energy required (steam and electricity) by the process, with a net electricity export of 48,476,797 kWh/y. However, due to the high temperature requirement for the pyrolysis, HDO and hydrogen production, some direct fire heating is required which may need to be provided by natural gas. The direct process heat required was 70,016 MJ/h of primary energy as natural gas, or 8.36 MJ/kg bio-oil.

5.2 *Life Cycle Assessment*

The main results of LCA are shown in Fig. 7 for the % mitigation of the AS using bio-oil in respect to the BS scenario using crude oil. In all cases, the AS has positive mitigation percentages which means that the base scenario has higher impacts than alternative scenario. Human toxicity has the highest mitigation followed by fresh water aquatic ecotoxicity with more than 50%. The global warming is reduced by 21%, and the other categories are also mitigated by a range between 15 and 30%. This is mainly due to the on-site hydrogen production which avoids using crude oil or natural gas, and due to the on-site energy generation and export which reduced impacts from fossil resource use to generate energy.

LCA was useful to evaluate environmental impacts of the biocrude oil versus fossil crude oil. The analysis quantified the major environmental hotspots of fossil crude oil. The best scenario for having less environmental impacts is with biocrude oil with most of the air quality indicators. However, at fossil crude oil scenario, a significant mitigation of environmental impacts related to toxicity can be achieved.

6 Discussion

Results have demonstrated that the economic and life cycle environmental sustainability assessment of the whole process for converting lignin pyrolysis oil into chemical can be competitive and cleaner with biocrude oil compared against fossil-based alternatives. The research for production of sustainable biofuels and chemicals from bio-oil is still under development. However, it is important to continue investigating

Table 3 Mass balance results of main streams

	Lignin	Evaporated water	Bio-oil	H2	Benzene	Toluene	Xylene	Blendstock	Feed to reforming
Type	Raw material	Emission	0.25124735	Raw material	Revenue	Revenue	Revenue	Revenue	
Total flow (kg/h)	33333.3	26178.7	8374.9	562.9	720.2	2636.9	1641.1	573.3	860.0
Temperature (°C)	100.0	562.7	65.5	40.0	81.3	111.2	137.3	150.0	150.0
Pressure (bar)	1.0	1.0	1.0	30.0	1.0	1.0	1.0	1.0	1.0
2,4-dimethylphe	0.0	0.0	818.0	0.0	0.0	0.0	0.0	0.0	0.0
Benzene	0.0	0.0	0.0	0.0	681.2	3.4	0.0	13.1	19.6
Carb. dioxide	0.0	0.0	0.0	24.0	0.0	0.0	0.0	0.0	0.0
Carbon monoxide	0.0	0.0	0.0	3.8	0.0	0.0	0.0	0.0	0.0
Catechol	0.0	0.0	818.0	0.0	0.0	0.0	0.0	0.0	0.0
Cresol	0.0	0.0	2490.5	0.0	0.0	0.0	0.0	0.0	0.0
Cyclohexane	0.0	0.0	0.0	0.0	3.9	0.0	0.0	146.6	219.9
Guaiacol	0.0	0.0	2738.1	0.0	0.0	0.0	0.0	0.0	0.0
Hydrogen	0.0	0.0	0.0	535.0	0.0	0.0	0.0	0.0	0.0
Methanol	0.0	0.0	0.0	0.0	0.0	0.0	0.0	2.6	3.9
Methyl Cyclohexane	0.0	0.0	0.0	0.0	8.7	0.0	0.0	328.8	493.3
Nitrogen	0.0	16735.1	0.0	0.0	0.0	0.0	0.0	0.0	0.0
Oxygen	0.0	5080.5	0.0	0.0	0.0	0.0	0.0	0.0	0.0
p-Xylene	0.0	0.0	0.0	0.0	0.0	66.2	1588.7	31.6	47.5
Soluble cellulose	1.8	0.0	0.0	0.0	0.0	0.0	0.0	0.0	0.0

(continued)

Table 3 (continued)

	Lignin	Evaporated water	Bio-oil	H2	Benzene	Toluene	Xylene	Blendstock	Feed to reforming
Soluble hemicellulose	0.5	0.0	0.0	0.0	0.0	0.0	0.0	0.0	0.0
Soluble lignin	24622.4	0.0	0.0	0.0	0.0	0.0	0.0	0.0	0.0
Syringol	0.0	0.0	1509.8	0.0	0.0	0.0	0.0	0.0	0.0
Toluene	0.0	0.0	0.0	0.0	26.5	2567.3	52.4	50.6	75.9
Water	8708.7	4363.1	0.5	0.1	0.0	0.0	0.0	0.0	0.0

Table 4 Mass balance results of other streams in process

	WW1	Steam to reform	Water recycle	Solvent	Solvent recycle	Air to boiler	BF water	Flue gas	Steam 2 process	Steam to turbine
Type	Aqueous waste	Raw material	Credit	Raw material	Credit	Raw material	Raw material	Emission	Credit	
Total flow (kg/h)	5616.7	2700.0	600.2	6800.0	8886.7	82313.5	59289.5	100108.7	10968.6	48321.0
Temperature (°C)	62.4	220.0	30.0	25.0	30.0	25.0	25.0	200.0	255.4	255.4
Pressure (bar)	1.0	25.0	25.0	1.0	1.0	1.0	1.0	1.0	45.0	45.0
2,4-dimethylphe										
Benzene	5616.7	2700.0	600.2	6800.0	8886.7	82313.5	59289.5	100108.7	10968.6	48321.0
Carb. dioxide	11.8	0.0	0.0	0.0	0.0	0.0	0.0	0.0	0.0	0.0
Carbon monoxide	0.0	0.0	0.0	0.0	6.9	0.0	0.0	0.0	0.0	0.0
Catechol	0.0	0.0	24.3	0.0	0.0	0.0	0.0	31847.6	0.0	0.0
Cresol	0.0	0.0	3.8	0.0	0.0	0.0	0.0	0.0	0.0	0.0
Cyclohexane	11.8	0.0	0.0	0.0	0.0	0.0	0.0	0.0	0.0	0.0
Guaiacol	6.2	0.0	0.0	0.0	0.0	0.0	0.0	0.0	0.0	0.0
Hydrogen	0.0	0.0	0.0	0.0	0.0	0.0	0.0	0.0	0.0	0.0
Methanol	39.3	0.0	0.0	0.0	0.0	0.0	0.0	0.0	0.0	0.0
Methyl Cyclohexane	0.0	0.0	0.0	0.0	0.0	0.0	0.0	0.0	0.0	0.0
Nitrogen	0.0	0.0	0.0	6800.0	6793.6	0.0	0.0	0.0	0.0	0.0
Oxygen	0.0	0.0	0.0	0.0	0.0	0.0	0.0	0.0	0.0	0.0
p-Xylene	0.0	0.0	0.0	0.0	0.0	63144.1	0.0	63144.1	0.0	0.0
Soluble cellulose	0.0	0.0	0.0	0.0	0.0	19169.3	0.0	2500.3	0.0	0.0

(continued)

Table 4 (continued)

	WW1	Steam to reform	Water recycle	Solvent	Solvent recycle	Air to boiler	BF water	Flue gas	Steam 2 process	Steam to turbine
Soluble hemicellulose	0.0	0.0	0.0	0.0	16.7	0.0	0.0	0.0	0.0	0.0
Soluble lignin	0.0	0.0	0.0	0.0	0.0	0.0	0.0	0.0	0.0	0.0
Syringol	0.0	0.0	0.0	0.0	0.0	0.0	0.0	0.0	0.0	0.0
Toluene	0.0	0.0	0.0	0.0	0.0	0.0	0.0	0.0	0.0	0.0
Water	21.7	0.0	0.0	0.0	0.0	0.0	0.0	0.0	0.0	0.0
Toluene	0.0	0.0	0.0	0.0	26.7	0.0	0.0	0.0	0.0	0.0
Water	5525.9	2700.0	572.1	0.0	2042.8	0.0	59289.5	2616.6	10968.6	48321.0

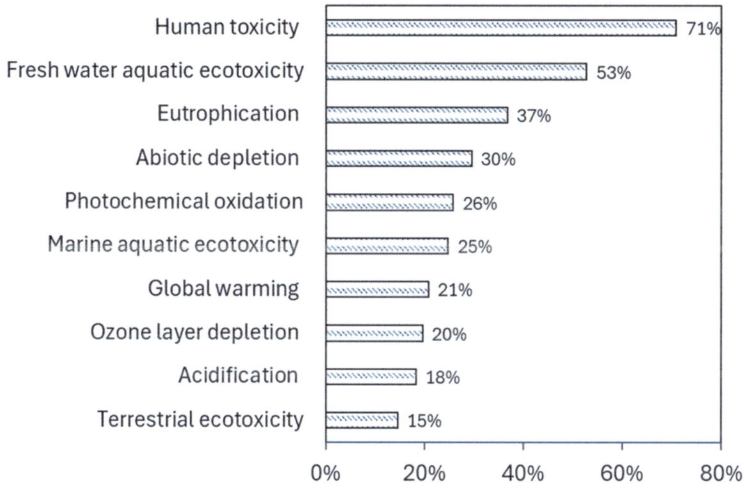

Fig. 7 Mitigation (%) per impact category for alternative scenario (AS) using pyrolysis oil versus base case scenario (BS) using crude oil

possible ways to optimize the process, reduce its cost, and protect the environment early since the experimental and conceptual stages. One of the best ways to reduce costs and mitigate the impact on the environment is using waste materials since some residues are considered renewable feedstocks with a high potential to produce energy through waste-to-energy technologies. With this, in recent years more research has been carried out at laboratory scale and with simulators, and favorable results have been obtained toward the production of biocrude oil. As shown in this brief case study, tight and smart integration of energy and material sources with the process demands is key to enable highly efficient biorefinery systems that can adapt and improve what conventional oil refineries achieve today.

7 Conclusions

This chapter reviewed the main aspects of the TEA and LCA of biocrude oil from pyrolysis to test its feasibility, profitability, and environmental impact. TE and LC assessments for bio-oil production are evolving as new feedstocks, conversion technologies, and production strategies are developed and optimized.

The production of biocrude oil can be economically viable, especially if it is integrated, internally, for utilization of all energy and material streams; and externally with other industries, such as waste management or agriculture. However, the profitability of biocrude production is still highly dependent on the price of crude oil and government policies, such as tax incentives and mandates. In terms of environmental impact, biocrude oil production can significantly reduce greenhouse gas emissions compared to conventional crude oil. But the environmental benefits of bio-oil production are highly dependent on the feedstock and production process used. The use of waste material is one of the best ways to reduce costs and mitigate the impact on the environment, with a high potential to produce energy and value-added chemicals through a combination of process technologies.

Acknowledgements The authors thank Mexican Institute of Petroleum for making this work possible.

Author Contributions All authors contributed to the study conception and design. Material preparation, data collection, and analysis for techno-economic analysis were performed by Elias Martinez-Hernandez. Analysis for life cycle assessment was performed by Myriam A. Amezcua-Allieri. The first draft of the manuscript was written by all the authors, and all of them commented on previous versions of the manuscript. All authors read and approved the final manuscript.

Funding No funding was received to assist with the preparation of this manuscript.

Data Availability Not applicable.

Declarations The authors have no financial or proprietary interests in any material discussed in this article.

References

1. Ahmed A, Abu Bakar MS, Azad AK, Sukri RS, Phusunti N (2018) Intermediate pyrolysis of *Acacia cincinnata* and *Acacia holosericea* species for bio-oil and biochar production. Energy Convers Manag 176:393–408. https://doi.org/10.1016/J.ENCONMAN.2018.09.041
2. Degn Jensen A, Yu Murzin D, Machado H, Cristino AF, Orišková S, Galhano dos Santos R (2022) Bio-oil: the next-generation source of chemicals. Reactions 3:118–137, 3:118–37. https://doi.org/10.3390/REACTIONS3010009
3. Lu X, Gu X (2022) A review on lignin pyrolysis: pyrolytic behavior, mechanism, and relevant upgrading for improving process efficiency. Biotechnol Biofuels Bioproducts 15:1, 15:1–43. https://doi.org/10.1186/S13068-022-02203-0
4. Ansari KB, Arora JS, Chew JW, Dauenhauer PJ, Mushrif SH (2019) Fast pyrolysis of cellulose, hemicellulose, and lignin: effect of operating temperature on bio-oil yield and composition and insights into the intrinsic pyrolysis chemistry. Ind Eng Chem Res 58:15838–15852. https://doi.org/10.1021/ACS.IECR.9B00920/SUPPL_FILE/IE9B00920_SI_001.PDF
5. Hosoya T, Kawamoto H, Saka S (2009) Role of methoxyl group in char formation from lignin-related compounds. J Anal Appl Pyrolysis 84:79–83. https://doi.org/10.1016/J.JAAP.2008.10.024
6. Guo H, Zhao Y, Chang JS, Lee DJ (2023) Lignin to value-added products: research updates and prospects. Bioresour Technol 384:129294. https://doi.org/10.1016/J.BIORTECH.2023.129294
7. Wang H, Male J, Wang Y (2013) Recent advances in hydrotreating of pyrolysis bio-oil and its oxygen-containing model compounds. ACS Catal 3:1047–1070. https://doi.org/10.1021/CS400069Z/ASSET/IMAGES/MEDIUM/CS-2013-00069Z_0030.GIF
8. Kallury RKMR, Restivo WM, Tidwell TT, Boocock DGB, Crimi A, Douglas J (1985) Hydrodeoxygenation of hydroxy, methoxy and methyl phenols with molybdenum oxide/nickel oxide/alumina catalyst. J Catal 96:535–543. https://doi.org/10.1016/0021-9517(85)90321-5
9. ISO. ISO 14040:2006—environmental management—life cycle assessment—principles and framework 2016. https://www.iso.org/standard/37456.html. Accessed 18 Mar 2020

10. ISO. ISO 14044:2006—environmental management—life cycle assessment—requirements and guidelines 2016. https://www.iso.org/standard/38498.html. Accessed 18 Mar 2020
11. Guinée JB (2002) Handbook on life cycle assessment: operational guide to the ISO standards. Kluwer Academic Publishers
12. Anaya-Reza O, Zermeño Eguía Lis JA, Aburto J, Amezcua-Allieri MA (2022) Environmental impact assessment of heat generation from residues: a forest-based industry case study. Bioenergy Res 15:1787–1796. https://doi.org/10.1007/S12155-021-10356-3/METRICS

Chapter 13
Biorefining of Biocrude Oil: Current Status and Future Developments

Raqeeba Aziz, Khalil Ul Rehman, Iftikhar Hussain, Ishfaq Majeed Mir, and Mohammad Aslam

Abstract The biorefining of biocrude oil has garnered significant attention as a sustainable alternative to conventional fossil fuels, bolstering global efforts to reduce carbon emissions and transition to renewable energy sources. The current and possible prospects of biocrude oil biorefining are examined in this chapter, with a focus on commercial opportunities, industry challenges, and technological advancements. One practical method for turning biomass into liquid fuels is the generation of biocrude oil through thermochemical processes like pyrolysis and hydrothermal liquefaction. However, because of its high oxygen concentration, instability, and corrosiveness, there are significant barriers to its refinement and application in the current fuel infrastructure. In order to improve the properties of biocrude oil and make it suitable for use in fuel applications, the chapter provides a comprehensive examination of potential biomass conversion techniques, such as fermentation, pyrolysis, and hydrothermal liquefaction. In addition, biocrude characteristics, sustainability, and environmental impact and metrics of sustainability as well as current status and future developments are emphasized as critical elements that can enhance the efficiency of biocrude oil upgrading processes.

Keywords Biorefining · Biocrude oil · Thermochemical conversion · Biocrude characteristics · Sustainability and environmental impact

1 Introduction

The global energy landscape has been significantly impacted by recurring crises and shortages. For instance, the energy crisis and oil embargo of the 1970s exposed the world's excessive dependence on limited petroleum supplies. The investigation and advancement of alternative energy sources became more important as a result of

R. Aziz · K. U. Rehman · I. Hussain · I. M. Mir · M. Aslam (✉)
Biofuels Laboratory, Department of Chemistry, National Institute of Technology, Srinagar, Jammu and Kashmir, India
e-mail: maslam@nitsri.ac.in

M. Aslam et al. (eds.), *Production and Biorefining of Biocrude Oil: Current Status and Future Developments*, Advances in Sustainability Science and Technology,
https://doi.org/10.1007/978-981-96-5198-6_13

these occurrences [1]. Interest in biomass as fuel has grown as a result of environmental concerns, energy security issues, and shifts in the price of crude oil. Biomass is a sustainable way to lower carbon emissions since it offers a closed carbon cycle, absorbing CO_2 during growth and releasing it during burning. Numerous organic resources, agricultural wastes, short-rotation woody crops, animal dung, and other waste goods can be converted into biomass fuels [2]. Biomass is a renewable resource that can be converted into energy in a variety of ways. For the production of liquid fuels, thermochemical procedures like pyrolysis and hydrothermal liquefaction are especially helpful. The resulting liquid byproducts are known as "bio-oils," and an increasing amount of research indicates that they may be a viable substitute for traditional petroleum fuels [3]. Bio-oils provide sustainable energy for heat, chemicals, and power. They are produced by flash pyrolysis and hydrothermal liquefaction (HTL). Flash pyrolysis rapidly heats biomass without oxygen to produce bio-oil, gas, and char. HTL processes biomass in water at high temperatures and pressures.

Biorefineries use these methods to convert biomass into useful products like polymers and gasoline, as well as fuels and energy. They can produce high-value, low-volume chemicals and huge quantities of lower-value fuels, such as biodiesel or bioenergy [4]. By adopting combined heat and power (CHP) to produce electricity for themselves and sell extra, biorefineries can reduce emissions and boost efficiency. Although biorefineries only provide a modest number of high-value products and fuels, fossil fuels remain crucial for meeting the high energy demands. Every year, almost 250 million tons of municipal solid trash are produced in our country alone, and the usage of fossil fuels raises pollution and greenhouse gas emissions. As the world's population grows, waste levels likewise rise. Of this, roughly 53% winds up in landfills, 13% is utilized to generate electricity, and 35% is recycled or composted [5]. By promoting biomass, such as pulp mill waste, municipal waste, and agricultural leftovers, as a sustainable source of energy and chemicals, this strategy tackles the environmental problems associated with fossil fuels. Biorefineries based on biomass promote environmentally responsible recycling and waste management, providing a more sustainable option to fossil fuels.

2 Biocrude Oil and Biorefining

Biocrude is a liquid biofuel produced by high-temperature; high-pressure liquid-phase thermal processing used to liquefy biomass. They are similar to bio-oil since they are both organic liquids produced by liquefying biomass [6]. But what distinguishes biocrude is its capacity to be gravity-separated from the aqueous phase created during hydrothermal liquefaction. Biocrude offers several benefits over raw biomass. It is easier to process, extract, or modify chemically, and it is easier to store, pump, and feed into processing equipment. Compared to biomass, biocrude is easier to pump, store, and incorporate into production processes. It is also more adaptable to processing, extraction, and chemical modifications. A complex mixture of organic molecules with different molecular weights that have been oxygenated makes up this substance. Biocrude is frequently viscous, but it eventually becomes thermally stable. Compared to conventional petroleum, biocrude has higher quantities of sulfur and nitrogen; levels for biocrudes made from organic wet wastes range from 3 to 8% by weight, which reflects the makeup of the original biomass [7]. Biocrude oil can be used as a major or intermediate ingredient in the production of gasoline, kerosene, heavy fuel oil, and biodiesel.

In order to convert biocrude oil into usable fuels, biorefining involves a number of procedures, including hydroprocessing, transesterification, and catalytic cracking. The co-processing of biocrudes in traditional oil refineries is another way that biorefining might transform heavy petroleum fractions into lucrative liquid and gaseous products [6]. Although a biorefinery uses biomass as its main raw material rather than crude oil, it functions similarly to a petroleum refinery. This biomass is transformed into a variety of goods, including materials, chemicals, heat, power, and transportation fuels. Biomass is a carbon–neutral alternative to fossil fuels because it is a renewable resource and does not raise atmospheric carbon dioxide levels. At the World Economic Forum in 2008 and 2009, industrial biorefineries were highlighted as a potential solution to the global climate change issues because of these benefits [6]. By using renewable biomass, biorefineries can reduce the environmental effect of traditional fossil fuel-based processes while helping to meet the world's growing demand for energy, fuels, chemicals, and materials.

Sustainable energy must be the foundation of a new energy paradigm in order to address the three primary global concerns of energy security, environmental protection, and socioeconomic growth [8]. To grasp the concept of sustainable energy, one must comprehend the connection between renewable and sustainable energy. Sustainable energy encompasses more than just renewable energy, despite the fact that the terms are commonly used interchangeably. Renewable energy is defined as energy sources that are dependent on non-depletable resources. Combining the concepts of sustainable and renewable energy can help us move toward a more forward-thinking energy model that takes into account both current and future demands. Because renewable energy sources are abundant and never run out, they offer a dependable alternative to fossil fuels, which are becoming more scarce and costly, which is why sustainable energy is so important [9]. By employing renewable energy, we help protect the planet's natural resources for coming generations. Sustainable energy encourages economic expansion and job development because of its stable and growing employment market.

Numerous renewable energy sources are easily accessible locally, which can strengthen local economies and reduce reliance on imported fuels. Energy from renewable resources, such as solar and wind, has demonstrated to be less expensive over time than energy from traditional fossil fuels. Renewable energy sources, as opposed to fossil fuels, are limitless and will never run out, ensuring a steady supply of energy for years to come. Biocrude oil and biorefining are essential for solving the environmental issues related to the chemical and petrochemical sectors, as well as providing a more sustainable alternative to conventional fossil fuels. Biorefining is the process of converting biomass into chemicals and fuels that are usually made from petroleum-based resources. Because biorefining uses renewable biomass sources, it has a smaller environmental impact than conventional petrochemical processes [10]. The manufacture of biocrude oil, particularly through HTL using biomass, provides a more environmentally benign option than many biofuels and traditional liquid fuels [11]. Biorefining can benefit from a broad range of biomass feedstocks, including forestry or agricultural wastes, algae, crop waste, and municipal solid waste [12].

3 Biomass Feedstocks

Biomass is the term used to describe organic material that comes from living organisms. It is utilized as a raw material for numerous chemical and industrial processes and poses fewer environmental hazards than more conventional sources like fossil fuels. Examples of organic materials that can be utilized as renewable feedstock for a range of industrial and energy-related applications include trash, algae, and crops. Different biomass generations exist. The term "food-based" biomass, which encompasses crops, such as soy, rapeseed, palm oil, sugarcane, wheat, and corn, is used to describe Generation 1 (Table 1). "Non-food" sources like wood, straw, and other organic waste are included in Generation 2. The biomass of Generation 3 is mostly derived from algae [13]. Thus, biomass-based materials can serve as carbon sinks. However, the conflict between food and fuel production is called into question because these feedstocks are derived from food sources. As a result, the emphasis is now on creating technologies that use feedstocks other than food.

4 Biomass Processing Technologies

Several specialized techniques are needed to convert biomass into biofuels, each of which aims to capture energy in a unique way. Fermentation is one of the main processes that are described in detail in the manufacture of biofuel. A well-known biochemical process known as fermentation is necessary to produce ethanol, a type of sustainable biofuel. Producing ethanol from crops that are high in sugars or starches, such as corn, sugarcane, and certain algae, is the primary objective of this process [14]. Microorganisms like yeast and specific bacterial strains are responsible for breaking down the sugars present in these crops. The microbes that consume the sugars during fermentation convert them into carbon dioxide and ethanol through a series of metabolic activities. Ethanol is separated and refined after production in order to be used as biofuel, as shown in Fig. 1.

Pyrolysis (Fig. 2) is an advanced thermochemical process that breaks down organic materials, especially biomass, by using high temperatures and no oxygen. By eliminating oxygen, this method stops burning and enables biomass to thermally decompose into smaller molecular pieces. The main products of pyrolysis are biochar, bio-oil, and syngas (synthetic gas). As an example, fast pyrolysis, which can produce up to 75% of the total yield, is made possible by high temperatures and

Table 1 Classifications of biomass feedstocks with their pros and cons [12, 13]

Generation	Feedstocks	Pros	Cons
First generation (1-G)	Edible oils, sugar starch	Decreased CO_2 emission, reduced fossil fuel consumption	Adverse effects on the availability of food, agricultural land, and food prices
Second generation (2-G)	Non-edible oils, waste cooking oil, forest residue, agricultural residue, aquatic plants	Low-cost feedstock, no threat to hunger, uses waste or non-food sources, potentially higher yields	Additional processing stages and the need for land
Third generation (3-G)	Microalgae and macroalgae	Less space and simple requirements for growth, ecofriendly, high oil content, high productivity	Expensive and unstable biofuel
Fourth generation (4-G)	Microalgae, genetically modified crops, and synthetic biology-derived organisms	High growth rate, high oil content, low structural complexity	Still in the early stages of development

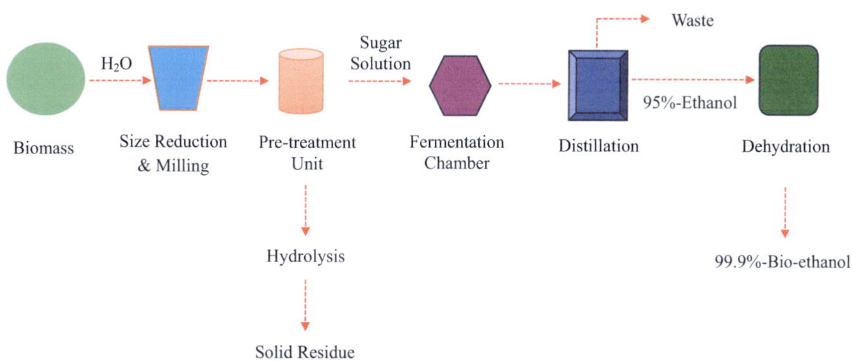

Fig. 1 Schematic diagram for fermentation process

short residence durations. Another significant result of pyrolysis is syngas, a blend of hydrogen, carbon monoxide, and other hydrocarbons that can be utilized as fuel to generate electricity or as a precursor for the synthesis of chemicals. Syngas is a valuable resource for decentralized power systems since it may be utilized to power gas turbines or internal combustion engines. Implementing policies, subsidies, or other incentives that encourage the production of biofuels and ensure their financial sustainability can significantly benefit in the development of pyrolysis technology.

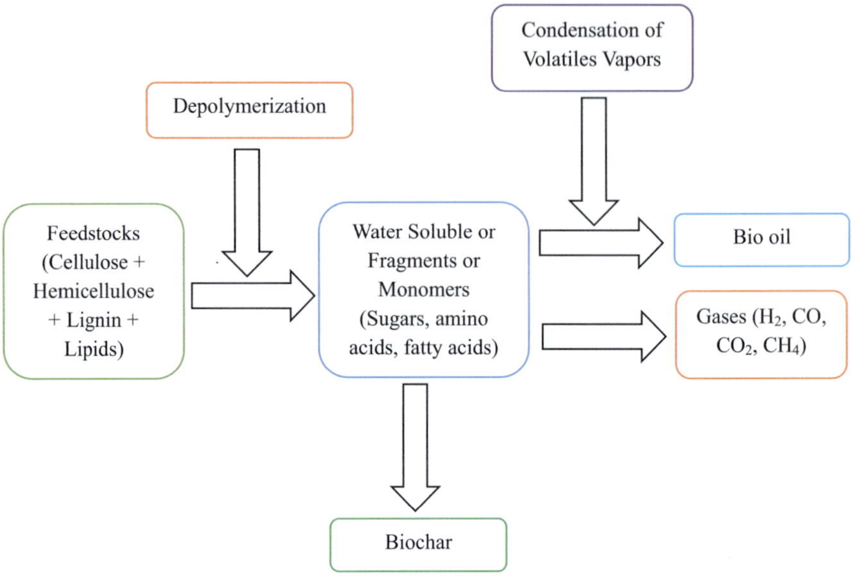

Fig. 2 Basic reaction pathways for biomass conversion through pyrolysis and liquefaction

One good alternative that can be further processed into other fuel types is biocrude made by hydrothermal liquefaction (Fig. 3). Biomass, or organic materials originating from recently or actively living organisms, such as plants and algae, is used to make biocrude oil. These fuels contribute to the fight against climate change and the reduction of greenhouse gas emissions by providing a sustainable substitute for fossil fuels. Because it is a cutting-edge method for turning wet biomass into bio-oil, it is especially helpful for feedstocks with a high-moisture content, such as algae and food waste. Water is used as a solvent in the basic HTL process to break down organic compounds at high pressures (up to 20 MPa or more) and temperatures (often between 200 and 400 °C) [15]. Hydrothermal liquefaction is a practical process for converting wet biomass into bio-oil and other beneficial products. Because it can handle high-moisture feedstocks without requiring a lot of preprocessing, it is a crucial technology for the sustainable exploitation of biomass. Through the manufacture of biofuels and the utilization of byproducts, such as charcoal and nutrient-rich aqueous phases, HTL enables the production of renewable energy, enhances agricultural practices, and promotes waste management.

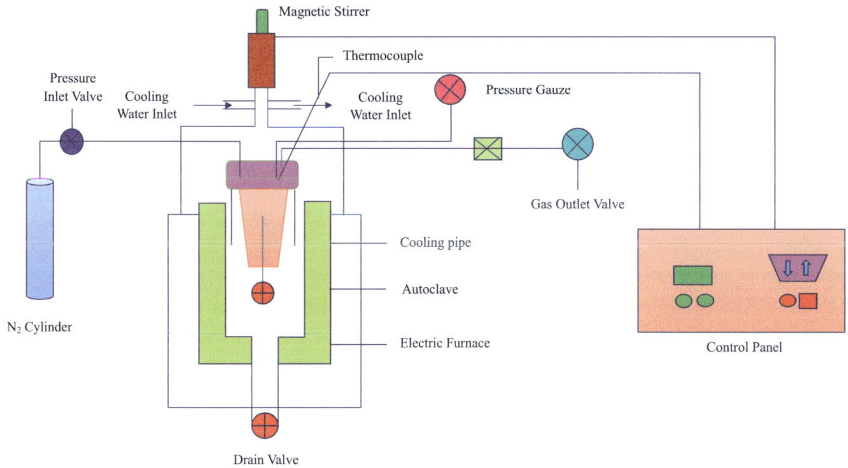

Fig. 3 Schematic diagram for hydrothermal liquefaction process

5 Characteristics of Biocrude

Biocrude is a renewable fuel derived from biomass that is produced through fermentation, HTL, or pyrolysis. It looks like oil even though it contains more nitrogen, water, and oxygen than ordinary crude. Understanding these features is essential in order to evaluate its potential as a supplement or replacement for fossil fuels [16]. These properties have an important impact on the refinement, processing, and use of biocrude. Density ranges from 950 to 1200 kg/m^3. The density of biocrude is generally higher than that of ordinary fossil crude oil, which typically ranges from 850 to 900 kg/m^3 [17]. Its increased density is a result of the raw material's high-water content and the incomplete breakdown of biomass, which results in the presence of more polar and oxygenated chemicals. At 40 °C, viscosity ranges from 0.01 to 1 Pa s. The viscosity of conventional crude oil, which normally has a viscosity of about 0.1 Pa s at 40 °C, is significantly lower than that of biocrude [18]. Biocrude has a higher viscosity because it includes more oxygen and has larger molecules that often form hydrogen bonds.

Additionally, depending on how it is manufactured, it might contain up to 30% water. For instance, biocrude from HTL, which is produced in an environment that is rich in water, typically contains more water than biocrude from pyrolysis [19]. The oxygen content of biocrude is 10–40% by weight, which is substantially higher than that of petroleum. Oxygen forms like phenols, acids, alcohols, and aldehydes are produced when biomass, which is rich in cellulose and lignin, breaks down. Usually, the range of sulfur content is less than 1%. Compared to conventional crude oil, which might have higher sulfur concentrations, biocrude has a reduced sulfur content [20]. Toxic sulfur oxides (SOx) are created when sulfur is burnt in fossil fuels. Compared to petroleum crude, which normally has a heating value of 42 to

45 MJ/kg, biocrude has a lower energy content (20–35 MJ/kg) [17]. The heating value, or the amount of energy accessible per kilogram of fuel, is impacted by the biocrude's high oxygen and water content. Acidic content ranges from 2 to 5. With a pH between 2 and 5, biocrude is often more acidic than conventional crude oil [21]. This enhanced acidity is caused by oxygenated molecules like carboxylic acids and phenols. The acidic nature of biocrude makes it challenging to manage and can cause corrosion issues in processing equipment, storage tanks, and pipelines.

The elemental composition of biocrude differs from that of fossil crude oil. Biocrude is composed of 50–70% carbon, 6–12% hydrogen, 10–40% oxygen, and less than 5% nitrogen. It has a lower energy density because it has more oxygen and less carbon, the components of biocrude have a wide boiling range, boiling between 150 and 600 °C [22]. Biocrude's diverse organic chemical composition gives it a wide range of qualities. Its oil phase separates; however, it is somewhat soluble in water due to oxygenated components. It dissolves in organic solvents like ether, hexane, and toluene just like regular crude oil does. Although it is compatible with current refining operations because to its solubility in organic solvents, it still has to be upgraded to match the performance of fossil fuels [6]. The distinct chemical properties of biocrude oil, which comes from renewable resources like wood and algae, include fatty acid esters and unsaturated double bonds that influence viscosity and biofuel potential [22]. Compared to petroleum, it contains more heteroatoms (nitrogen, sulfur, and oxygen), which affects combustion and calls for further processing. Research is ongoing to improve biocrude stability and reduce unwanted chemicals to promote fuel viability [23].

6 Biocrude Oil Applications

There are several uses for biocrude oil (Fig. 4), which is a renewable energy source. It is produced from biomass using pyrolysis and hydrothermal liquefaction [24]. One of its most well-known applications is turning biocrude oil into transportation fuel for vehicles. Two products that are crucial for reducing greenhouse gas emissions are biodiesel and gasoline, which can be produced from biocrude. Because it needs little to no modification to be used in modern diesel engines, biodiesel is a feasible alternative [17]. In addition to being used as a feedstock for chemicals, biofuels, biodegradable oils, and lubricants, biocrude is also used to generate industrial heat and energy. It decarbonizes transportation, encourages renewable energy, and helps slow down climate change because it has the potential to be carbon–neutral.

Waste reduction is greatly aided by the manufacture of biocrude from organic materials, forestry waste, and agricultural wastes [25]. Biocrude production reduces landfill loads and encourages a circular economy by turning waste into energy. By giving crop residues and other organic wastes a way to be disposed of, this procedure promotes sustainable agricultural methods. Research into biocrude manufacturing technology is a rapidly evolving topic. Increasing the yield and efficiency of biocrude extraction processes is the aim of scientists and engineers. Techniques, such

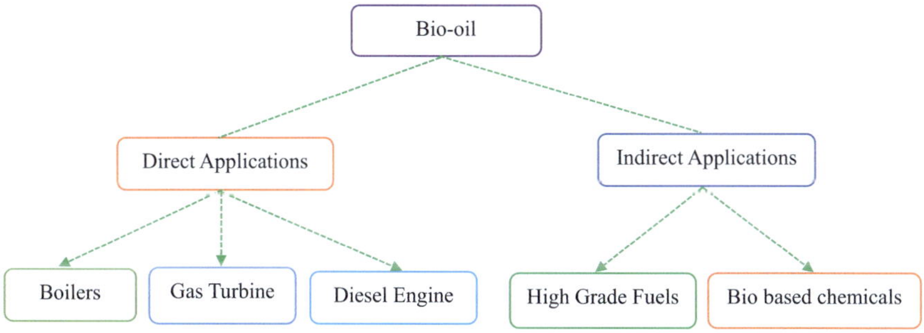

Fig. 4 Applications of biocrude oil

as catalytic pyrolysis and enhanced hydrothermal liquefaction, are being researched to optimize the conversion of biomass into high-quality biocrude. Developments in these areas of technology might significantly reduce production costs and make biocrude a more viable renewable energy source overall. The long-term sustainability of the biocrude sector depends on the development of sustainable biomass procurement practices [26]. Integrated biorefineries increase efficiency by decreasing waste. The biocrude sector has the potential to boost rural economies, create jobs, and increase energy security by stabilizing energy costs and reducing reliance on imported fossil fuels.

7 Current Status and Future Developments

Biocrude oil is often produced using pyrolysis, gasification, hydrothermal liquefaction (HTL), and catalytic depolymerization. Interest in biocrude is primarily driven by the desire to reduce reliance on fossil fuels, energy security, and climate change. Biocrude production can lower greenhouse gas emissions, boost domestic energy, recycle waste, and promote sustainability using renewable resources with the aid of fuel restrictions, carbon pricing, and bioenergy subsidies. Biocrude Research and Innovation: By 2024, efforts are underway to improve the economy and efficiency of refining processes, as well as the output of biocrude oil [27]. Advanced refining is costly, and conversion uses a lot of energy. Production costs are still expensive in comparison with fossil fuels, despite advancements in algae biofuels, hybrid processes, and additional biomass sources improving biocrude economics. Despite their potential, technologies like pyrolysis and HTL are still in the early phases of widespread commercialization [28]. The manufacturing of biocrude is continuously improved, and output is increased. Existing refineries must be substantially updated or constructed from the ground up for biocrude processing, which relies on sustainable biomass sources. Competition with food production and specific growth requirements are additional challenges.

In recent years, several companies and educational organizations have taken steps to boost biocrude production. For instance, commercial biocrude refineries are being developed by companies, such as Neste, Diamond, and Ensyn [29]. Biocrude oil can reduce lifecycle greenhouse gas emissions by using waste and renewable biomass, but the impact depends on feedstock and production efficiency. While other feedstocks may need a lot of water or compete with food, algae can thrive on land that cannot be used. Sustainable development requires a balance between these factors. By using waste materials like food scraps, agricultural wastes, and even municipal sewage, biocrude manufacturing is positioned as a key player in circular economy models [30]. Government regulations, especially in the US and the EU, are essential to biocrude development in order to prevent environmental impact, such as deforestation. These rules promote the growth of sustainable biomass, biofuel, and renewable energy.

8 Sustainability and Environmental Impact

The sustainability of biocrude oil depends on a number of factors, including the feedstock source, manufacturing techniques, and end product. Limiting the use of resources (such as energy, water, and land), reducing greenhouse gas emissions, and avoiding negative impacts on ecosystems and biodiversity are all essential components of a sustainable biocrude production pathway [31]. Biocrude oil reduces greenhouse gas emissions and promotes sustainability since biomass absorbs CO_2 during growth, creating a cycle that is almost carbon–neutral. With carbon capture and storage (CCS), biocrude can potentially become carbon negative, reducing atmospheric CO_2 concentrations by storing more CO_2 than it releases. Selecting feedstocks from non-food or marginal land, including algae or agricultural residues, can lower the risks related to indirect land use change (ILUC) [32]. The amount of land required for the production of biocrude is determined by the feedstocks. Using non-food crops like switchgrass or waste biomass preserves arable land and lessens the impact on the ecology. To maintain biodiversity, feedstocks like algae, forestry waste, and agricultural wastes are recommended.

When producing biocrude, energy, water, and land use must be balanced. For instance, HTL processes require a lot of energy and water for pre-treatment, liquefaction, and upgrading, especially when using wet feedstocks. Reducing this energy consumption through better reactor designs, process integration (e.g., combined heat and power systems), and the use of renewable energy sources can increase the sustainability of biocrude production. Life cycle analysis (LCA) is an environmental impact assessment tool that assesses a process or product from "cradle to grave." From feedstock cultivation to final combustion or end-of-life, life cycle assessment helps evaluate the sustainability of the entire biocrude oil production chain [6, 33, 34].

9 Metrics of Sustainability—An Important LCA Stage

In feedstock cultivation, biomass growth or gathering results in emissions and the consumption of resources, such as energy, water, and nutrients. The energy and chemicals required in conversion processes, such as pyrolysis and HTL, can affect emissions. The energy balance and environmental impact may also be impacted by advanced catalytic processes [17]. Energy consumption and emissions are associated with catalytic hydroprocessing and other refining techniques that transform raw biocrude into useable fuels. The LCA results are significantly influenced by the kinds of catalysts employed and the effectiveness of the process [35]. LCA incorporates emissions associated with the transportation of biomass, biocrude oil, and finished fuel products [36]. The carbon footprint can be increased by using non-local feedstocks or by long transportation distances. By employing sustainable feedstocks, CO_2 emissions from the last step of the life cycle-fuel combustion can be reduced.

Metrics like global warming potential (GWP), which is a measure of the total quantity of greenhouse gases produced and consumed, are used to assess the environmental impact of biocrude. The reduced greenhouse gas of biocrude is a key indicator of its sustainability when compared to conventional fossil fuels [37]. Acid rain is mostly caused by chemicals produced during the upgrading or combustion of sulfur- and nitrogen-containing biocrude materials. This effect can be lessened by appropriately controlling the amounts of sulfur and nitrogen in biocrude. The overall amount of water utilized throughout the biocrude production process is a significant factor to take into account when considering the sustainability of wet feedstocks like algae [38, 39]. Improving the sustainability profile can be achieved by using recycled water or by consuming less water. The production of biocrude, particularly from agricultural feedstocks, has the potential to increase nutrient runoff, specifically nitrogen and phosphorus, which can eutrophicate water bodies [40].

Biocrude sustainability is measured using metrics like carbon intensity (CI), which is measured in grams of CO_2 equivalent per megajoule (g CO_2e/MJ). A smaller CI indicates greater sustainability. Biocrude, especially when made from waste or algae, usually has a lower CI than fossil fuels. Regulations like California's Low Carbon Fuel Standard (LCFS) give credit for lower CI fuels. In markets where carbon emissions are controlled, low-CI biocrude oil would be more advantageous [41]. Energy efficiency measures how well biomass converts to biocrude fuel, balancing energy input and output. Improving this yield makes the process more sustainable. Advanced catalysts and integrated systems that lower energy usage during conversion and upgrading are necessary to improve this statistic [42]. This statistic calculates how much water is required to generate a certain amount of energy. Water-saving technology, such hydrothermal processes or recycling water for algal production, can boost this efficiency. The land use per unit of energy metric calculates the area of land needed to generate a specific quantity of biocrude energy [43]. Compared to traditional crops, algae require less space and can be cultivated on non-arable terrain. The cost of manufacturing biocrude ($/MJ) can be reduced by refinery integration, economies of scale, and process improvements [1].

10 Conclusion

The biorefining of biocrude oil is a rapidly evolving field in a bioenergy sector. It creates sustainable fuel alternatives by using waste, algae, and lignocellulosic materials as renewable biomass sources. Nowadays, hydrothermal liquefaction (HTL) and catalytic upgrading are two crucial processes that improve the yield and quality of biocrude. Recent developments have concentrated on utilizing sophisticated catalysts, co-processing with fossil fuels, and integrating combined heat and power (CHP) systems in order to improve energy efficiency. Future plans call for the use of novel feedstocks, genetic engineering, and enhanced life cycle evaluations to reduce environmental impacts and make biocrude production sustainable and lucrative. This development lends credence to the idea that biocrude could replace conventional fossil fuels.

Acknowledgements The authors thank to J&K State Science, Technology and Innovation Council (JKSTIC), Department of Science and Technology (DST), Government of India, for financial support of this research through JKST&IC/SRE/426-30. The first author also thanks to JKSTIC (DST) for providing the research fellowship.

Declarations Mohammad Aslam is an editorial board member for the book tittle "Production and Biorefining of Biocrude Oil: Current Status and Future Development" and was not involved in the editorial review, or the decision to publish, this article. All authors declare that there are no competing interests.

References

1. Ahmed SF, Rafa SJ, Mehjabin A, Tasannum N, Ahmed S, Mofijur M et al (2023) Bio-oil from microalgae: materials, production, technique, and future. Energy Reports [Internet] 10:3297–314. https://www.sciencedirect.com/science/article/pii/S2352484723013070
2. Sahoo G, Sharma A, Chandra Dash A (2022) Biomass from trees for bioenergy and biofuels—a briefing paper. Mater Today Proc [Internet] 65:461–7. https://www.sciencedirect.com/science/article/pii/S2214785322012846
3. Paramasivam P, Balasubramanian A, Adeleke AA, Ikubanni PP, Kumar S, Prakash C et al (2024) Exploring alternative fuel solutions: lemon grass oil biodiesel blend with dibutyl ether additive for VCR diesel engines—an experimental analysis. Sci Rep 1:14(1)
4. Alsultan AG, Asikin-Mijan N, Obeas LK, Isalam A, Mansir N, Nassar MF et al (2022) Biofuel and biorefinery technologies. In: Bartoli M, Giorcelli M, Tagliaferro A (eds). Biochar [Internet]. Rijeka: IntechOpen. https://doi.org/10.5772/intechopen.104984
5. Chaturvedi T, Torres AI, Stephanopoulos G, Thomsen MH, Schmidt JE (2020) Developing process designs for biorefineries-definitions, categories, and unit operations. Energies (Basel) 1:13(6)
6. Usman M, Cheng S, Boonyubol S, Cross JS (2024) From biomass to biocrude: innovations in hydrothermal liquefaction and upgrading. Energy Convers Manag [Internet] 302:118093. https://www.sciencedirect.com/science/article/pii/S0196890424000347
7. Goswami G, Makut BB, Das D (2019) Sustainable production of bio-crude oil via hydrothermal liquefaction of symbiotically grown biomass of microalgae-bacteria coupled with effective wastewater treatment. Sci Rep 1:9(1)

8. Ubando AT, Felix CB, Chen WH (2020) Biorefineries in circular bioeconomy: a comprehensive review. Bioresour Technol [Internet] 299:122585. https://www.sciencedirect.com/science/article/pii/S0960852419318152
9. Lindfors C, Elliott DC, Prins W, Oasmaa A, Lehtonen J (2023) Co-processing of biocrudes in oil refineries. Energy Fuels [Internet] 37(2):799–804. https://doi.org/10.1021/acs.energyfuels.2c04238
10. Jaiswal KK, Chowdhury CR, Yadav D, Verma R, Dutta S, Jaiswal KS et al (2022) Renewable and sustainable clean energy development and impact on social, economic, and environmental health. Energy Nexus [Internet] 7:100118. https://www.sciencedirect.com/science/article/pii/S2772427122000687
11. Dey S, Sreenivasulu A, Veerendra GTN, Rao KV, Babu PSSA (2022) Renewable energy present status and future potentials in India: an overview. Innov Green Dev [Internet] 1(1):100006. https://www.sciencedirect.com/science/article/pii/S2949753122000066
12. Clauser NM, González G, Mendieta CM, Kruyeniski J, Area MC, Vallejos ME (2021) Biomass waste as sustainable raw material for energy and fuels. Sustainability (Switzerland) 13(2):1–21
13. Saengsuriwong R, Onsree T, Phromphithak S, Tippayawong N (2021) Biocrude oil production via hydrothermal liquefaction of food waste in a simplified high-throughput reactor. Bioresour Technol [Internet] 341:125750. https://www.sciencedirect.com/science/article/pii/S0960852421010919
14. Khoo CG, Dasan YK, Lam MK, Lee KT (2019) Algae biorefinery: review on a broad spectrum of downstream processes and products. Bioresour Technol [Internet] 292:121964. https://www.sciencedirect.com/science/article/pii/S0960852419311940
15. Padder SA, Khan R, Rather RA (2024) Biofuel generations: new insights into challenges and opportunities in their microbe-derived industrial production. Biomass Bioenergy [Internet] 185:107220. https://www.sciencedirect.com/science/article/pii/S0961953424001739
16. Watson J, Swoboda M, Aierzhati A, Wang T, Si B, Zhang Y (2021) Biocrude oil from algal bloom microalgae: a novel integration of biological and thermochemical techniques. Environ Sci Technol 55(3):1973–1983
17. Mishra RK, kumar V, Kumar P, Mohanty K (2022) Hydrothermal liquefaction of biomass for bio-crude production: a review on feedstocks, chemical compositions, operating parameters, reaction kinetics, techno-economic study, and life cycle assessment. Fuel [Internet] 316:123377. https://www.sciencedirect.com/science/article/pii/S0016236122002459
18. Ranjbar S, Malcata FX (2023) Hydrothermal liquefaction: how the holistic approach by nature will help solve the environmental conundrum, Molecules, vol 28. Multidisciplinary Digital Publishing Institute (MDPI)
19. Cao M, Wang Y, Li H, Jia J, Li R, Liu Z (2023) Renewable biocrude blends boost tribological properties and maintain fuel performance of diesel. Chem Eng J [Internet] 474:145696. https://www.sciencedirect.com/science/article/pii/S1385894723044273
20. Cavalli MC, Wu W, Poulikakos L (2024) Bio-based rejuvenators in asphalt pavements: a comprehensive review and analytical study. J Road Eng [Internet] 4(3):282–91. https://www.sciencedirect.com/science/article/pii/S2097049824000325
21. Barrère-Mangote C, Roubaud A, Bouyssiere B, Maillard J, Hertzog J, Le MJ et al (2021) Study of biocrudes obtained via hydrothermal liquefaction (Htl) of wild alga consortium under different conditions. Processes 1:9(9)
22. Di Lauro F, Balsamo M, Solimene R, Alfieri ML, Manini P, Migliaccio R et al (2024) Characterization of biocrude produced by hydrothermal liquefaction of municipal sewage sludge in a 500 mL batch reactor. Ind Eng Chem Res [Internet] 63(2):955–967. https://doi.org/10.1021/acs.iecr.3c03058
23. Guirguis PM, Seshasayee MS, Motavaf B, Savage PE (2024) Review and assessment of models for predicting biocrude yields from hydrothermal liquefaction of biomass. vol 2. RSC Sustainability. Royal Society of Chemistry, pp 736–756
24. Venkatachalam CD, Sekar S, Ravichandran SR, Sengottian M, Sukumar KC, Chenniappan D et al (2023) A review on bio-crude production from algal biomass using catalytic hydrothermal liquefaction process. Environ Eng Res 1:28(3)

25. Varjani S, Shahbeig H, Popat K, Patel Z, Vyas S, Shah AV et al (2022) Sustainable management of municipal solid waste through waste-to-energy technologies. Bioresour Technol [Internet] 355:127247. https://www.sciencedirect.com/science/article/pii/S0960852422005764
26. Chaudhary V, Awadakkam S, Churchill JGB, Borugadda VB, Dalai AK (2024) Production of sustainable biocrude from Canadian agricultural biomass: process optimization and product characterization. Fuel Process Technol [Internet] 261:108108. https://www.sciencedirect.com/science/article/pii/S037838202400078X
27. Inayat A, Ahmed A, Tariq R, Waris A, Jamil F, Ahmed SF et al (2022) Techno-economical evaluation of bio-oil production via biomass fast pyrolysis process: a review, vol 9. Frontiers in Energy Research. Frontiers Media S.A.
28. Deepika C, Mrinal, Pohrmen CB, Jaiswal KS, Sangmesh B, Jaiswal KK et al (2024) Hydrothermal liquefaction of wet microalgal biomass for biofuels and platform chemicals: advances and future prospects, vol 6, Discover Applied Sciences. Springer Nature
29. Panwar NL, Paul AS (2021) An overview of recent development in bio-oil upgrading and separation techniques. Environ Eng Res [Internet] 26(5):200382. http://www.eeer.org/journal/view.php?number=1217
30. Osman AI, Mehta N, Elgarahy AM, Al-Hinai A, Al-Muhtaseb AH, Rooney DW (2021) Conversion of biomass to biofuels and life cycle assessment: a review, vol 19, Environmental Chemistry Letters. Springer Science and Business Media Deutschland GmbH, pp 4075–4118
31. Mohanty A, Ajmera S, Chinnam S, Kumar V, Mishra RK, Acharya B (2024) Pyrolysis of waste oils for biofuel production: an economic and life cycle assessment. Fuel Commun [Internet] 18:100108. https://www.sciencedirect.com/science/article/pii/S2666052024000037
32. Bishop G, Styles D, Lens PNL (2022) Land-use change and valorisation of feedstock side-streams determine the climate mitigation potential of bioplastics. Resour Conserv Recycl [Internet] 180:106185. https://www.sciencedirect.com/science/article/pii/S0921344922000337
33. Patel MR, Panwar NL (2023) Biochar from agricultural crop residues: environmental, production, and life cycle assessment overview. Resour Conserv Recycling Adv [Internet] 19:200173. https://www.sciencedirect.com/science/article/pii/S2667378923000457
34. Bishop G, Styles D, Lens PNL (2021) Environmental performance comparison of bioplastics and petrochemical plastics: a review of life cycle assessment (LCA) methodological decisions. Resour Conserv Recycl [Internet] 168:105451. https://www.sciencedirect.com/science/article/pii/S0921344921000586
35. Zhou Y, Remón J, Pang X, Jiang Z, Liu H, Ding W (2023) Hydrothermal conversion of biomass to fuels, chemicals and materials: a review holistically connecting product properties and marketable applications. Sci Total Environ [Internet] 886:163920. https://www.sciencedirect.com/science/article/pii/S004896972302541X
36. Liu F, Shafique M, Luo X (2023) Literature review on life cycle assessment of transportation alternative fuels. Environ Technol Innov [Internet] 32:103343. https://www.sciencedirect.com/science/article/pii/S2352186423003395
37. Rahimi Z, Anand A, Gautam S (2022) An overview on thermochemical conversion and potential evaluation of biofuels derived from agricultural wastes. Energy Nexus [Internet] 7:100125. https://www.sciencedirect.com/science/article/pii/S2772427122000808
38. Mohajan H, Mohajan HK (2018) Acid rain is a local environment pollution but global concern, vol 3. Open Science Journal of Analytical Chemistry
39. Prakash J, Agrawal SB, Agrawal M (2023) Global trends of acidity in rainfall and its impact on plants and soil, vol 23. Journal of Soil Science and Plant Nutrition. Springer Science and Business Media Deutschland GmbH, pp 398–419
40. Grennfelt P, Engleryd A, Forsius M, Hov Ø, Rodhe H, Cowling E (2020) Acid rain and air pollution: 50 years of progress in environmental science and policy, vol 49, Ambio. Springer, pp 849–864
41. Narayana Sarma R, Vinu R (2023) An assessment of sustainability metrics for waste-to-liquid fuel pathways for a low carbon circular economy. Energy Nexus [Internet] 12:100254. https://www.sciencedirect.com/science/article/pii/S2772427123000840

42. Lee SY, Sankaran R, Chew KW, Tan CH, Krishnamoorthy R, Chu DT et al (2019) Waste to bioenergy: a review on the recent conversion technologies. BMC Energy 1(1)
43. Eboibi BEO, Lewis DM, Ashman PJ, Chinnasamy S (2014) Hydrothermal liquefaction of microalgae for biocrude production: improving the biocrude properties with vacuum distillation. Bioresour Technol 174:212

Chapter 14
Sustainable Production of Biofuels and Value-Added Products from Biorefinery

Shreya Anand and Padmini Padmanabhan

Abstract Since the enormous energy demand unavoidably exacerbates the depletion of non-renewable resources and ecological and social problems, renewable energy has emerged as a key component of a sustainable approach. In order to develop a circular bioeconomy that focuses on the effective and sustainable valorization of biomass and encourages resource regeneration and restoration, biorefinery emerged as a sustainable strategy and recognised viable transformation platforms for products. Using various conversion technologies, biorefinery is a sustainable process for producing biofuels and other bio-products from biomass feedstock. A key element of renewable energy sources that helps to create a carbon–neutral energy system is biofuel. By 2023, according to studies, over 90% of petroleum products might be generated globally from renewable resources, along with 33% of chemicals and 50% of the pharmaceutical sector. The latest research includes combining numerous unit processes to create a variety of bio-based products, including biochemical, thermochemical, physical, and catalytic conversion. This chapter will discuss about the strategies of biorefinery integration to produce biofuels and value-added products.

Keywords Biorefinery · Biofules · Value-added products · Sustainable production · Circular bioeconomy

1 Introduction

The environment and natural resources are directly impacted by the substantial increase in energy consumption brought about by the Industrial Revolution and advancement [1]. Two of the energy sector's main issues are the rising greenhouse gas (GHG) emissions and reducing fossil fuel reserves [2, 3]. The earth's temperature has risen by 1.9 °C due to overuse of fossil fuels, which has a negative impact

S. Anand · P. Padmanabhan (✉)
Department of Bioengineering and Biotechnology, Birla Institute of Technology, Mesra, Ranchi, India
e-mail: padmini@bitmesra.ac.in

M. Aslam et al. (eds.), *Production and Biorefining of Biocrude Oil: Current Status and Future Developments*, Advances in Sustainability Science and Technology,
https://doi.org/10.1007/978-981-96-5198-6_14

on biodiversity and life [4]. In order to reduce GHG emissions, policymakers are placing a strong emphasis on sustainable development strategies. States must work together extensively to address climate change and its negative repercussions.

As a result, 196 nations decided to manage climate change and reduce GHG emissions at the Conference of the Parties (COP21–Paris, December 2015) [4]. Scientists and researchers are also focusing on biofuels and other alternative renewable energy sources [5, 6]. According to a global status report on energy, energy generation from renewable resources and non-renewable fossil fuels is 19% and 78.4%, respectively [7]. Conventional biomass, biofuels, hydropower, wind, geothermal heat, solar energy, and geothermal energy are examples of renewable energy sources [8, 9]. The biofuel industry is anticipated to expand at a compound annual growth rate (CAGR) of 8% from 2019 to 2024.

It is projected that by 2030, 50% of the global pharmaceutical industry may be bio-based, 33% of chemicals and various materials could be produced from renewable resources, and over 90% of petroleum-based commodities might be produced from renewable resources [10]. According to Ajanovic [11] and Tsita et al. [12], biofuels are classified according to their chemical characteristics, nature, biomass feedstock, conversion methods, and technology. According to Zhou et al. [13], biofuels come in a variety of states, including liquid, gaseous, solid, and liquid. There are two primary types of biofuels that are commonly recognised: the traditional biofuels that come from edible crops, and the advanced biofuels that come from non-food feedstock and agricultural wastes [14].

Two primary biofuels are commonly recognised: traditional biofuels from edible crops and advanced biofuels from non-food feedstock and agricultural wastes [14]. Depending on the starting material, biofuels are classified as first, second, third, or fourth-generation products [15]. According to Bhatia et al. [16], the term "first-generation biofuels" mostly refers to conventional biofuels, while "second, third, and fourth-generation biofuels" are classified as advanced biofuels. First-generation biofuels, such as ethanol and biodiesel, account for most of the biofuel market share [17]. While plant material and animal fats are often esterified and transesterified during the first generation of biodiesel synthesis, bioethanol is produced by fermentation of plant carbohydrates. The only energy source for second-generation biofuels is non-food crops, such as sugar cane bagasse [16]. The process of producing second-generation biofuels entails the hydrolysis and fermentation of lignocellulosic material. Gasification and pyrolysis are further processes used to produce biofuels based on hydrocarbons. This kind of product, however, requires additional catalytic procedures. The main feedstock for the biochemical and thermochemical procedures used to produce third-generation biofuel is algae and some other microorganisms.

Utilising cutting-edge biological technologies, fourth-generation biofuel is essentially a continuation of third generation [18, 19]. Fourth-generation biofuels, for example, include biofuels made from genetically modified algae feedstock. Enhancing photosynthetic efficiency, decreasing photoinhibition, and increasing light penetration are standard methods for genetically modifying algae [20]. Notably, metabolic engineering dramatically increases the amount of lipids or carbs in algae

[21]. This chapter focuses on the strategies of biorefinery integration to produce biofuels and value-added products.

2 Biorefinery and Its Integration Strategies of Production

A biorefinery is a facility or processing unit that uses biomass as a substrate to create a wide variety of products. According to Alonso et al. [22], biomass can be made of wood, straw, starch, sugars, trash, or algae. It can be processed into a number of economic goods, such as chemicals and biomaterials. Since biomass is used in place of crude oil in the biorefinery model, it is comparable to petroleum plants [23]. Four industrial trends gave rise to the term "biorefinery" in the 1990s: (a) increased paper mill profitability and competitiveness, (b) ethanol production from wood and pulp; (c) biomass gasification and growing interest in starch production for energy application; and (d) growing interest in upgrading low-quality lignocellulose to valuable products [24]. To effectively produce sustainable bio-based products, such as bioenergy, biochemicals, and biofuels, biorefineries integrate many conversion processes, including biochemical, thermochemical, combustion, and microbial growth platforms [25]. Products from biorefineries are renewable carbon sources that reduce the risk of climate change and satisfy the endless need for fuels, materials, and energy [26].

The raw material used in a traditional oil refinery is a complicated blend of hydrocarbons. Fractional distillation, a highly energy-intensive process, is used to separate the oil [27]. In general, to generate fuel with a higher octane number or boiling point, the purified oil fractions must be upgraded using a variety of techniques, such as fluid catalytic cracking in the case of tiny chain products or carbon–carbon linkage [28, 29]. Biorefineries involve the opposite process. The pretreatment phase is first used to reduce the crystallinity, which is followed by the biopolymers' recalcitrance [30].

The biorefineries are divided into three phases based on the type of substrate, production method, potential product, and yield. These phases are as follows: (a) Phase I biorefineries, which have a single feedstock, a single process, and a single major product; (b) Phase II biorefineries, which involve a single feedstock, multiple processes, and multiple major products; and (c) Phase III biorefineries, which have multiple feedstocks, multiple processes, and multiple major products.

According to Bozell and Petersen [31], phase I biorefineries are restricted to processing only one feedstock in order to produce one key product. Biodiesel is created by transesterification, which comes after the seed oil is first removed by crushing [32]. In pulp and paper mills and processing facilities, the phase I biorefinery has been used to convert maize grains into ethanol and has proven to be a financially viable solution [33]. In Europe, phase I biorefineries have been used to produce biodiesel [34].

The phase II biorefinery employs a single feedstock, just as the phase I biorefinery. Phase II biorefineries, on the other hand, employ a number of processes that produce a number of important products, such as vitamins, hydrocolloids, amino acids, polyols, biofuels, and polymers with a variety of uses in the chemical, material, and energy sectors [35]. According to surveys, biorefineries currently meet 30% of the demand for biofuel. Starch-derived thermoplastics and biodegradable polyesters were manufactured at a phase II Novamont Plant in Italy using maize starch as feedstock. The Roquett site of Lastrem is an example of a phase II biorefinery. It uses a biochemical catalysis process to transform cereal grain feedstock into a variety of key products, including proteins, modified starches, and organic acid.

Phase III biorefineries are an advanced category that use a variety of feedstocks and a combination of processes, including extraction, separation, and chemical and biological transformations, to generate a wide range of products [36]. According to the type of feedstock, the methods used, and intermediates like the sugar platform and syngas platform, phase III biorefineries are currently divided into seven distinct systems [37].

3 Biofuels, Value-Added Products, and Their Production

Three decades' worth of technological developments and precursor materials are used to produce several generations of biofuels [38]. A sizable amount of the feedstock used in the large-scale manufacture of biofuels has come from plants. According to Naik et al. [39], the first generation of biofuels was made from food crops cultivated over a number of arable land areas, such as corn, wheat, and sugar beetroot. The main procedure used to process the plant carbohydrates was fermentation, which resulted in bioethanol, and transesterification of vegetable oil, which produced biodiesel. However, the exclusive use of food crops for energy production puts them in direct competition with the food market and agricultural resources (land and water), undermining their significance as a dependable energy source on a global scale [40, 41].

Second-generation biofuels solved these drawbacks. In order to liberate stored carbon sources and sugar, which experienced chemical modifications to make biofuel, pretreatment processes were devised for wood and non-edible plant parts (lignocellulose components) [42]. According to Yokoyama and Matsumura [43], a sizable percentage of non-feed crops were included as raw materials during the manufacturing cycle. The third generation of biofuels was identified by the bioenergy derived from algae. According to Davis et al. [44], algae are explicitly significant because they play a non-competitive function in the main food chain and provide a platform for the production of several end products, including bioethanol, biobutanol, biogas, and biodiesel. Because of the extra steps involved, the products became more expensive, which prevented developing and underdeveloped nations from using them as their main source of fuel.

Biodegradable organic matter derived from plants, animals, and microorganisms is referred to as biomass. There are two additional categories for biomass: virgin and waste materials. "Virgin biomass" refers to biomass from forests, grasslands, energy crops from terrestrial habitats, and algae from aquatic habitats that are directly used in the production process. "Waste biomass" is the term used to describe the waste material from industrial, agricultural, and municipal settings that is utilised to produce biofuel [45]. Plants still make up the majority of biomass, despite algae offering extremely sophisticated and sustainable platforms for the production of biofuels. Both forms of biomasses can be used to produce various kinds of biofuel through a variety of methods.

According to Neshat et al. [46], plants are photosynthetic living things with a higher proportion of biological molecules that are carbohydrates. Triglycerides, oligosaccharides, and polysaccharides are the three types of carbohydrates, which are carbon molecules that have been hydrated. Wood logs and dried plant stalks have been utilised as a direct energy source since ancient times [47]. Fuel was produced by processing plant material as the chemical and industrial sectors advanced. During fermentation, lignocellulosic substances such as cellulose, hemicellulose, and lignin are the main reactants. To produce ethanol, molasses that was collected from *Saccharum officinarum* was fermented using the Acetone-Butanol-Ethanol pathway [48]. The most prevalent polysaccharide in plants is cellulose, which is made up of 800–17,000 glucose units joined by β-1, 4-glucosidic bonds. Hemicellulose is a structural polymer made up of heterogeneous 5-C and 6-C monosaccharide subunits. The branching structure has partial permeability in water due to an abundance of xylose and a varied amount of glucose, galactose, mannose, and arabinose. It strengthens the plant cell wall by binding cellulose and lignin [49].

Plants provide the most biomass, followed by microbes. The costs associated with producing plant-based organic matter also include those associated with agriculture and transportation to the final destination. A plant's capacity as an industrial feedstock is determined by a number of factors, including its availability as a consistent fuel source, high yield per hectare, resistance to pests and salinity, sustainability of uniform composition, transportability of desirable material, conditions, compaction, and strength [50].

The production cycle incorporates waste biomass (unneeded forestry, agricultural, and residential materials), cellulose plants (trees and grasses), starch crops (wheat, potatoes, and corn), and sucrose crops (cane, beetroot root) [51]. 80% of the world's output of maize is shared by the USA, China, Brazil, Argentina, Ukraine, and India, where it is grown as a significant food crop. Due to particular climatic circumstances, sugarcane is the most widely produced crop and is primarily collected in Brazil and India [51]. Switch grass shows promise as a possible substitute for this interruption in the food supply [49]. In addition, it grows primarily in marginal places, is a non-feed crop, and prevents erosion of the soil [50].

Additionally, the main ingredients in the manufacturing of biofuel include animal fat and waste manure [52]. Animal fat methyl ester (AFME) is produced by processing animal fats and tallows [53]. Free fatty acids found in raw animal fats impair the quality of biodiesel, for this reason, they must be actively eliminated during the pretreatment stage. There has only been claimed to be a 2% active contribution to the creation of biodiesel, since refineries and biodiesel production primarily process biomass derived from plants. When animal dung is digested anaerobically, the primary by-product of animal origin produced is biogas [46]. Animal manure's carbon to nitrogen ratio is unsuitable for anaerobic digestion since animal waste is a carbon-poor and nitrogen-rich source.

Algae are quickly expanding photosynthetic creatures that can live in various climatic situations. They are also known to grow in both aquatic and terrestrial environments. Regardless of the soil type or land availability, algae are being researched as a powerful, sustainable biomass with high productivity and a non-specific global distribution [54]. Because different strains of algae contain large amounts of lipids, which are necessary for obtaining biodiesel and cellulosic cell walls, algal biomass produces multiple bio-products with significant economic and industrial values. It is also used as a biomass for the production of bioethanol and biobutanol [55].

Various feedstocks, such as beet, oils, organic waste, cereals, and biomass treatment, are used to manufacture biofuels. Lignocellulose is the most preferred starting material for manufacturing biofuels out of all the substrates [56]. However, lignocellulosic material is very difficult to degrade because of its intricate structure. Enzymatic and thermochemical treatment is necessary to break down lignocellulosic macromolecules into smaller subunits that can be processed further. Generally, hydrolysis—which can be acidic or enzymatic—is required to convert macromolecules into monomers. A few drawbacks of acid hydrolysis are its low yield, need for higher temperatures, and requirement for neutralisation following hydrolysis. The range of enzymes called cellulases are involved in the enzymatic hydrolysis process [57].

4 Future Developments in Production of Bio-crude Oil

In addition to producing additional products with added value, microalgae is a prospective source of biodiesel synthesis. This microbe is being used to investigate a wide range of process strategies to improve biofuel yield and lowering harvesting costs. Depending on the desired final product, there are two ways to utilise microalgae-based biofuel: thermochemical and biochemical. Anaerobic digestion, fermentation, transesterification, and other processes are the main categories of biochemical conversion. Anaerobic digestion yields methane as its principal result, whereas transesterification primarily deals with synthesising biodiesel from microalgal lipids. Ethanol is produced during fermentation. In the meantime, biohydrogen is produced through photo-fermentation. By thermal converting the entire biomass to biofuel and bioenergy, the thermochemical method often converts biomass from algae robustly while shortening the processing time [58].

Reducing reliance on conventional fossil fuels is necessary due to the depletion of fossil fuel reserves and the policies being implemented to generate net-zero release of carbon dioxide by 2050. The growing global population leads to a rise in the energy demand, and thus, it is critical to create novel strategies for utilising nontraditional, sustainable, and renewable energy sources (including solar, wind, and lignocellulosic biomass) to fulfil the rising energy demands. With a focus on hydrothermal liquefaction, the current study investigates different methods for turning bark, a lignocellulosic biomass, into biofuels. These processes include feedstock pretreatment, the creation of bio-crude, and product upgradation to make the product appropriate for use as fuel for vehicles. There are few studies on synthesising bio-crude from bark, particularly when employing hydrothermal liquefaction, and several information gaps need to be filled.

The pyrolysis of biomass sustainably produces chemicals and is appealing. This is important to prevent excessive reliance on fossil fuels. However, there is still a significant amount of water in the liquid by-product known as raw bio-crude oil following biomass pyrolysis. This study produced light distillate by batch vacuum distillation of RBCO using empty bunches of fruit pyrolysis at 80–95 °C and 30–45 kPa. This is further supported by the data from RBCO's Karl-Fischer, which show that water content was 66.77%, while acetic acid and phenol concentrations were 16.02% and 26.53%, respectively, according to RBCO's GC–MS. The light distillate's physical characteristics included a light brown hue, a smoky smell, and a density and viscosity 0.99 g/ml and 0.93 cP, respectively. This investigation indicates that the water content of light distillate ranges from 70.57 to 98.81%. The two main components of RBCO, acetic acid and phenol, are also dissolved in the water phase during vacuum distillation, with the greatest quantities of 0.28% and 0.09%, respectively. This light distillate may contain agricultural biopesticides or food preservatives based on its chemical makeup [59].

5 Conclusion

Reliance upon non-detrimental renewable energy sources is necessary in light of the worrying circumstances brought about by exploiting petroleum-based fuel goods. Algae, plants, and mammals are the main sources of biofuels. Starch and both cellulosic and non-cellulosic biomolecules are produced by plants and are widely used in the cycles that produce bioethanol and biobutanol. The fats and oils found in plants, algae, and animals are the basic materials used to produce biogas and biodiesel. Biorefineries' construction and technological advancement played a pivotal role in the energy sector's shift from petroleum to biochemical energy. It is critical to assess when developing highly efficient, effective, and sustainable processes. Clean technology and environmentally friendly processes can be achieved based on integrated biorefineries.

References

1. Cheng Y, Awan U, Ahmad S, Tan Z (2021) How do technological innovation and fiscal decentralization affect the environment? A story of the fourth industrial revolution and sustainable

growth. Technol Forecast Soc Change 162:120398. https://doi.org/10.1016/j.techfore.2020.120398
2. Neukirchen F, Ries G (2020) The world of mineral deposits. Springer. https://doi.org/10.1007/978-3-030-34346-0
3. Peters GP, Andrew RM, Canadell JG, Friedlingstein P, Jackson RB, Korsbakken JI, Le Qu´er´e C, Peregon A (2020) Carbon dioxide emissions continue to grow amidst slowly emerging climate policies. Nat Clim Change 10:3–6. https://doi.org/10.1038/s41558-019-0659-6
4. Umar M, Ji X, Kirikkaleli D, Xu Q (2020) COP21 roadmap: do innovation, financial development, and transportation infrastructure matter for environmental sustainability in China? J Environ Manag 271:111026. https://doi.org/10.1016/j.jenvman.2020.111026
5. Adeniyi OM, Azimov U, Burluka A (2018) Algae biofuel: current status and future applications. Renew Sustain Energy Rev 90:316–335. https://doi.org/10.1016/j.rser.2018.03.067
6. Zeppini P, van den Bergh JCJM (2020) Global competition dynamics of fossil fuels and renewable energy under climate policies and peak oil: a behavioural model. Energy Pol 136:110907. https://doi.org/10.1016/j.enpol.2019.110907
7. Milano J, Ong HC, Masjuki HH, Chong WT, Lam MK, Loh PK, Vellayan V (2016) Microalgae biofuels as an alternative to fossil fuel for power generation. Renew Sustain Energy Rev 58:180–197. https://doi.org/10.1016/j.rser.2015.12.150
8. Vakulchuk R, Overland I, Scholten D (2020) Renewable energy and geopolitics: a review. Renew Sustain Energy Rev 122:109547. https://doi.org/10.1016/j.rser.2019.109547
9. Pal K, Yadav P, Tyagi SK (2017) Renewable sources in India and their applications. In: Sustainable biofuels development in India. Springer, pp 39–71. https://doi.org/10.1007/978-3-319-50219-9_3
10. Clauser NM, Felissia FE, Area MC, Vallejos ME (2021) A framework for the design and analysis of integrated multi-product biorefineries from agricultural and forestry wastes. Renew Sustain Energy Rev 139:110687. https://doi.org/10.1016/j.rser.2020.110687
11. Ajanovic A (2011) Biofuels versus food production: does biofuels production increase food prices? Energy 36:2070–2076. https://doi.org/10.1016/j.energy.2010.05.019
12. Tsita KG, Kiartzis SJ, Ntavos NK, Pilavachi PA (2020) Next generation biofuels derived from thermal and chemical conversion of the Greek transport sector. Therm. Sci. Eng. Prog. 17:100387. https://doi.org/10.1016/j.tsep.2019.100387
13. Zhou Y, Zhang Z, Zhang Y, Wang Y, Yu Y, Ji F, Ahmad R, Dong R (2016) A comprehensive review on densified solid biofuel industry in China. Renew Sustain Energy Rev 54:1412–1428. https://doi.org/10.1016/j.rser.2015.09.096
14. Ak´e SC (2017) The nonlinear relation between biofuels and food prices. Invest Econ 76:3–50. https://doi.org/10.1016/j.inveco.2016.12.001
15. Callegari A, Bolognesi S, Cecconet D, Capodaglio AG (2020) Production technologies, current role, and future prospects of biofuels feedstocks: a state-of-theart review. Crit Rev Environ Sci Technol 50:384–436. https://doi.org/10.1080/10643389.2019.1629801
16. Bhatia SK, Kim SH, Yoon JJ, Yang YH (2017) Current status and strategies for second generation biofuel production using microbial systems. Energy Convers Manag 148:1142–1156. https://doi.org/10.1016/j.enconman.2017.06.073
17. Araújo K, Mahajan D, Kerr R, Silva MD (2017) Global biofuels at the crossroads: an overview of technical, policy, and investment complexities in the sustainability of biofuel development. Agric For 7:32. https://doi.org/10.3390/agriculture7040032
18. Kumar R, Ghosh AK, Pal P (2020) Synergy of biofuel production with waste remediation along with value-added co-products recovery through microalgae cultivation: a review of membrane-integrated green approach. Sci Total Environ 698:134169. https://doi.org/10.1016/j.scitotenv.2019.134169
19. Zhou Y, Chen Y, Li M, Hu C (2020) Production of high-quality biofuel via ethanol liquefaction of pretreated natural microalgae. Renew Energy 147:293–301. https://doi.org/10.1016/j.renene.2019.08.136
20. Tandon P, Jin Q (2017) Microalgae culture enhancement through key microbial approaches. Renew Sustain Energy Rev 80:1089–1099. https://doi.org/10.1016/j.rser.2017.05.260

21. Abdullah B, Syed Muhammad SAF, Shokravi Z, Ismail S, Kassim KA, Mahmood AN, Aziz MMA (2019) Fourth generation biofuel: a review on risks and mitigation strategies. Renew Sustain Energy Rev 107:37–50. https://doi.org/10.1016/j.rser.2019.02.018
22. Alonso DM, Bond JQ, Dumesic JA (2010) Catalytic conversion of biomass to biofuels. Green Chem 12:1493–1513. https://doi.org/10.1039/c004654j
23. Esteban J, Ladero M (2018) Food waste as a source of value-added chemicals and materials: a biorefinery perspective. Int J Food Sci Technol 53:1095–1108. https://doi.org/10.1111/ijfs.13726
24. Ohlrogge J, Allen D, Berguson B, Della Penna D, Shachar-Hill Y, Stymne S (2009) Driving on biomass. Science 324(80):1019–1020. https://doi.org/10.1126/science.1171740
25. Ong HC, Chen WH, Singh Y, Gan YY, Chen CY, Show PL (2020) A state-of-theart review on thermochemical conversion of biomass for biofuel production: a TGFTIR approach. Energy Convers Manag 209:112634. https://doi.org/10.1016/j.enconman.2020.112634
26. Esposito D, Antonietti M (2015) Redefining biorefinery: the search for unconventional building blocks for materials. Chem Soc Rev 44:5821–5835. https://doi.org/10.1039/c4cs00368c
27. Hülsey MJ (2018) Shell biorefinery: a comprehensive introduction. Green Energy Environ 3:318–327. https://doi.org/10.1016/j.gee.2018.07.007
28. Mortensen PM, Grunwaldt JD, Jensen PA, Knudsen KG, Jensen AD (2011) A review of catalytic upgrading of bio-oil to engine fuels. Appl Catal Gen 407:1–19. https://doi.org/10.1016/j.apcata.2011.08.046
29. Xiu S, Shahbazi A (2012) Bio-oil production and upgrading research: a review. Renew Sustain Energy Rev 16:4406–4414. https://doi.org/10.1016/j.rser.2012.04.028
30. Kamm B, Kamm M (2004) Principles of biorefineries. Appl Microbiol Biotechnol 64:137–145. https://doi.org/10.1007/s00253-003-1537-7
31. Bozell JJ, Petersen GR (2010) Technology development for the production of biobased products from biorefinery carbohydrates—the US Department of Energy's "top 10" revisited. Green Chem 12:539–554. https://doi.org/10.1039/b922014c
32. Ed de Jong, Higson A, Walsh P, Wellisch M (eds) (2012) Bio-based chemicals value added products from biorefineries. IEA Bioenergy. Task 42 Biorefinery
33. Gravitis J (2007) Zero techniques and systems—ZETS strength and weakness. J Clean Prod 15:1190–1197. https://doi.org/10.1016/j.jclepro.2006.07.038
34. Wellisch M, Jungmeier G, Karbowski A, Patel MK, Rogulska M (2010) Biorefinery systems—potential contributors to sustainable innovation. Biofuels Bioprod Biorefining 4:275–286. https://doi.org/10.1002/bbb.217
35. de Jong E, van Ree R, van Tuil R, Elbersen W (2008) Biorefineries for the chemical industry—a Dutch point of view. In: Biorefineries-industrial processes and products: status quo and future directions, pp 85–111. 9783527619849.ch4
36. Diep NQ, Sakanishi K, Nakagoshi N, Fujimoto S, Minowa T, Tran XD (2012) Biorefinery: concepts, current status, and development trends. Int J Biomass Renewables 1:1–8
37. Wright M, Brown RC (2007) Establishing the optimal sizes of different kinds of biorefineries. Biofuels Bioprod Biorefining 1:191–200. https://doi.org/10.1002/bbb.25
38. Lee RA, Lavoie J-M (2013) From first- to third-generation biofuels: challenges of producing a commodity from a biomass of increasing complexity. Anim Front 3:6–11. https://doi.org/10.2527/af.2013-0010
39. Naik SN, Goud VV, Rout PK, Dalai AK (2010) Production of first and second generation biofuels: a comprehensive review. Renew Sustain Energy Rev 14:578–597. https://doi.org/10.1016/j.rser.2009.10.003
40. Basu P (2010) Biomass gasification and pyrolysis practical design and theory. Elsevier
41. Wang Y, Ho SH, Yen HW, Nagarajan D, Ren NQ, Li S, Hu Z, Lee DJ, Kondo A, Chang JS (2017) Current advances on fermentative biobutanol production using third generation feedstock. Biotechnol Adv 35:1049–1059. https://doi.org/10.1016/j.biotechadv.2017.06.001
42. Soltanian S, Aghbashlo M, Almasi F, Hosseinzadeh-Bandbafha H, Nizami A-S, Ok YS, Lam SS, Tabatabaei M (2020) A critical review of the effects of pretreatment methods on the exergetic aspects of lignocellulosic biofuels. Energy Convers Manag 212:112792. https://doi.org/10.1016/j.enconman.2020.112792

43. Yokoyama S, Matsumura Y (2008) The Asian biomass handbook: a guide for biomass production and utilization. Japan Instit Energy 1:61–62
44. Davis R, Kataria R, Cerrone F, Woods T, Kenny S, O'Donovan A, Guzik M, Shaikh H, Duane G, Gupta VK, Tuohy MG, Padamatti RB, Casey E, O'Connor KE (2013) Conversion of grass biomass into fermentable sugars and its utilization for medium chain length polyhydroxyalkanoate (mcl-PHA) production by Pseudomonas strains. Bioresour Technol 150:202–209. https://doi.org/10.1016/j.biortech.2013.10.001
45. Harmsen P, Huijgen W, Bermudez L, Bakker R (2010) Literature review of physical and chemical pretreatment processes for lignocellulosic biomass. Wageningen UR Food and Biobased Research
46. Neshat SA, Mohammadi M, Najafpour GD, Lahijani P (2017) Anaerobic codigestion of animal manures and lignocellulosic residues as a potent approach for sustainable biogas production. Renew Sustain Energy Rev 79:308–322. https://doi.org/10.1016/j.rser.2017.05.137
47. Falarz LJ, Deyholos MK, Chen G (2018) Plant carbohydrates and production of renewable biofuel from starch, sugar, and cellulose. In: Plant bioproducts. Springer, pp 87–107. https://doi.org/10.1007/978-1-4939-8616-3_6
48. Dahman Y, Syed K, Begum S, Roy P, Mohtasebi B (2019) Biofuels: their characteristics and analysis. In: Biomass, biopolymer-based materials, and bioenergy: construction, biomedical, and other industrial applications. Elsevier Ltd, pp 277–325. https://doi.org/10.1016/B978-0-08-102426-3.00014-X
49. Vohra M, Manwar J, Manmode R, Padgilwar S, Patil S (2014) Bioethanol production: feedstock and current technologies. J Environ Chem Eng 2:573–584. https://doi.org/10.1016/j.jece.2013.10.013
50. Parrish DJ, Casler MD, Monti A (2012) The evolution of switchgrass as an energy crop. In: Switchgrass: a valuable biomass crop for energy. https://doi.org/10.1007/978-1-4471-2903-5
51. Medipally SR, Yusoff F, Banerjee S, Shariff M (2015) Microalgae as sustainable renewable energy feedstock for biofuel production. BioMed Res Int 2015:1–13. https://doi.org/10.1155/2015/519513
52. Karmakar A, Karmakar S, Mukherjee S (2010) Properties of various plants and animals feedstocks for biodiesel production. Bioresour Technol 101:7201–7210. https://doi.org/10.1016/j.biortech.2010.04.079
53. Alptekin E, Canakci M, Sanli H (2014) Biodiesel production from vegetable oil and waste animal fats in a pilot plant. Waste Manag 34:2146–2154. https://doi.org/10.1016/j.wasman.2014.07.019
54. Tran DT, Yeh KL, Chen CL, Chang JS (2012) Enzymatic transesterification of microalgal oil from Chlorella vulgaris ESP-31 for biodiesel synthesis using immobilized Burkholderia lipase. Bioresour Technol 108:119–127. https://doi.org/10.1016/j.biortech.2011.12.145
55. Cesário MT, da Fonseca MMR, Marques MM, de Almeida MCMD (2018) Marine algal carbohydrates as carbon sources for the production of biochemicals and biomaterials. Biotechnol Adv 36:798–817. https://doi.org/10.1016/jbiotechadv.2018.02.006
56. Kucharska K, Słupek E, Ciéslinski H, Kaminski M (2019) Advantageous conditions of saccharification of lignocellulosic biomass for biofuels generation via fermentation processes. Chem Pap 74:1199–1209. https://doi.org/10.1007/s11696-019-00960-1
57. Clark JH, Luque R, Matharu AS (2012) Green chemistry, biofuels, and biorefinery. Annu Rev Chem Biomol Eng 3:183–207. https://doi.org/10.1146/annurevchembioeng-062011-081014
58. Bhowmick S, Mitra R, Kumar R, Yadav S, Tripathi S, Mukherjee S, Sen R (2024) An insight into the recent developments of biofuel production from microalgae. Biofuels 311–334
59. Oktavia FD, Steven S, Putri AHI, Hernowo P, Restiawaty E, Bindar Y (2024) Vacuum distillation of raw bio-crude oil (RBCO) from empty fruit bunches (EFB) pyrolysis to produce light distillate containing acetic acid and phenol for sustainable chemical purposes. Brazilian J Chem Eng 1–18. https://doi.org/10.1007/s43153-024-00448-w

Chapter 15
Biocrude Upgrading Technology and Challenges

Patrick da Silva Sousa, José Roberto de Matos Filho, Letícia da Costa Silva, Rodrigo da Silva Vieira, Mauricio Quintas, and José C. S. dos Santos

Abstract The world's growing energy needs drive the search for alternative sources. Fossil fuels, despite their importance, are unsustainable due to emissions and limited supply. Therefore, renewable sources, such as biomass energy, are gaining attention. Biomass, derived from organic materials, including plants and animals, can be converted into fuels like biocrude through processes, such as hydrothermal liquefaction. Biocrude, comparable to diesel, offers a sustainable fuel alternative. Bibliometric analysis, a tool for quantitative and qualitative assessment of scientific output, was used in this study to investigate contributions from countries, authors, institutions, and highly cited articles, as well as the most frequently occurring keywords in articles, reviews, and conference papers on biocrude oil production and biorefining. The analysis revealed that the United States leads in publications with 343 documents, followed by China with 340 and India with 101. Promising collaborations were noted between the United States and China. The University of Illinois System, University of Illinois Urbana-Champaign, and Aalborg University were the top contributing institutions. The significance of hydrothermal liquefaction in bioproduct synthesis and the development of catalysts to enhance efficiency and reduce waste were highlighted, promoting sustainable development.

Keywords Biocrude · Crude bio-oil · Upgrade bio-oil · Hydrothermal liquefaction

---

P. da Silva Sousa
Departamento de Engenharia Química, Universidade Federal do Ceará, Fortaleza, Ceará, Brazil

J. R. de Matos Filho · L. da Costa Silva · R. da Silva Vieira · M. Quintas · J. C. S. dos Santos (✉)
Instituto de Engenharias e Desenvolvimento Sustentável, Universidade da Integração Internacional da Lusofonia Afro-Brasileira, Redenção, Ceará, Brazil
e-mail: jcs@unilab.edu.br

M. Aslam et al. (eds.), *Production and Biorefining of Biocrude Oil: Current Status and Future Developments*, Advances in Sustainability Science and Technology,
https://doi.org/10.1007/978-981-96-5198-6_15

1 Introduction

The continued global demand for energy has propelled the concurrent pursuit and exploration of alternative sources [1, 2]. Fossil fuels, while historically crucial for energy production, are increasingly facing criticism due to their significant greenhouse gas emissions and the finite nature of their reserves [3, 4]. Therefore, renewable energy sources have attracted wide attention, especially biomass energy, which researchers widely recognize. The organic components of biomass can be extracted and converted into fuels through physical and chemical methods [5]. Biomass, a biological material that includes fauna and flora [6, 7], offers a wide variety of resources, such as wood [8] and wood waste [9], energy crops, aquatic plants [4, 10], and agricultural residues [11, 12].

This diversity of biomass is essential to produce biocrude, a liquid biofuel obtained through the thermochemical conversion of biomass, often via processes, such as hydrothermal liquefaction (HTL) [1–14]. Studies demonstrate the effectiveness of biocrude as a substitute for fossil fuels. For example, biocrude obtained from orange peels exhibited properties comparable to diesel and biodiesel standards [15].

In this context, the biorefining of biocrude gains prominence, being interconnected in converting biomass into biofuels [4, 6–12]. This process involves processing technologies [13, 14, 16, 17], waste valorization [9, 11, 12], sustainability, and economic feasibility [18], contributing significantly to a more sustainable and efficient energy future [5].

Bibliometric analysis is widely used in various research areas [6, 19–21]. This method allows quantitative and qualitative activities to identify research trends over time or in specific periods and provide detailed analysis based on consistent data [22]. In recent years, bibliometric studies have significantly advanced, revealing emerging trends in various fields and collecting essential data, such as the number of publications, citations, keywords, leading countries, and institutions [23].

In this context, bibliometric analysis was used to develop and discuss the proposed topic. This analysis is crucial for understanding innovative technologies and schemes in biocrude biorefining, refining global interest, and identifying highly relevant points for the biorefinery industry, highlighting strengths and weaknesses. The obtained results make it possible to relate analyses and practical applications, highlighting various possibilities and their concepts.

The main objective of this article is to conduct an advanced bibliometric analysis of studies on biocrude biorefining published between 2000 and 2024. This study aims to benefit industries, researchers, and other interested parties by demonstrating trends and valuable insights. The analysis included an extensive literature review on biocrude, with data obtained from the Web of Science database. Tools, such as VOSviewer software, were used to construct and discuss the analyses and generate figures and graphs throughout the work.

2 Methodology

2.1 Data Source

Bibliometric analysis is a tool for gathering and assessing data related to scientific research [22, 24]. It examines authors, citations, journals, institutions, and emerging research areas. The dataset used in this study was obtained from the Web of Science website, which contains around 74.8 million academic publications [25, 26].

2.2 Data Collection

During the research study, a meticulous analysis was conducted covering the period from 2000 to 2024, focusing extensively on keywords, such as "Biocrude," "Crude bio-oil," and "Upgrade bio-oil." Only materials in the English language were considered to ensure a comprehensive evaluation of the present scenario and minimize the potential impact of future updates. The research database encompassed diverse review articles and conference papers to support this endeavor.

2.3 Data Extraction

The current research focuses on analyzing the scientific output and identifying the countries actively involved in the production and biorefining of biocrude oil. In pursuit of this, we gathered a comprehensive database, which was subsequently imported into Microsoft Excel 2023 for processing and analysis. Statistical results were based on the frequency of citations, authors, annual publication numbers, agencies financing the projects, and the involved institutions. The data was extracted using the Journal website Citation Reports (JCR) (https://www.thomsonreuters.com/journal-citationreports/) and manually entered and processed in Excel during the analysis.

2.4 Data Visualization

This study utilized Excel 2023 software to conduct a detailed descriptive database analysis, accounting for various restrictions, such as author, institution, and statistical criteria. The analysis provided a reliable assessment of publication performance in the study area, enhancing the interpretation of results and facilitating informed decision-making. The database considered the quality of the publications, including their impact factor (IF) and the category (Q1, Q2, Q3, and Q4) in which each magazine and article falls. These indexes highlight the most relevant publications in a particular area of knowledge.

3 Results and Discussions

3.1 *Bibliometric Analysis*

3.1.1 Results of Publication: Overall

A search conducted in the Web of Science (WoS) database from 2000 to May 2024 resulted in data that allowed the creation of various figures, including network maps, tables, and charts. This data facilitated the quantification and creation of the topics below, such as the contribution of countries/regions, the contribution of institutions, the analysis of journals, and the most cited articles.

The United States stands out with 343 publications and 14,118 citations. The University of Illinois System and the University of Illinois Urbana-Champaign lead in the number of publications, with sixty-two papers each. Despite the leadership of the United States in these topics, the United Kingdom excels in the potential quality of its work. This country leads the analysis of journals and most cited articles, with the journal Bioresource Technology contributing 7.6% of the publications. Additionally, three UK-based journals listed in Table 2 have the highest impact factor (IF), all with an IF above 11.0. Among the most cited articles, the paper titled "Competitive liquid biofuels from biomass" [27] has accumulated 576 citations, and three other articles from the United Kingdom are also among the top four most cited. Therefore, although the United States leads in the number of publications, the impact factors and the ranking of journals and most cited articles suggest a significant influence of publications from the United Kingdom.

Figure 1 shows a significant and exponential growth in publications and citations when comparing 2000 with 2020. Despite the elevated level of publications and citations in 2020, a decrease in subsequent years is noticeable, demonstrating a drop of approximately 50–53%. It is essential to highlight that despite the large volume of work, the research was based exclusively on the Web of Science (WoS) database, and there are several other relevant databases, such as Scopus and SciELO, which could also provide complementary and potentially different information.

3.1.2 Contributions of Countries/regions

Figure 2 represents the contribution of countries to the biorefining of biocrude oil from 2000 to May 2024. The "Countries with the highest number of publications" bar chart highlights the number of documents published by researchers from various countries. The United States and China have the most publications, with 343 and 340

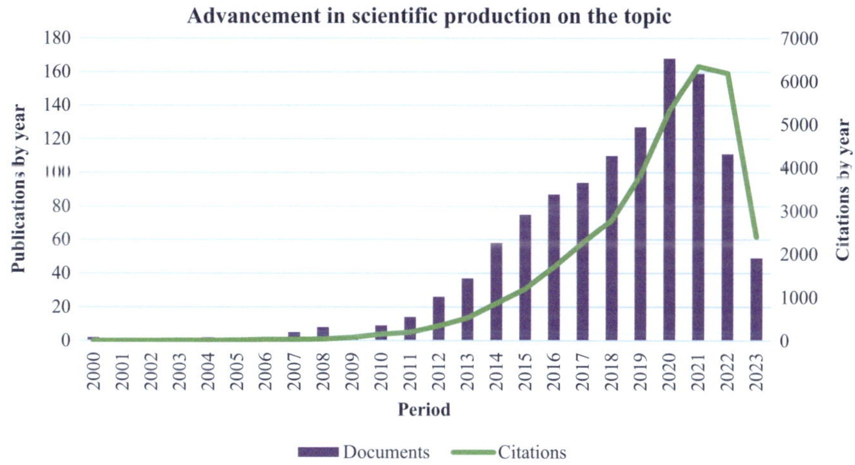

Fig. 1 Progress in annual publication of scientific documents during period analyzed, accompanied by number of recurring citations

papers, respectively. Following them, India appears with about 101 publications. Canada has ninety-one publications, while Denmark, Australia, and South Korea have approximately 89, 68, and 53 publications each. Italy, England, and Germany have the lowest publications among the listed countries, each contributing around 44, 43, and 40 documents. Figure 2 highlights the dominance of the United States and China in research production, followed by a significant gap before India and the other countries, providing a view of global research trends on the subject and the distribution of scientific output among different regions.

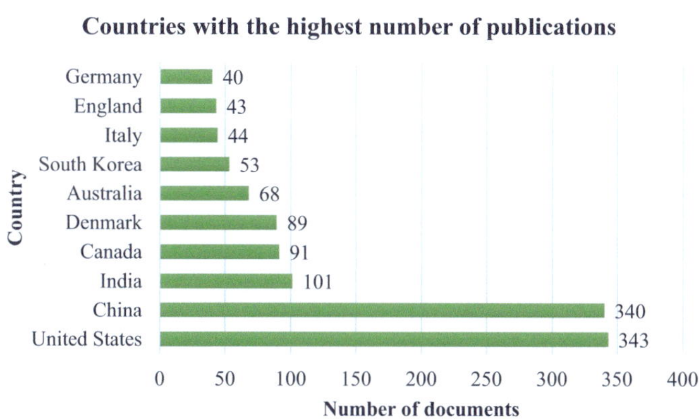

Fig. 2 Ten countries with highest quantities of scientific documents published within range investigated

Figure 3 shows the number of citations received by documents published by researchers from various countries. The United States has the highest number of citations, accumulating around 14,118. China follows in second place with approximately 9,990 citations. Denmark has about 2322 citations, while Canada and India have approximately 1808 and 1974 citations each. This suggests that despite India and Canada having more publications than Denmark, their works significantly influence the studied topic.

Although the United States is the most significant contributor in the number of publications and citations, Fig. 4 shows China occupies a central position regarding cooperation and partnership between countries. Another noteworthy aspect is the presence of Brazil, which, although being the only Latin American country represented, has a modest contribution and connects only with the USA.

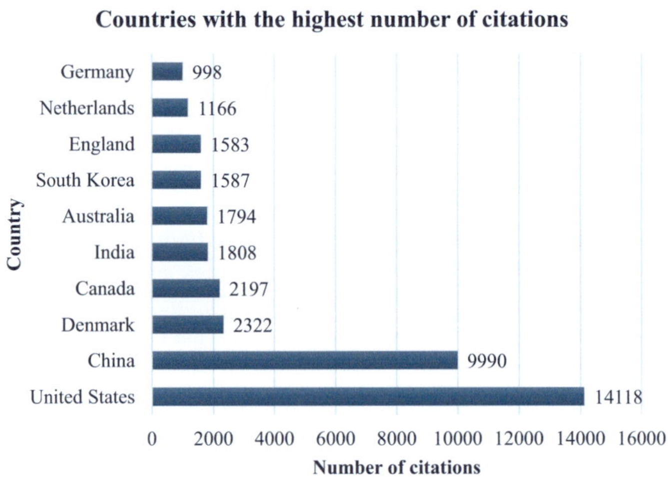

Fig. 3 These are top 10 countries with highest citations for their published articles

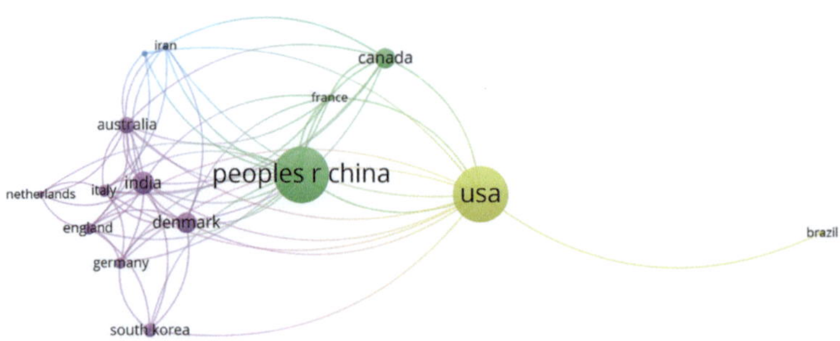

Fig. 4 Assessment of partnership among countries that contributed most to article publications

Table 1 Ranking of 10 institutions that have published most scientific papers on biocrude production

Rank	Institution	Country/Region	Number of publications
1	University of Illinois System	United States	62
2	University of Illinois Urbana-Champaign	United States	62
3	Aalborg University	Denmark	59
4	United States Department of Energy	United States	54
5	China Agricultural University	China	52
6	Pennsylvania Commonwealth System of Higher Education	United States	43
7	Pennsylvania State University	United States	43
8	Pennsylvania State University, University Park	United States	40
9	Xi'an Jiaotong University	China	39
10	Ministry of Agriculture Rural Affairs	China	38

3.1.3 Contributions of Institutions

Among the various institutions surveyed, only ten met the minimum criterion of thirty-nine publications, as proposed in the ranking in Table 1. This cutoff reveals the concentration of academic production in a few prominent institutions. The United States shows significant interest in the topic. The University of Illinois System and the University of Illinois Urbana-Champaign occupy the 1st and 2nd places in the ranking, with sixty-two publications each, highlighting the significant role of these institutions in the research. Third is Aalborg University from Denmark, which has fifty-nine works, also indicating the relevance of European institutions on the global stage. Considering the perspective of countries, the United States maintains the lead with 166 publications, confirming its dominant position in scientific production on the topic. China follows in second place, with 129 works, reflecting its growing involvement in the research area. These numbers highlight the disparity between countries, evidencing the concentration of academic production on the topic in a few nations and the need for greater inclusion and international collaboration.

3.1.4 Contributions of Authors

Figure 5 highlights the ten authors who published the most on the topic from 2000 to May 2024, while Fig. 6 shows the authors with the highest number of citations during the same period. Zhang Yuanhui leads in publications with fifty-one works, 3020 citations, and a Total Link Strength (TLS) of 706. Following him, Savage Phillip

Table 2 Scientific journals with highest number of published works on biocrude production

Rank	Journal title	Country	Percentage (%)	IF
1	Bioresource Technology	United Kingdom	7.6	11.889
2	Energy Fuels	United States	7.34	8.857
3	Fuel	Netherlands	6.83	8.035
4	Algal Research Biomass, Biofuels and Bioproducts	Netherlands	3.75	5.276
5	Renewable Energy	United Kingdom	3.58	8.634
6	Biomass Energy	United Kingdom	3.5	6.0
7	Energy Conversion and Management	United Kingdom	3.5	11.533
8	ACS Sustainable Chemistry Engineering	United States	3.41	8.4
9	Applied Energy	United Kingdom	3.07	11.446
10	Energies	Switzerland	2.9	3.252

E. occupies the second position with fifty publications, 2401 citations, and a TLS of 410. Liu Zhidan is in third place with thirty-three publications, 1776 citations, and a TLS of 518. It is worth noting that Lu Jianwen, although ranked eighth in several publications in Fig. 5, appears in fourth place in Fig. 6 with 956 citations and a TLS of 518 from his twenty works. This highlights the considerable influence of his publications. These data emphasize the number of works published and the relevance and impact of the authors in the research field.

Figure 7 presents a network graph illustrating the connections and collaborations among different authors in the research field of biorefining of biocrude oil. Each node represents an author, while the lines indicate co-authorships or collaborations

Fig. 5 Ten authors have published most scientific works on biocrude

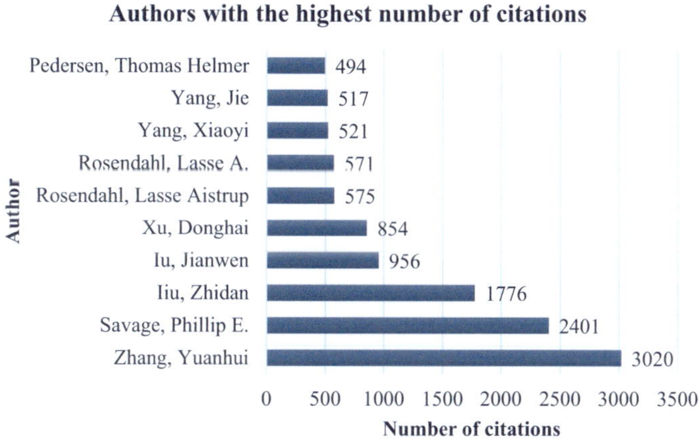

Fig. 6 Ranking of 10 most cited authors in scientific literature on biocrude production technology

in publications. The nodes' and edges' color and size reflect different metrics, such as the number of scientific technologies and challenges that have been produced. The graph reveals several clusters of distinct colors, indicating groups of authors who frequently collaborate. The largest cluster is the group of authors in the green area, led by Savage Phillip E., who is connected to other authors like Wang Shuzhong and Xu Donghai. Another significant group is the purple one, with authors like Lu Jianwen, Liu Zhidan, and Zhang Yuanhui. Savage Phillip E. is a central author in the graph, with several strong connections, indicating frequent collaborations within his cluster. Zhang Yuanhui, Liu Zhidan, and Lu Jianwen are also well connected within their group, showing significant collaborations.

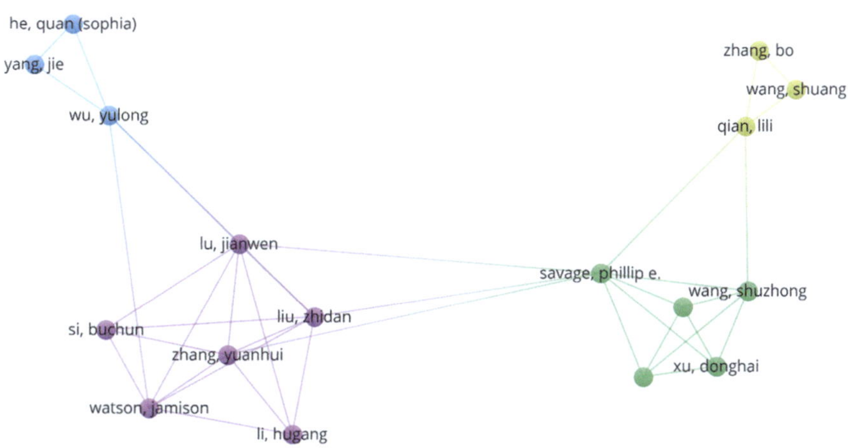

Fig. 7 Collaboration network among authors who contributed significantly to scientific research of biocrude production technologies and challenges

Some authors form isolated small clusters, such as He Quan, Yang Jie, and Wu Yulong (blue cluster), and Zhang Bo, Wang Shuang, and Qian Lili (yellow cluster), showing more restricted collaborations. There are interconnections between different clusters, such as the line connecting Lu Jianwen (purple cluster) to Savage Phillip E. (green cluster), indicating collaboration between these two main groups. Images were created using VOSviewer software.

3.1.5 Journal Analysis

The top ten journals that published the most on the production of biocrude oil and biorefinery are shown in Table 2. It also details each journal's impact factor (IF), the percentage of papers published on the subject, and the nation of origin of these publications. The United Kingdom's Bioresource Technology stands out with an impact factor of 11,889 and 7.6% of publications. Next, with an impact factor of 8.857 and 7.34% of the scientific papers examined, is energy fuels from the United States. Fuel, a Dutch publication, provides 6.83% of the assessed scientific papers and has an impact factor 8.035. This analysis is essential to evaluate the quality of journals that publish works on biocrude oil biorefinery. Furthermore, it guides future researchers interested in which journals are most suitable for submitting their scientific articles.

3.1.6 Most Articles Cited

Bibliometric analysis enables a quantitative and qualitative analysis of published works on a given topic. Therefore, Table 3 presents the ranking of the most cited articles on biocrude oil biorefinery, accompanied by the number of citations and year of publication. Consequently, it becomes possible to identify the most influential works within the analyzed period. It also indicates relevance metrics and helps recognize emerging trends and topics.

The most cited article, "Competitive liquid biofuels from biomass," published in 2011 by Applied Energy magazine, has 576 citations. The review aims to analyze the competitiveness of the production of liquid biofuels (biodiesel, bioethanol, and biocrude) and explain biofuel generation's characteristics, advantages, and disadvantages. In this way, it contributes to its dissemination in the academic community and industrial sector, as it corroborates with significant numbers regarding production costs [27]. Next, "Hydrothermal liquefaction (HTL) of microalgae for biofuel production: State of the art review and prospects," published in 2013 by the magazine Biomass and Bioenergy, has 495 citations. The article explains, through a literature review, the pertinent concepts regarding the hydrothermal liquefaction of microalgae and identifies in emerging research areas [28].

"Chemical properties of biocrude oil from hydrothermal liquefaction of Spirulina algae, swine manure, and digested anaerobic sludge," published in 2011 by the magazine Bioresource Technology, totals 487 citations. Therefore, this is an experimental

Table 3 Ten most cited articles on development of biocrude production technologies

Rank	Title	Journal	Year published	Citations
1	Competitive liquid biofuels from biomass	Applied Energy	2011	576
2	Hydrothermal liquefaction (HTL) of microalgae for biofuel production: state of the art review and future prospects	Biomass and Bioenergy	2013	495
3	Chemical properties of biocrude oil from the hydrothermal liquefaction of Spirulina algae, swine manure, and digested anaerobic sludge	Bioresource Technology	2011	487
4	Effect of operating conditions of thermochemical liquefaction on biocrude production from Spirulina platensis	Bioresource Technology	2011	368
5	Process development for hydrothermal liquefaction of algae feedstocks in a continuous-flow reactor	Algal Research Biomass Biofuels and Bioproducts	2013	356
6	Hydrothermal liquefaction of various biomass and waste feedstocks for biocrude production: a state of the art review	Renewable and Sustainable Energy Reviews	2017	347
7	Additives to lower and stabilize the viscosity of pyrolysis oils during storage	Energy and Fuels	1997	298
8	Hydrothermal liquefaction of *Nannochloropsis* sp.: systematic study of process variables and analysis of the product fractions	Biomass and Bioenergy	2012	289
9	Hydrothermal liquefaction for algal biorefinery: a critical review	Renewable and Sustainable Energy Reviews	2014	274
10	Complete utilization of spent coffee grounds to produce biodiesel, bio-oil, and biochar	ACS Sustainable Chemistry and Engineering	2014	235

article where the authors investigated the influence of the composition of the raw material on the properties of biocrude. Consequently, they concluded that it was necessary to select and create profiles of essential raw materials to guarantee the safety and quality of the product [29].

3.1.7 Research Area

Figure 8 highlights the prominent research areas in biocrude oil biorefinery. It is observed that "Energy Fuels" leads with 690 publications, underscoring the relevance of renewable liquid fuel synthesis within the academic community and, consequently, the industrial sector. In contrast, "Materials Science" has only 40 publications, indicating an area with fewer studies and offering guidance for future researchers to

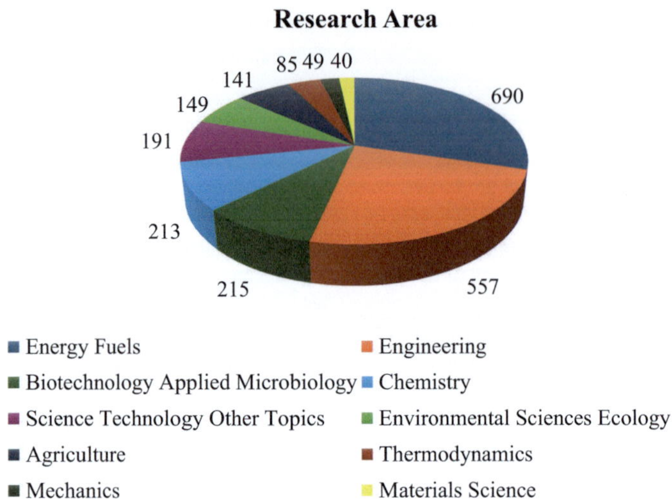

Fig. 8 Prominent research areas in development of biocrude synthesis technologies

develop emerging research and address this scientific gap. "Engineering" has 557 publications demonstrating the importance of investing in developing methods for producing alternative fuels to promote sustainability. "Environmental Sciences and Ecology" is another promising area with 215 publications, reflecting the impact of using waste to encourage the circular economy on the environment. Finally, Fig. 8 and the present study help guide the development of emerging research areas and recognize prominent trends.

4 Keyword Analysis

With this bibliometric analysis, it was also possible to investigate the most relevant keywords in the scope of the research in question. Figure 9 presents a collection of words in purple, highlighting "Biocrude" as the main word, strongly connected with terms, such as "biomass," "conversion," "temperature," "pyrolysis," "liquefaction," "biofuels," "oil," and "water." Pyrolysis and liquefaction, for example, are technologies that use biomass as raw material to produce biofuels, bio-oil, or biocrude under high temperatures. Often, biocrude production processes use water as a reaction medium [30–32].

It is also possible to observe in Fig. 9 a green cluster network, with the central word "hydrothermal liquefaction" strongly connected with terms, such as "microalgae," "biocrude," "co-liquefaction," and "sewage sludge." Hydrothermal liquefaction is a thermal process technology that stands out because it can use wet biomass without the need for primary processes, such as drying, optimizing, and reducing the costs

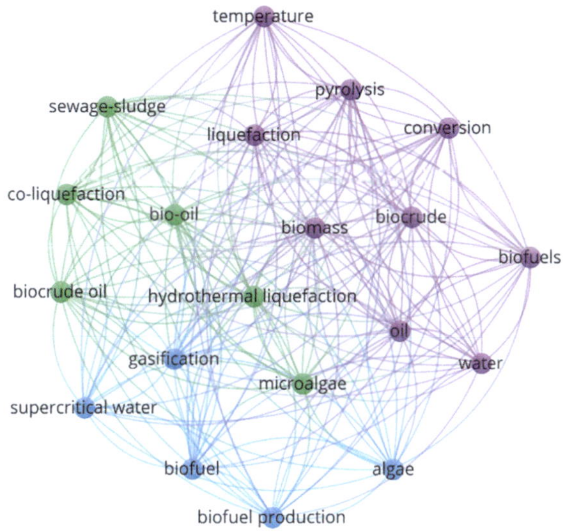

Fig. 9 Connection network among 20 key terms in assessed scientific papers on biocrude production

of this technology [33, 34]. A mixture of biomass in the liquefaction process (co-liquefaction) is often used to increase the yield of this technology. Many studies point out microalgae as a viable and sustainable source for biocrude production [35–37].

However, another connection in blue can be verified with words, such as "gasification," "supercritical water," "biofuel production," and "algae." In Fig. 10, we can observe the unfolding of keywords on biorefinery technologies over time. Initially, the keywords were more related to "gasification," "liquefaction," and "supercritical water." However, in early 2019, the focus shifted to "biofuel production," "conversion," "biomass," and "biocrude." In the second half of 2019, the analysis shows that the research was more focused on terms, such as "pyrolysis," "bio-oil," "biocrude oil," "microalgae," and "hydrothermal liquefaction." From 2020 to the date of the research, studies have focused on words, such as "co-liquefaction," "temperature," and "sewage sludge." It is worth noting that hydrogenation is an essential process for the treatment of biocrude, using a catalyst to make it suitable for use as fuel [38, 39].

Table 4 highlights the ranking of the 20 most prominent keywords identified in the articles, their respective Total Link Strength and the frequency with which they were found. Although these words are also highlighted in Figs. 9 and 10, it is essential to emphasize the role of "temperature," which appears in 11th place with a Total Link Strength of 462, in converting biomass or any other waste into biocrude. High temperatures are present in the main methods of converting raw materials into biocrude, such as hydrothermal liquefaction, pyrolysis, and steam reforming [40–42].

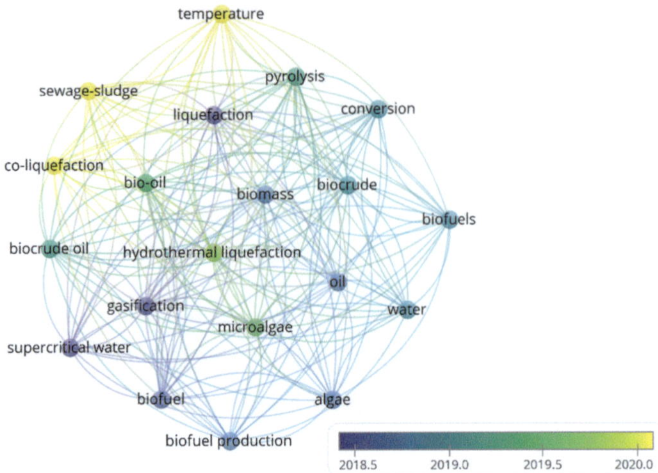

Fig. 10 Investigating 20 recurring keywords in scientific literature on biorefinery technologies

Table 4 Ranking of top 20 keywords with highest occurrence in evaluated scientific documents on biorefinery development technologies

Rank	Keyword	F	Total link strength	Rank	Keyword	F	Total link strength
1	Hydrothermal liquefaction	560	1,975	**11**	Temperature	120	462
2	Biomass	500	1,693	**12**	Biocrude oil	120	421
3	Bio-oil	413	1,391	**13**	Biofuel production	116	449
4	Biocrude	392	1,325	**14**	Sewage-sludge	108	405
5	Microalgae	388	1,546	**15**	Co-liquefaction	101	371
6	Conversion	214	847	**16**	Liquefaction	97	351
7	Algae	188	769	**17**	Biofuels	90	322
8	Pyrolysis	178	630	**18**	Gasification	86	357
9	Water	135	526	**19**	Biofuel	86	301
10	Oil	135	472	**20**	Supercritical water	83	327

5 Thermal and Catalytic Processes

For sustainable energy production, industries have spared no effort in converting biomass and other organic and inorganic waste into high-value products. However, thermal and catalytic processes are widely used in biomass conversion into biocrude [43–46]. Plastic waste, which does not degrade under normal environmental conditions and causes damage to human health and the environment, is of great use to industries that employ these thermal and catalytic processes to convert it into valuable products, such as oil. This way, the environmental impacts of plastic waste accumulation in landfills are minimized [47].

Put, thermal processes involve high temperatures, such as pyrolysis, hydrothermal liquefaction (HTL), steam reforming, gasification, and thermal cracking, to preferentially convert biomass into products like biochar, bio-oil, gases, and fuels [48–50]. Among these technologies, HTL has stood out considerably, as it uses microalgae as raw material due to its rapid growth and ability to absorb harmful environmental gases [41]. Other technologies, such as steam reforming, have also gained visibility in global research. For example, Yun J. et al., in their study on "Performance improvement of methanol steam reforming system with auxiliary heat recovery units," compare three distinct steam reforming configurations. They claim that the internally evaporated methanol steam reformer more efficiently converts methanol into a hydrogen-rich mixture, as the exhaust gases are used to heat the methanol/water mixture input [41].

Technologies that use thermal processes, such as cracking, which breaks larger hydrocarbon molecules into smaller ones, and pyrolysis, which transforms biomass into bio-oil, gas, or fuel, are also widely used in various industries, betting on a more sustainable energy matrix.

On the other hand, catalytic processes use catalysts to convert various raw materials into high-value products. This process is so necessary that it presents itself as a fundamental pillar of the chemical industry, being present in most processes related to converting waste into high-value-added fuels [51–55]. Moreover, catalytic processes can convert CO_2 to produce products, such as syngas, ethanol, and other higher-value products, which benefit the energy matrix and contribute positively to the environment [56, 57]. However, although fuels obtained through processes, such as catalytic cracking, catalytic reforming, hydrogenation, and catalytic oxidation, are more environmentally friendly compared to fossil fuels, they also present challenges, such as the high costs of catalysts, which may be subject to deactivation [58–61].

A study that focuses on replacing fossil fuels through fluid catalytic cracking tested five types of oils derived from municipal solid waste and biogenic materials in a reactor, using a catalyst based on ultra-stable Y zeolite. The objective was to test the co-processing of fluid catalytic cracking. The study states that although these oils exhibited some unique properties, such as high Conradson carbon (up to 19.41 wt%), water (up to 5.7 wt%), and contaminants (up to 227 ppm Cl), they showed yields similar to that of vacuum gas oil, including gasoline, LPG, olefinicity, light cycle oils (LCO), coke selectivity, among others [62]. This study examines the importance of catalyst selectivity, solid waste, and biogenic oils for refineries.

Studies also point to electromagnetic induction heating as an alternative for the heat control reaction in catalytic processes, presenting advantages, such as process intensification, energy efficiency, and simplification of reactor configuration. This approach also offers solutions to such problems as slow heating and cooling rates, variable heating environments, and low energy efficiency [51, 63–65]. However, much research has been developed in catalytic processes, whether improving product yield or discovering new raw materials or catalysts that make the process more efficient. Artificial intelligence has been incorporated into modeling studies aimed at prediction and optimization, making the operation brighter [66–70]. Tai et al., in their research "Multi-objective optimization with a hybrid machine learning strategy for complex catalytic processes," state that the physics-based model in the data-driven artificial neural network framework provides a lower quadratic error of 0.01 compared to the physics-based simulation, promoting hybrid machine learning for aiding various catalytic processes [71].

Still in the quest to improve catalytic processes, taking into account pillars, such as biomass availability, economy, efficient catalysts, and process sustainability, Xu et al. present alternatives and valuable insights for the future with their study on "Targeted Catalytic Cracking to Olefins (TCO): Reaction Mechanism, Production Scheme, and Process Perspectives." The study highlights that this process's main innovations include multiple raw material supplies, mid-sized catalysts, and a fluidized bed reactor with different feed schemes. They state that the generation and conversion of olefins are essential for expanding upstream and downstream processes. TCO is presented to meet current demands for oil refining transformation and ensure low NO_x levels in chemical processes [72].

6 Challenges in Impurity Removal

To study the challenges in removing impurities in the production and biorefining of biocrude oil, several significant challenges were observed. The uncertainties related to the variability of the raw material (i.e., composition of the residues), equipment design and cost, and other knowledge gaps in the process can also have a significant impact [73]. Challenges that remain a barrier to industrial-scale dissemination include unreliable cultivation methods, large nutrient requirements, low energy return on investment, high capital and operational costs, the complexity of bio-oil composition, and challenges in separating products and producing high-value co-products [74]. An analysis of a study on improving biocrude production [75] stated that the structural differences between raw materials can govern catalytic reaction pathways.

Being a flexible technology in raw materials, hydrothermal liquefaction (HTL) is highly dependent on each biomass's chemical and biochemical characteristics as they affect the physical, chemical, and thermochemical properties of biocrude [76]. Since it is biomass rich in proteins and lipids, the effect of different alkaline catalysts in the HTL process of pulp or its equivalents has been little discussed. However, the selectivity and specific role of alkaline catalysts during the catalytic HTL operation of the organic fraction of MSW are still unknown. The organic and nutrient-loaded aqueous phase is among the challenges that have not been adequately addressed, especially in treating wet biomass. Depending on the type of raw material, the aqueous phase contains a considerable amount of water-soluble organic and inorganic products, increasing the substantial need for downstream water treatment [75]. Another study investigated the impact of activated carbon as an adsorbent material for treating the aqueous phase produced from HTL of SS and its recycling effect on the properties of biocrude.

The results revealed that activated carbon significantly affected moving organics, while the inorganics of the aqueous phase remained unchanged [77]. Thus, it was evident that the aqueous phase, which maintains the presence of inorganic inputs, can be characterized as another challenge in impurity removal, given that after the experiments, there was still the presence of inorganic material. The organic aqueous phase is nutrient-loaded and is among the challenges that have not been adequately addressed, especially in treating wet biomass. Depending on the type of raw material, the aqueous phase contains a considerable amount of water-soluble organic and inorganic products, increasing the substantial need for downstream water treatment. Recently, some researchers reported recirculating the aqueous phase as an advantageous method to recapture light hydrophilic organics and increase process economics, reducing the need for wastewater treatment. Klemmer et al. investigated the effect of aqueous phase recirculation in continuous subcritical HTL [78]. After an extensive analysis of studies and research related to the topic, it was noticeable that there was a small amount of research related to the subject. Therefore, these challenges highlight the paramount importance of technological advancements and more in-depth research to develop more efficient and cost-effective purification methods to make biocrude a viable and sustainable alternative to fossil fuels.

7 Development of Efficient Catalysts

Hydrothermal liquefaction (HTL) is the latest innovative technology for converting waste into energy [79]. It has also been identified as a creative technique to convert aquatic or wet biomass, such as algae, into biofuels [80]. Mimicking the geological conditions for crude oil generation, HTL uses pressurized water to transform organic matter into a liquid biofuel, called biocrude, under moderate temperature (280–374 °C) and autogenous pressure (8–22 MPa) [79]. Currently, catalytic hydrothermal liquefaction of biomass has been widely studied, including homogeneous catalysts [81] and heterogeneous catalysts [82]. Catalysts can be added to HTL to control rates and select chemical reactions. Both alkaline salts and metals can be employed as catalysts and are referred to as homogeneous and heterogeneous catalysts, respectively [83].

In general, catalysts are used to increase the yield or quality of biocrude. Homogeneous catalysts have been more commonly used for HTL than heterogeneous catalysts, although homogeneous catalysts are often more challenging to separate from the reaction products [84]. Additionally, catalysis is essential in transforming HTL biocrude into final products, a field in which recent developments have been observed [85]. However, no apparent catalytic effect in HTL can be found due to its complex composition, high ash content, and high heavy metal distribution. Typically, solid heterogeneous catalysts have more robust selectivity, are easy to recover, and can avoid the corrosive effect of liquid acid catalysts on equipment [86]. Heterogeneous catalysts, such as metal nanoparticles and supported materials, have been widely studied for their ease of separation and reuse. Noble metal nanoparticles, such as Pt, Pd, and Ni, are particularly effective in the hydrodesulfurization and deoxygenation of bio-oils, essential processes for producing high-quality biocrude [87].

However, substantial obstacles, such as high cost and sintering propensity, must be addressed by lowering synthesis costs and enhancing nanoparticle stability [87]. Supported catalysts are favored because of their ease of separation from end products and thermal stability, such as zeolites and activated carbon. They work well in processes involving deoxygenation and transesterification [87]. The solubility of homogeneous catalysts in the reaction media allows for intimate contact with reactants and provides excellent selectivity and efficiency, particularly for transition metal complexes. They are employed in the crucial processes of hydrogenation and oxidation, which eliminate oxygen and other contaminants from bio-oils [87]. Sodium carbonate (Na_2CO_3) is the most commonly used homogeneous catalyst for microalgae liquefaction [84, 88]. The development of more efficient catalysts for the production and biorefining of biocrude oil is progressing, focusing on improving the processes' efficiency, stability, and sustainability. Continuous research and technological innovation are essential to overcome current challenges and promote biofuels' commercial and environmental viability. Therefore, a considerable increase in studies in this area is necessary to enhance and expand the variety of publications addressing the proposed topic (Table 5).

Table 5 Characteristics of selected articles on most efficient catalysts

Type of catalyst	Main application	Advantages	Disadvantage	Reference
Homogeneous Transition metal complexes (Mo, W, Co, Ni)	Hydrogenation, oxidation	High selectivity, high efficiency	Difficult separation and reuse	[96, 97]
Heterogeneous Metallic nanoparticles (Pt, Pd, Ni), zeolites, activated carbon	Hydrodesulfurization, deoxygenation, transesterification	High active surface area, thermal stability, ease of separation	High cost, deactivation by coke, sintering	[98–101]
Enzymatic lipases	Transesterification, bioconversion of biomass	High selectivity, mild conditions	Limited stability, high cost	[102, 103]
Biocatalysts Enzymes derived from microorganisms	Biomass conversion	Sustainability specificity	Stability and cost	[104, 105]

8 Sustainability and Environmental Impact

Since biocrude can be processed into various liquid fuels based on hydrocarbons, it may be viewed as a renewable substitute for fossil crude oil [89]. A possible method for creating biocrude is hydrothermal liquefaction (HTL), which dispenses with the requirement for energy-intensive drying by directly converting wet biomass—which has a moisture content of over 80%—into biocrude, which has a high energy density [90].

The use of chemicals, energy, and fossil fuels in these processes raises concerns regarding their carbon impact. Nevertheless, implementing novel technologies for biocrude enhancement offers a viable solution. The production of biofuels and green hydrogen, along with the utilization of waste for fertilizer development, presents a substantial opportunity for mitigating greenhouse gas emissions and reducing the carbon footprint [91].

A viable substitute for fossil fuels, biocrude, presents a sustainable approach to energy generation and transportation. Its adaptability goes beyond fuels; it can be developed into various essential goods that support a more circular and sustainable economy, including biofuels, biochemicals, and specialized chemicals [92]. Considering this, boosting biocrude output is crucial to developing sustainable biofuels. However, a balance between yield increase and economic and environmental issues must be struck. For the aim of producing biofuels to be in line with the overall sustainability goals, a thorough and rigorous analysis of the trade-offs involved is necessary [93].

9 Integration with Existing Refining Infrastructures

Crude oil is the primary fuel used by oil refineries. However, short- and long-term problems need to be fixed to reduce this dependency. Two of these challenges are the depletion of crude oil reserves and the unpredictability of oil prices. Furthermore, the oil refining sector is under investigation and criticism for its detrimental environmental effects, particularly concerning greenhouse gas (GHG) emissions [94].

Because they lower greenhouse gas emissions and open the door for a gradual transition away from fossil fuels, biofuels are essential in reducing the effects of climate change. They do this by establishing themselves as highly sustainable and low-carbon substitutes [95]. Decarbonizing the oil and transportation sectors is crucial to overcome the abovementioned challenges.

About 150 billion gallons of traditional biofuels, such as ethanol and biodiesel, are produced worldwide each year, covering 4% of the world's transportation fuel needs. Conventional biofuels with low-carbon content do not fit the "drop-in" category and can only be utilized in specific mixes with current engines and infrastructure [96].

Drop-in biofuels are functionally equivalent to petroleum-derived transport fuels, boasting complete compatibility with existing petroleum infrastructure [97].

Lipids should be used for co-processing in the beginning, according to Van Dyk et al. (2019), as they are easily incorporated into petroleum refining procedures and are widely accessible in huge numbers. However, lignocellulose-derived biocrudes and bio-oils are not yet widely accessible and need more involved co-processing in refineries. The reason for this unique biocrude processing is the notable variation in oxygen content [96, 98]. Thus, pre-treatment of biocrudes would be required before their insertion into a refinery unit [97], and specific policies to reduce production costs [99].

10 Conclusions

The crude oil biorefinery is relevant in the economic, social, and environmental sectors, as it is responsible for valuing waste, generating jobs, and promoting sustainable development. Given this, the bibliometric analysis made it possible to observe the significant contribution of countries, such as the United States, China, and India in publishing scientific documents (articles, review articles, and conference articles). Furthermore, emerging demands were visible, such as the obstacles faced in removing impurities that compromise the quality and safety of the bioproduct. We also target promising areas, such as the development of efficient improvements to boost production, improve efficiency, and contribute to the reduction of waste generated during the process. Therefore, the impact on the environment and sustainability was evaluated as well as the integration with existing refinery infrastructures. Consequently, this study contributes to scientific dissemination in the academic and industrial community, ensuring sustainable development.

Acknowledgements The authors would like to express their gratitude for the support that was provided through Programa de Pós Graduação em Engenharia Química at the Universidade Federal do Ceará and the Instituto de Engenharias e Desenvolvimento Sustentável (IEDS) at the Universidade da Integração Internacional da Lusofonia Afro-Brasileira (UNILAB). We gratefully acknowledge the following Brazilian Agencies for Scientific and Technological Development: Fundação Cearense de Apoio ao Desenvolvimento Científico e Tecnológico (FUNCAP) (PS1-0186-00216.01.00/21; 31052.000417/2023; UNI-0210-00537.01.00/23), Conselho Nacional de Desenvolvimento Científico e Tecnológico (CNPq) (311062/2019-9; 440891/2020-5; 307454/2022-3), and Coordenação de Aperfeiçoamento de Ensino Superior (CAPES) (finance code 001).

Declaration of Competing Interest The authors declare that they have no known competing financial interests or personal relationships that could have appeared to influence the work reported in this paper.

Data Availability Data will be made available on request.

References

1. Mishra RK, Kumar V, Kumar P, Mohanty K (2022) Hydrothermal liquefaction of biomass for bio-crude production: a review on feedstocks, chemical compositions, operating parameters, reaction kinetics, techno-economic study, and life cycle assessment. Fuel 316:123377. https://doi.org/10.1016/j.fuel.2022.123377
2. Kumar M, Olajire Oyedun A, Kumar A (2018) A review on the current status of various hydrothermal technologies on biomass feedstock. Renew Sustain Energy Rev 81:1742–1770. https://doi.org/10.1016/j.rser.2017.05.270
3. Gollakota ARK, Kishore N, Gu S (2018) A review on hydrothermal liquefaction of biomass. Renew Sustain Energy Rev 81:1378–1392. https://doi.org/10.1016/j.rser.2017.05.178
4. He Y, Liang X, Jazrawi C, Montoya A, Yuen A, Cole AJ et al (2016) Continuous hydrothermal liquefaction of macroalgae in the presence of organic co-solvents. Algal Res 17:185–195. https://doi.org/10.1016/j.algal.2016.05.010
5. Xu D, Lin G, Guo S, Wang S, Guo Y, Jing Z (2018) Catalytic hydrothermal liquefaction of algae and upgrading of biocrude: a critical review. Renew Sustain Energy Rev 97:103–118. https://doi.org/10.1016/j.rser.2018.08.042
6. Akhtar J, Amin NAS (2011) A review on process conditions for optimum bio-oil yield in hydrothermal liquefaction of biomass. Renew Sustain Energy Rev 15:1615–1624. https://doi.org/10.1016/j.rser.2010.11.054
7. Castello D, Pedersen TH, Rosendahl LA (2018) Continuous hydrothermal liquefaction of biomass: a critical review. Energies (Basel) 11:3165. https://doi.org/10.3390/en11113165
8. Zhu Y, Biddy MJ, Jones SB, Elliott DC, Schmidt AJ (2014) Techno-economic analysis of liquid fuel production from woody biomass via hydrothermal liquefaction (HTL) and upgrading. Appl Energy 129:384–394. https://doi.org/10.1016/j.apenergy.2014.03.053
9. Jensen CU, Guerrero JKR, Karatzos S, Olofsson G, Iversen SB (2018) Hydrofaction^{TM} of forestry residues to drop-in renewable transportation fuels. In: Direct thermochemical liquefaction for energy applications. Elsevier, pp 319–345. https://doi.org/10.1016/B978-0-08-101029-7.00009-6
10. Elliott DC, Hart TR, Schmidt AJ, Neuenschwander GG, Rotness LJ, Olarte MV et al (2013) Process development for hydrothermal liquefaction of algae feedstocks in a continuous-flow reactor. Algal Res 2:445–454. https://doi.org/10.1016/j.algal.2013.08.005
11. Elliott DC, Schmidt AJ, Hart TR, Billing JM (2017) Conversion of a wet waste feedstock to biocrude by hydrothermal processing in a continuous-flow reactor: grape pomace. Biomass Convers Biorefin 7:455–465. https://doi.org/10.1007/s13399-017-0264-8
12. Seehar TH, Toor SS, Shah AA, Pedersen TH, Rosendahl LA (2020) Biocrude production from wheat straw at sub and supercritical hydrothermal liquefaction. Energies (Basel) 13:3114. https://doi.org/10.3390/en13123114
13. Biller P, Sharma BK, Kunwar B, Ross AB (2015) Hydroprocessing of bio-crude from continuous hydrothermal liquefaction of microalgae. Fuel 159:197–205. https://doi.org/10.1016/j.fuel.2015.06.077
14. Elliott DC, Biller P, Ross AB, Schmidt AJ, Jones SB (2015) Hydrothermal liquefaction of biomass: Developments from batch to continuous process. Bioresour Technol 178:147–156. https://doi.org/10.1016/j.biortech.2014.09.132
15. Divyabharathi R, Subramanian P (2022) Biocrude production from orange (*Citrus reticulata*) peel by hydrothermal liquefaction and process optimization. Biomass Convers Biorefin 12:183–194. https://doi.org/10.1007/s13399-021-01383-3
16. Valadão L, Duarte C, Andrade G, Santos D de los, Sanches Filho P (2021) Estudo comparativo dos bio-óleos obtidos por pirólises rápida e lenta do caroço de pêssego. Engenharia Sanitaria e Ambiental 26:757–764. https://doi.org/10.1590/s1413-415220200077
17. Hao B, Xu D, Jiang G, Sabri TA, Jing Z, Guo Y (2021) Chemical reactions in the hydrothermal liquefaction of biomass and in the catalytic hydrogenation upgrading of biocrude. Green Chem 23:1562–1583. https://doi.org/10.1039/D0GC02893B

18. Shemfe M, Gu S, Fidalgo B (2017) Techno-economic analysis of biofuel production via bio-oil zeolite upgrading: an evaluation of two catalyst regeneration systems. Biomass Bioenergy 98:182–193. https://doi.org/10.1016/j.biombioe.2017.01.020
19. Tamala JK, Maramag EI, Simeon KA, Ignacio JJ (2022) A bibliometric analysis of sustainable oil and gas production research using VOSviewer. Clean Eng Technol 7:100437. https://doi.org/10.1016/j.clet.2022.100437
20. Negris L, Santos M de FP dos, Vicente MA (2023) Análise bibliométrica e tendências da pesquisa sobre determinação de cloretos na indústria do petróleo. Brazilian J Prod Eng 9:01–22. https://doi.org/10.47456/bjpe.v9i3.40905
21. da Silva Aires FI, Dari DN, Freitas IS, da Silva JL, de Matos Filho JR, dos Santos KM et al (2024) Advanced and prospects in phenol wastewater treatment technologies: unveiling opportunities and trends. Discover Water 4:20. https://doi.org/10.1007/s43832-024-00076-y
22. de Castro Bizerra V, Sales MB, Fernandes Melo RL, Andrade do Nascimento JG, Junior JB, França Silva MP et al (2024) Opportunities for cleaner leather processing based on protease enzyme: current evidence from an advanced bibliometric analysis. Renew Sustain Energy Rev 191:114162. https://doi.org/10.1016/j.rser.2023.114162
23. Donthu N, Kumar S, Mukherjee D, Pandey N, Lim WM (2021) How to conduct a bibliometric analysis: an overview and guidelines. J Bus Res 133:285–296. https://doi.org/10.1016/j.jbusres.2021.04.070
24. de Castro Bizerra V, Leandro Fernandes Melo R, Simão Neto F, Carlos de Castro E, Bessa Sales M, Gonçalves de Sousa Junior P et al (2024) Exploring the synergy of ionic liquids and lipase in sustainable enzymatic engineering. J Mol Liq 399:124373. https://doi.org/10.1016/J.MOLLIQ.2024.124373
25. Melo RLF, Neto FS, Dari DN, Fernandes BCC, Freire TM, Fechine PBA et al (2024) A comprehensive review on enzyme-based biosensors: advanced analysis and emerging applications in nanomaterial-enzyme linkage. Int J Biol Macromol 264:130817. https://doi.org/10.1016/J.IJBIOMAC.2024.130817
26. Cavalcante IO, Simão Neto F, Sousa P da S, Aires FI da S, Dari DN, Chaves de Lima RK et al (2024) Evolving sustainable energy technologies and assessments through global research networks: advancing the role of blue hydrogen for a cleaner future. RSC Sustain 2:348–368. https://doi.org/10.1039/D3SU00309D
27. Demirbas A (2011) Competitive liquid biofuels from biomass. Appl Energy 88:17–28. https://doi.org/10.1016/j.apenergy.2010.07.016
28. López Barreiro D, Prins W, Ronsse F, Brilman W (2013) Hydrothermal liquefaction (HTL) of microalgae for biofuel production: State of the art review and future prospects. Biomass Bioenergy 53:113–127. https://doi.org/10.1016/j.biombioe.2012.12.029
29. Vardon DR, Sharma BK, Scott J, Yu G, Wang Z, Schideman L et al (2011) Chemical properties of biocrude oil from the hydrothermal liquefaction of Spirulina algae, swine manure, and digested anaerobic sludge. Bioresour Technol 102:8295–8303. https://doi.org/10.1016/j.biortech.2011.06.041
30. Wu Y, Wang H, Li H, Han X, Zhang M, Sun Y et al (2022) Applications of catalysts in thermochemical conversion of biomass (pyrolysis, hydrothermal liquefaction and gasification): a critical review. Renew Energy 196. https://doi.org/10.1016/j.renene.2022.07.031
31. Xu Y, Hu Y, Peng Y, Yao L, Dong Y, Yang B et al (2020) Catalytic pyrolysis and liquefaction behavior of microalgae for bio-oil production. Bioresour Technol 300. https://doi.org/10.1016/j.biortech.2019.122665
32. Yang M, Song XR, Deng PF, Yang HR (2000) Pyrolysis and liquefaction of biomass. Linchan Huaxue Yu Gongye/Chemistry and Industry of Forest Products 20
33. Hao B, Xu D, Wei Y, Diao Y, Yang L, Fan L et al (2023) Mathematical models application in optimization of hydrothermal liquefaction of biomass. Fuel Process Technol 243. https://doi.org/10.1016/j.fuproc.2023.107673
34. Fan Y, Hornung U, Dahmen N (2022) Hydrothermal liquefaction of sewage sludge for biofuel application: a review on fundamentals, current challenges and strategies. Biomass Bioenergy 165. https://doi.org/10.1016/j.biombioe.2022.106570

35. Yang J, He Q, Yang L (2019) A review on hydrothermal co-liquefaction of biomass. Appl Energy 250. https://doi.org/10.1016/j.apenergy.2019.05.033
36. Sharma K, Shah AA, Toor SS, Seehar TH, Pedersen TH, Rosendahl LA (2021) Co-hydrothermal liquefaction of lignocellulosic biomass in supercritical water. Energies (Basel) 14. https://doi.org/10.3390/en14061708
37. Li Q, Yuan X, Hu X, Meers E, Ong HC, Chen WH et al (2022) Co-liquefaction of mixed biomass feedstocks for bio-oil production: a critical review. Renew Sustain Energy Rev 154. https://doi.org/10.1016/j.rser.2021.111814
38. Ren M, Zhang Y, Wang X, Qiu H (2022) Catalytic hydrogenation of CO_2 to methanol: a review. Catalysts 12. https://doi.org/10.3390/catal12040403
39. Lau S, Gasperini D, Webster RL (2021) Amine–Boranes as transfer hydrogenation and hydrogenation reagents: a mechanistic perspective. Angewandte Chemie—International Edition 60. https://doi.org/10.1002/anie.202010835
40. Brindhadevi K, Anto S, Rene ER, Sekar M, Mathimani T, Thuy Lan Chi N et al (2021) Effect of reaction temperature on the conversion of algal biomass to bio-oil and biochar through pyrolysis and hydrothermal liquefaction. Fuel 285. https://doi.org/10.1016/j.fuel.2020.119106
41. Venkatachalam CD, Ravichandran SR, Sengottian M (2022) Lignocellulosic and algal biomass for bio-crude production using hydrothermal liquefaction: conversion techniques, mechanism and process conditions: a review. Environ Eng Res 27. https://doi.org/10.4491/eer.2020.555
42. Saengsuriwong R, Onsree T, Phromphithak S, Tippayawong N (2023) Conversion of tobacco processing waste to biocrude oil via hydrothermal liquefaction in a multiple batch reactor. Clean Technol Environ Policy 25. https://doi.org/10.1007/s10098-021-02132-w
43. Thermal and catalytic processes in petroleum refining. Focus on Catalysts 2004. https://doi.org/10.1016/s1351-4180(04)00326-5
44. Aguado J, Serrano DP, Escola JM (2008) Fuels from waste plastics by thermal and catalytic processes: a review. Ind Eng Chem Res 47. https://doi.org/10.1021/ie800393w
45. Nigam M, Rajoriya S, Singh SR, Kumar P (2020) Thermal catalytic treatment (thermolysis): an effective process for the removal of COD and color from industrial wastewater. J Environ Treat Techn 8
46. Grebe M, Ruland M (2022) Influence of mechanical, thermal, oxidative and catalytic processes on thickener structure and thus on the service life of rolling bearings. Lubricants 10. https://doi.org/10.3390/lubricants10050077
47. Al-Salem SM, Antelava A, Constantinou A, Manos G, Dutta A (2017) A review on thermal and catalytic pyrolysis of plastic solid waste (PSW). J Environ Manage 197. https://doi.org/10.1016/j.jenvman.2017.03.084
48. Maiwald M (2020) Integrated and networked systems and processes—a perspective for digital transformation in thermal process engineering. Chem Eng 4. https://doi.org/10.3390/chemengineering4010015
49. Price CR, Nimbalkar SU, Thirumaran K, Cresko J (2023) Smart manufacturing pathways for industrial decarbonization and thermal process intensification. Smart Sustain Manuf Syst 7. https://doi.org/10.1520/SSMS20220027
50. Hu F, Chen G, Lin Y, Wang H, Zhu Z (2024) Numerical and experimental study on the thermal process during additive friction stir deposition. CIRP J Manuf Sci Technol 48. https://doi.org/10.1016/j.cirpj.2023.12.002
51. Wang W, Tuci G, Duong-Viet C, Liu Y, Rossin A, Luconi L et al Induction heating: an enabling technology for the heat management in catalytic processes. ACS Catal 9. https://doi.org/10.1021/acscatal.9b02471
52. Ambrosetti M (2022) A perspective on power-to-heat in catalytic processes for decarbonization. Chem Eng Process Process Intensification 182. https://doi.org/10.1016/j.cep.2022.109187
53. Chew TL, Bhatia S (2008) Catalytic processes towards the production of biofuels in a palm oil and oil palm biomass-based biorefinery. Bioresour Technol 99. https://doi.org/10.1016/j.biortech.2008.03.009

54. Kumar R, Singh R, Dutta S (2024) Review and outlook of hydrogen production through catalytic processes. Energy Fuels 38. https://doi.org/10.1021/acs.energyfuels.3c04026
55. Mackintosh AF, Shin T, Yang H, Choe K (2020) Hydrothermal polymerization catalytic process effect of various organicwastes on reaction time, yield, and temperature. Processes 8. https://doi.org/10.3390/pr8030303
56. Asadullah M, Tomishige K, Fujimoto K (2001) A novel catalytic process for cellulose gasification to synthesis gas. Catal Commun 2. https://doi.org/10.1016/S1566-7367(01)00011-5
57. Aasberg-Petersen K, Dybkjær I, Ovesen CV, Schjødt NC, Sehested J, Thomsen SG (2011) Natural gas to synthesis gas—catalysts and catalytic processes. J Nat Gas Sci Eng 3. https://doi.org/10.1016/j.jngse.2011.03.004
58. Patiño Y, Faba L, Peláez R, Cueto J, Marín P, Díaz E et al (2023) The role of ion exchange resins for solving biorefinery catalytic processes challenges. Catalysts 13. https://doi.org/10.3390/catal13060999
59. Sousa-Aguiar EF, Noronha FB, Faro A (2011) The main catalytic challenges in GTL (gas-to-liquids) processes. Catal Sci Technol 1. https://doi.org/10.1039/c1cy00116g
60. Zhou J, Zhao J, Zhang J, Zhang T, Ye M, Liu Z (2020) Regeneration of catalysts deactivated by coke deposition: a review. Chin J Catal 41. https://doi.org/10.1016/S1872-2067(20)63552-5
61. Dresp S, Ngo Thanh T, Klingenhof M, Brückner S, Hauke P, Strasser P (2020) Efficient direct seawater electrolysers using selective alkaline NiFe-LDH as OER catalyst in asymmetric electrolyte feeds. Energy Environ Sci 13. https://doi.org/10.1039/d0ee01125h
62. Mastry MC, Dorazio L, Fu JC, Gómez JP, Sedano S, Ail SS et al (2023) Processing renewable and waste-based feedstocks with fluid catalytic cracking: impact on catalytic performance and considerations for improved catalyst design. Front Chem 11. https://doi.org/10.3389/fchem.2023.1067488
63. Kim YT, Lee JJ, Lee J (2023) Electricity-driven reactors that promote thermochemical catalytic reactions via joule and induction heating. Chem Eng J 470. https://doi.org/10.1016/j.cej.2023.144333
64. Wong SL, Armenise S, Nyakuma BB, Bogush A, Towers S, Lee CH et al (2023) Plastic pyrolysis over HZSM-5 zeolite and fluid catalytic cracking catalyst under ultra-fast heating. J Anal Appl Pyrolysis 169. https://doi.org/10.1016/j.jaap.2022.105793
65. Essyed A, Pham XH, Truong-Phuoc L, Romero T, Nhut JM, Duong-Viet C et al (2024) High-efficiency graphene-coated macroscopic composite for catalytic methane decomposition operated with induction heating. Chem Eng J 485. https://doi.org/10.1016/j.cej.2024.150006
66. Yang F, Xu M, Lei W, Lv J (2023) Artificial intelligence methods applied to catalytic cracking processes. Big Data Mining Anal 6. https://doi.org/10.26599/BDMA.2023.9020002
67. Khaldi MK, Al-Dhaifallah M, Taha O (2023) Artificial intelligence perspectives: a systematic literature review on modeling, control, and optimization of fluid catalytic cracking. Alexandria Eng J 80. https://doi.org/10.1016/j.aej.2023.08.066
68. Foppa L, Ghiringhelli LM, Girgsdies F, Hashagen M, Kube P, Hävecker M et al (2021) Materials genes of heterogeneous catalysis from clean experiments and artificial intelligence. MRS Bull 46. https://doi.org/10.1557/s43577-021-00165-6
69. Al-Jamimi HA, BinMakhashen GM, Saleh TA (2022) Artificial intelligence approach for modeling petroleum refinery catalytic desulfurization process. Neural Comput Appl 34. https://doi.org/10.1007/s00521-022-07423-x
70. Moses OA, Chen W, Adam ML, Wang Z, Liu K, Shao J et al (2021) Integration of data-intensive, machine learning and robotic experimental approaches for accelerated discovery of catalysts in renewable energy-related reactions. Mater Rep Energy 1. https://doi.org/10.1016/j.matre.2021.100049
71. Tai XY, Ocone R, Christie SDR, Xuan J (2022) Multi-objective optimisation with hybrid machine learning strategy for complex catalytic processes. Energy 7. https://doi.org/10.1016/j.egyai.2021.100134

72. Xu Y, Zuo Y, Yang W, Shu X, Chen W, Zheng A (2023) Targeted catalytic cracking to olefins (TCO): reaction mechanism, production scheme, and process perspectives. Engineering 30. https://doi.org/10.1016/j.eng.2023.02.018
73. Li S, Jiang Y, Snowden-Swan LJ, Askander JA, Schmidt AJ, Billing JM (2021) Techno-economic uncertainty analysis of wet waste-to-biocrude via hydrothermal liquefaction. Appl Energy 283:116340. https://doi.org/10.1016/j.apenergy.2020.116340
74. Gu X, Martinez-Fernandez JS, Pang N, Fu X, Chen S (2020) Recent development of hydrothermal liquefaction for algal biorefinery. Renew Sustain Energy Rev 121:109707. https://doi.org/10.1016/j.rser.2020.109707
75. Kohansal K, Sharma K, Toor SS, Sanchez EL, Zimmermann J, Aistrup Rosendahl L et al (2021) Bio-crude production improvement during hydrothermal liquefaction of biopulp by simultaneous application of alkali catalysts and aqueous phase recirculation. Energies (Basel) 14:4492. https://doi.org/10.3390/en14154492
76. Blessing E, Jena U, Chinnasamy S (2019) Laboratory conversion of cultivated oleaginous organisms into biocrude for biofuel applications, pp 183–193. https://doi.org/10.1007/978-1-4939-9484-7_12
77. Shah AA, Toor SS, Nielsen AH, Pedersen TH, Rosendahl LA (2021) Bio-crude production through recycling of pretreated aqueous phase via activated carbon. Energies (Basel) 14:3488. https://doi.org/10.3390/en14123488
78. Klemmer M, Madsen RB, Houlberg K, Mørup AJ, Christensen PS, Becker J et al (2016) Effect of aqueous phase recycling in continuous hydrothermal liquefaction. Ind Eng Chem Res 55:12317–12325. https://doi.org/10.1021/acs.iecr.6b03414
79. Liu H, Basar IA, Eskicioglu C (2023) Hydrothermal liquefaction for sludge-to-energy conversion: an evaluation of biocrude production and management of waste streams. Energy 281:128268. https://doi.org/10.1016/j.energy.2023.128268
80. Shakya R, Whelen J, Adhikari S, Mahadevan R, Neupane S (2015) Effect of temperature and Na_2CO_3 catalyst on hydrothermal liquefaction of algae. Algal Res 12:80–90. https://doi.org/10.1016/j.algal.2015.08.006
81. Ferreira L, Ribeiro R, Madriaga V, Vasconcelos S, Shimabukuro E, Rossa V et al (2021) Uma breve revisão sobre a catálise por átomos isolados: conceitos e aplicações. Quim Nova. https://doi.org/10.21577/0100-4042.20170822
82. Braga AL, Lüdtke DS, Schneider PH, Andrade LH, Paixão MW (2013) Catálise assimétrica no Brasil: desenvolvimento e potencialidades para o avanço da indústria química brasileira. Quim Nova 36:1591–1599. https://doi.org/10.1590/S0100-40422013001000017
83. Bassoli SC, da Fonseca YA, Wandurraga HJL, Baeta BEL, de Souza AM (2023) Research progress, trends, and future prospects on hydrothermal liquefaction of algae for biocrude production: a bibliometric analysis. Biomass Convers Biorefin. https://doi.org/10.1007/s13399-023-03905-7
84. Yeh TM, Dickinson JG, Franck A, Linic S, Thompson LT, Savage PE (2013) Hydrothermal catalytic production of fuels and chemicals from aquatic biomass. J Chem Technol Biotechnol 88:13–24. https://doi.org/10.1002/jctb.3933
85. Shah AA, Sharma K, Haider MS, Toor SS, Rosendahl LA, Pedersen TH et al (2022) The role of catalysts in biomass hydrothermal liquefaction and biocrude upgrading. Processes 10:207. https://doi.org/10.3390/pr10020207
86. Hao B, Xu D, Wang Y, Wang Y, Kapusta K, Guo Y (2023) Catalytic hydrothermal liquefaction of municipal sludge for biocrude production over non-noble bimetallic catalyst in ethanol solvent. Fuel 331:125812. https://doi.org/10.1016/j.fuel.2022.125812
87. de Meneses JM (2018) Utilização de catalisadores heterogêneos suportados em carvão ativado: síntese de biodiesel de óleo de algodão. Environ Smoke 1:173. https://doi.org/10.32435/envsmoke.201811173
88. Chen W-H, Lin B-J, Huang M-Y, Chang J-S (2015) Thermochemical conversion of microalgal biomass into biofuels: a review. Bioresour Technol 184:314–327. https://doi.org/10.1016/j.biortech.2014.11.050

89. Heracleous E, Vassou M, Lappas AA, Rodriguez JK, Chiaberge S, Bianchi D (2022) Understanding the upgrading of sewage sludge-derived hydrothermal liquefaction biocrude via advanced characterization. Energy Fuels 36:12010–12020. https://doi.org/10.1021/acs.energyfuels.2c01746
90. Xu D, Wang Y, Lin G, Guo S, Wang S, Wu Z (2019) Co-hydrothermal liquefaction of microalgae and sewage sludge in subcritical water: ash effects on bio-oil production. Renew Energy 138:1143–1151. https://doi.org/10.1016/J.RENENE.2019.02.020
91. Zhao Y, Yang Z, Niu J, Du Z, Federica C, Zhu Z et al (2023) Systematical analysis of sludge treatment and disposal technologies for carbon footprint reduction. J Environ Sci 128:224–249. https://doi.org/10.1016/J.JES.2022.07.038
92. Usman M, Cheng S, Boonyubol S, Cross JS (2024) From biomass to biocrude: innovations in hydrothermal liquefaction and upgrading. Energy Convers Manag 302. https://doi.org/10.1016/j.enconman.2024.118093
93. Scarsella M, de Caprariis B, Damizia M, De Filippis P (2020) Heterogeneous catalysts for hydrothermal liquefaction of lignocellulosic biomass: a review. Biomass Bioenergy 140. https://doi.org/10.1016/j.biombioe.2020.105662
94. Prasetyo WD, Putra ZA, Bilad MR, Meurah T, Mahlia I, Wibisono Y et al (2020) Polymers Insight into the sustainable integration of bio-and petroleum refineries for the production of fuels and chemicals. Polymers (Basel) 12:1091. https://doi.org/10.3390/polym12051091
95. Bezergianni S, Dimitriadis A, Kikhtyanin O, Kubička D (2018) Refinery co-processing of renewable feeds. Prog Energy Combust Sci 68:29–64. https://doi.org/10.1016/j.pecs.2018.04.002
96. van Dyk S, Su J, Ebadian M, Saddler J (2022) Production of lower carbon-intensity fuels by co-processing biogenic feedstocks: potential and challenges for refineries. Fuel 324:124636. https://doi.org/10.1016/j.fuel.2022.124636
97. van Dyk S, Su J, Mcmillan JD, Saddler J (2019) Potential synergies of drop-in biofuel production with further co-processing at oil refineries. Biofuels Bioproducts Biorefining 13:760–775. https://doi.org/10.1002/bbb.1974
98. Stefanidis SD, Kalogiannis KG, Lappas AA (2018) Co-processing bio-oil in the refinery for drop-in biofuels via fluid catalytic cracking. Wiley Interdiscip Rev Energy Environ 7:e281. https://doi.org/10.1002/WENE.281
99. Karka P, Johnsson F, Papadokonstantakis S (2021) Perspectives for greening European fossil-fuel infrastructures through use of biomass: the case of liquid biofuels based on lignocellulosic resources. Front Energy Res 9. https://doi.org/10.3389/fenrg.2021.636782
100. Na J-G, Yi BE, Han JK, Oh Y-K, Park J-H, Jung TS, Han SS, Yoon HC, Kim J-N, Lee H, Ko CH (2012) Deoxygenation of microalgal oil into hydrocarbon with precious metal catalysts: optimization of reaction conditions and supports. Energy 47(1):25–30, ISSN 0360-5442. https://www.sciencedirect.com/science/article/pii/S0360544212005348. https://doi.org/10.1016/j.energy.2012.07.004
101. Coumans AE, Hensen EJM (2017) A real support effect on the hydrodeoxygenation of methyl oleate by sulfided NiMo catalysts. Catal Today 298:181–189, ISSN 0920-5861. https://www.sciencedirect.com/science/article/pii/S0920586117302833. https://doi.org/10.1016/j.cattod.2017.04.051
102. Lewandrowski J, Rosenfeld J, Pape D, Hendrickson T, Jaglo K, Moffroid K (2019) The greenhouse gas benefits of corn ethanol—assessing recent evidence. Biofuels 11(3):361–375. https://doi.org/10.1080/17597269.2018.1546488
103. Xia S et al (2023) Lipase-catalyzed production of biodiesel: a critical review on feedstock, enzyme carrier and process factors. Biofuel Bioprod Bior. https://doi.org/10.1002/bbb.2561
104. Sheldon RA, Brady D, Bode ML (2020) The Hitchhiker's guide to biocatalysis: recent advances in the use of enzymes in organic synthesis. Chem Sci 11(10): 2587–2605. https://doi.org/10.1039/C9SC05746C
105. Fusco S, Siebers B (2024) Extreme biocatalysts: from basic research to biotechnological applications. Int J Mol Sci 25(12):6492–6492. https://doi.org/10.3390/ijms25126492

MIX
Papier aus verantwortungsvollen Quellen
Paper from responsible sources
FSC® C105338

If you have any concerns about our products,
you can contact us on
ProductSafety@springernature.com

In case Publisher is established outside the EU,
the EU authorized representative is:
Springer Nature Customer Service Center GmbH
Europaplatz 3, 69115 Heidelberg, Germany

Printed by Libri Plureos GmbH
in Hamburg, Germany